On the Shoulders of Giants

On the Revolutions of Heavenly Spheres

On the Shoulders of Giants

On the Revolutions of Heavenly Spheres

by Nicolaus Copernicus
Edited, with Commentary, by Stephen Hawking

RUNNING PRESS
PHILADELPHIA · LONDON

9   8   7   6   5   4   3   2   1

Digit on the right indicates the number of this printing

Library of Congress Control Number: 2004094111

ISBN 0-7624-2021-9

Author photo courtesy of Book Laboratory

Cover Design: Doogie Horner

Typography: Scribe

Cover images: Portrait of Nicolaus Copernicus: Hulton/Archive by Getty Images; (bottom) Cat's Eye Nebula STScI-1995-01, courtesy of J.P. Harrington and K.J. Borkowski (University of Maryland), and NASA

Text of *On the Revolutions of Heavenly Spheres* courtesy of Annapolis: The St. John's Bookstore, © 1939.

This book may be ordered by mail from the publisher.

Please include $2.50 for postage and handling.

But try your bookstore first!

Running Press Book Publishers

125 South Twenty-second Street

Philadelphia, PA 19103-4399

Visit us on the web!

www.runningpress.com

Contents

A Note to the Reader

On the Revolutions of Heavenly Spheres is the first in a series of five scientific classics that Running Press has chosen to publish in a new format with commentary by renowned physicist, Stephen Hawking. Hawking chose these five particular essays to illuminate the evolution of modern physics and astronomy, as well as the process through which scientific knowledge evolves.

Also included in this series are:

- *Dialogues Concerning Two New Sciences* by Galileo Galilei
- *Harmonies of the World* by Johannes Kepler
- *Principia* by Sir Isaac Newton
- *Selections from* The Principle of Relativity by Albert Einstein

Look for all five essays combined in one volume in:

- *On the Shoulders of Giants* edited, with commentary, by Stephen Hawking
- *The Illustrated On the Shoulders of Giants* edited, with commentary, by Stephen Hawking

The text for *On the Revolutions of Heavenly Spheres* is based on a translation of the original, printed edition. We have made no attempt to modernize the author's own distinct usage, spelling, or punctuation. The details of this essay are as follows:

On the Revolutions of Heavenly Spheres, by Nicolaus Copernicus was first published in 1543 under the title *De revolutionibus orbium colestium.* This translation is by Charles Glen Wallis.

The Editors

Introduction

If I have seen farther, it is by standing on the shoulders of giants, wrote Isaac Newton in a letter to Robert Hooke in 1676. Although Newton was referring to his discoveries in optics rather than his more important work on gravity and the laws of motion, it is an apt comment on how science, and indeed the whole of civilization, is a series of incremental advances, each building on what went before. This is the theme of this fascinating series, which uses the original texts to trace the evolution of our picture of the heavens from the revolutionary claim of Nicolaus Copernicus that the earth orbits the sun to the equally revolutionary proposal of Albert Einstein that space and time are curved and warped by mass and energy. It is a compelling story because both Copernicus and Einstein have brought about profound changes in what we see as our position in the order of things. Gone is our privileged place at the center of the universe, gone are eternity and certainty, and gone are Absolute Space and Time to be replaced by rubber sheets.

It is no wonder both theories encountered violent opposition: the Inquisition in the case of the Copernican theory and the Nazis in the case of Relativity. We now have a tendency to dismiss as primitive the earlier world picture of Aristotle and Ptolemy in which the Earth was at the center and the Sun went round it. However we should not be too scornful of their model, which was anything but simpleminded. It incorporated Aristotle's deduction that the Earth is a round ball rather than a flat plate and it was reasonably accurate in its main function, that of predicting the apparent positions of the heavenly bodies in the sky for astrological purposes. In fact, it was about as accurate as the heretical suggestion put forward in 1543 by Copernicus that the Earth and the planets moved in circular orbits around the Sun.

Galileo found Copernicus's proposal convincing not because it better fit the observations of planetary positions but because of its simplicity and elegance, in contrast to the complicated epicycles of the Ptolemaic model. In *Dialogues Concerning Two New Sciences,* Galileo's characters, Salviati and Sagredo, put forward persuasive arguments in support of Copernicus. Yet, it was still possible for his third character, Simplicio, to defend Aristotle and Ptolemy and to maintain that in reality the Earth was at rest and the Sun went round the Earth.

It was not until Kepler's work made the Sun-centered model more accurate and Newton gave it laws of motion that the Earth-centered picture finally lost all credibility. It was quite a shift in our view of the universe: If we are not at the center, is our existence of any importance? Why should God or the Laws of Nature care about what happens on the third rock from the Sun, which is where Copernicus has left us? Modern scientists have out-Copernicused Copernicus by seeking an account of the universe in which Man (in the old pre-politically correct sense) played no role. Although this approach has succeeded in finding objective impersonal laws that govern the universe, it has not (so far at least) explained why the universe is the way it is rather than being one of the many other possible universes that would also be consistent with the laws.

Some scientists would claim that this failure is only provisional, that when we find the ultimate unified theory, it will uniquely prescribe the state of the universe, the strength of gravity, the mass and charge of the electron and so on. However, many features of the universe (like the fact that we are on the third rock, rather than the second or fourth) seem arbitrary and accidental and not the predictions of a master equation. Many people (myself included) feel that the appearance of such a complex and structured universe from simple laws requires the invocation of something called the anthropic principle, which restores us to the central position we have been too modest to claim since the time of Copernicus. The anthropic principle is based on the self-evident fact that we wouldn't be asking questions about the nature of the universe if the universe hadn't contained stars, planets and stable chemical compounds, among other prerequisites of (intelligent?) life as we know it. If the ultimate theory made a unique prediction for the state of the universe and its contents, it would be a remarkable coincidence that this state was in the small subset that allow life.

However the work of the last thinker in this series, Albert Einstein, raises a new possibility. Einstein played an important role in the development of quantum theory which says that a system doesn't just have a single history as one might have thought. Rather it has every possible history with some probability. Einstein was also almost solely responsible for the general theory of relativity in which space and time are curved and become dynamic. This means that they are subject to quantum theory and that the universe itself has every possible shape and history. Most of these histories will be quite unsuitable for the development of life but a very few have all the conditions needed. It doesn't matter if these few have a very low probability relative to the others: the lifeless universes will have no one to observe them. It is sufficient that there is at least one history in which life develops, and we ourselves are evidence for that, though maybe not for intelligence. Newton said he was *"standing on the shoulders of giants."* But as this series illustrates so well, our understanding doesn't advance just by slow and steady building on previous work. Sometimes as with Copernicus and Einstein, we have to make the intellectual leap to a new world picture. Maybe Newton should have said, "*I used the shoulders of giants as a springboard.*"

Nicolaus Copernicus

(1473–1543)
HIS LIFE AND WORK

Nicolaus Copernicus, a sixteenth-century Polish priest and mathematician, is often referred to as the founder of modern astronomy. That credit goes to him because he was the first to conclude that the planets and Sun did not revolve around the earth. Certainly there was speculation that a heliocentric—or sun-centered—universe had existed as far back as Aristarchus (d. 230 B.C.), but the idea was not seriously considered before Copernicus. Yet to understand the contributions of Copernicus, it is important to consider the religious and cultural implications of scientific discovery in his time.

As far back as the fourth century B.C., the Greek thinker and philosopher Aristotle (384–322 B.C.) devised a planetary system in his book, *On the Heavens*, (*De Caelo*) and concluded that because the Earth's shadow on the Moon during eclipses was always round, the world was spherical in shape rather than flat. He also surmised the Earth was round because when one watched a ship sailed out to sea one noticed that the hull disappeared over the horizon before the sails did.

In Aristotle's geocentric vision, the earth was stationary and the planets Mercury, Venus, Mars, Jupiter, and Saturn, as well as the sun and the moon performed circular orbits around the earth. Aristotle also believed the stars were fixed to the celestial sphere, and his scale of the universe purported these fixed stars to be not much further beyond the orbit of Saturn. He believed in perfect circular motions and had good evidence to believe the earth to be at rest. A stone dropped from a tower fell straight down. It did not fall to the west, as we would expect it to do if the earth rotated from west to east. (Aristotle did not consider that the stone might partake in the Earth's rotation). In an attempt to combine physics with the metaphysical, Aristotle devised his theory of a "prime mover," which held that a mystical force behind the fixed stars caused the circular motions he observed. This model of the universe was accepted and embraced by theologians, who often interpreted prime movers as angels, and Aristotle's vision endured for centuries. Many modern scholars believe universal acceptance of this theory by religious authorities hindered the progress of science, as to challenge Aristotle's theories was to call into question the authority of the church itself.

Five centuries after Aristotle's death, an Egyptian named Claudius Ptolemaeus (Ptolemy, A.D. 87–150), created a model for the universe that more accurately predicted the movements and actions of spheres in the heavens. Like Aristotle, Ptolemy believed the earth was stationary. Objects fell to the center of the earth, he reasoned, because the earth must be fixed at the center of the universe. Ptolemy ultimately elaborated a system in which the celestial bodies moved around the circumference of their own epicycles (a circle in which a planet moves and which has a center that is itself carried around at the same time on the circumference of a larger circle. To accomplish this, he put the Earth slightly off center of the universe and called this new center the "equant"—an imaginary point that helped him account for observable planetary movements. By custom designing the sizes of circles, Ptolemy was better able to predict the motions of celestial bodies. Western Christendom had little quarrel with Ptolemy's geocentric system, which left room in the universe behind the fixed stars to accommodate a heaven and a hell, and so the church adopted the Ptolemaic model of the universe as truth.

Aristotle and Ptolemy's picture of the cosmos reigned, with few significant modifications, for well over a thousand years. It wasn't until 1514 that the Polish priest Nicolaus Copernicus revived the heliocentric model of the universe. Copernicus proposed it merely as a model for calculating planetary positions because he was concerned that the church might label him a heretic if he proposed it as a description of reality. Copernicus became convinced, through his own study of planetary motions, that the earth was merely another planet and the sun was the center of the universe. This hypothesis became known as a heliocentric model. Copernicus's breakthrough marked one of the greatest paradigm shifts in world history, opening the way to modern astronomy and broadly affecting science, philosophy, and religion. The elderly priest was hesitant to divulge his theory, lest it provoke church authorities to any angry response, and so he withheld his work from all but a few astronomers. Copernicus's landmark *De Revolutionibus* was published while he was on his deathbed, in 1543. He did not live long enough to witness the chaos his heliocentric theory would cause.

Copernicus was born on February 19, 1473 in Torun, Poland, into a family of merchants and municipal officials who placed a high priority on education. His uncle, Lukasz Watzenrode, prince-bishop of Ermland, ensured that his nephew received the best academic training available in Poland. In 1491, Copernicus enrolled at Cracow University, where he pursued a course of general studies for four years before traveling to Italy to study law and medicine, as was common practice among Polish elites at the time. While studying at the University of Bologna (where he would eventually become a professor of astronomy), Copernicus boarded at the home of Domenico Maria de Novara, the renowned mathematician of whom Copernicus would ultimately become a disciple. Novara was a critic of Ptolemy, whose second-century astronomy he regarded with skepticism. In November 1500, Copernicus observed a lunar eclipse in Rome. Although he spent the next few years in Italy studying medicine, he never lost his passion for astronomy.

After receiving the degree of Doctor of Canon Law, Copernicus practiced medicine at the episcopal court of Heilsberg, where his uncle lived. Royalty and high clergy requested his medical services, but Copernicus spent most of his time in service of the poor. In 1503, he returned to Poland and moved into his uncle's bishopric palace in Lidzbark Warminski. There he tended to the administrative matters of the diocese, as well as serving as an advisor to his uncle. After his uncle's death in 1512, Copernicus moved permanently to Frauenburg and would spend the rest of his life in priestly service. But the man who was a scholar in mathematics, medicine and theology was only beginning the work for which he would become best known.

In March of 1513, Copernicus purchased 800 building stones, and a barrel of lime from his chapter so that he could build an observation tower. There, he made use of astronomical instruments such as quadrants, parallactics and astrolabes to observe the sun, moon and stars. The following year, he wrote a brief *Commentary on the Theories of the Motions of Heavenly Objects from Their Arrangements* (*De hypothesibus motuum coelestium a se constitutis commentariolus*), but he refused to publish the manuscript and only discreetly circulated it among his most trusted friends. The *Commentary* was a first attempt to propound an astronomical theory that the earth moves and the sun remains at rest. Copernicus had become dissatisfied with the Aristotelian-Ptolemaic astronomical system that had dominated Western thought for centuries. The center of the earth, he thought, was not the center of the universe, but merely the center of the Moon's orbit. Copernicus had come to believe that apparent perturbations in the observable motion of the planets was a result of the earth's own rotation around its axis and of its travel in orbit. "We revolve around the Sun," he concluded in *Commentary*, "like any other planet."

Despite speculation about a sun-centered universe as far back as the third century B.C. by Aristarchus, theologians and intellectuals felt more comfortable with a geocentric theory, and the premise was barely challenged in earnest. Copernicus prudently abstained from disclosing any of his views in public, preferring to develop his ideas quietly by exploring mathematical calculations and drawing elaborate diagrams, and to keep his theories from circulating outside of a select group of friends. When, in 1514, Pope Leo X summoned Bishop Paul of Fossombrone to recruit Copernicus to offer an opinion on reforming the ecclesiastical calendar, the Polish astronomer replied that knowledge of the motions of the sun and moon in relation to the length of the year was insufficient to have any bearing on reform. The challenge must have preoccupied Copernicus, however, for he later wrote to Pope Paul III, the same Pope who commissioned Michaelangelo to paint the Sistine Chapel, with some relevant observations, which later served to form the foundation of the Gregorian calendar seventy years later.

Still, Copernicus feared exposing himself to the contempt of the populace and the church, and he spent years working privately to amend and expand the *Commentary.* The result was *On the Revolutions of Heavenly Spheres* (*De Revolutionibus Orbium Coelestium*) which he completed in 1530, but withheld from publication for thirteen years. The risk of the church's condemnation was not, however, the only reason for Copernicus's hesitancy to publish. Copernicus was a perfectionist and considered his observations in constant need of verification and revision. He continued to lecture on these principles of his planetary theory, even appearing before Pope Clement VII, who approved of his work. In 1536, Clement formally requested that Copernicus publish his theories. But it took a former pupil, 25-year-old Georg Joachim Rheticus of Germany, who relinquished his chair in mathematics in Wittenberg so that he could study under Copernicus, to persuade his master to publish *On the Revolutions.* In 1540, Rheticus assisted in the editing of the work and presented the manuscript to a Lutheran printer in Nuremberg, ultimately giving birth to the Copernican Revolution.

When *On the Revolutions* appeared in 1543, it was attacked by Protestant theologians who held the premise of a heliocentric universe to be unbiblical. Copernicus's theories, they reasoned, might lead people to believe that they are simply part of a natural order, and not the masters of nature, the center around which nature was ordered. Because of this clerical opposition, and perhaps also general incredulity at the prospect of a non-geocentric universe, between 1543 and 1600, fewer than a dozen scientists embraced Copernican theory. Still, Copernicus had done nothing to resolve the major problem facing any system in which the earth rotated on its axis (and revolved around the sun), namely, how it is that terrestrial bodies stay with the rotating Earth. The answer was proposed by Giordano Bruno, an Italian scientist and avowed Copernican, who suggested that space might have no boundaries and that the solar system might be one of many such systems in the universe. Bruno also expanded on some purely speculative areas of astronomy that Copernicus did not explore in *On the Revolutions.* In his writings and lectures, the Italian scientist held that there were infinite worlds in the universe with intelligent life, some perhaps with beings superior to humans. Such audacity brought Bruno to the attention of the Inquisition, which tried and condemned him for his heretical beliefs. He was burned at the stake in 1600.

On the whole, however, the book did not have an immediate impact on modern astronomic study. In *On the Revolutions,* Copernicus did not actually put forth a heliocentric system, but rather a heliostatic one. He considered the Sun to be not precisely at the center of the universe, but only close to it, so as to account for variations in observable retrogression and brightness. The earth, he asserted, made one full rotation on its axis daily, and orbited around the sun once yearly. In the first section of the book's six sections, he took issue with the Ptolemaic system, which placed all heavenly bodies in orbit around the earth, and established the correct heliocentric order: Mercury, Venus, Mars, Jupiter and Saturn (the six planets known at the time). In the second section, Copernicus used mathematics (namely epicycles and equants) to explain the motions of the stars and planets, and reasoned that the sun's motion coincided with that of the earth. The third section gives a mathematical explanation of the precession of the equinoxes, which Copernicus attributes to the Earth's gyration around its axis. The remaining sections of *On the Revolutions* focus on the motions of the planets and the moon.

Copernicus was the first to position Venus and Mercury correctly, establishing with remarkable accuracy the order and distance of the known planets. He saw these two planets (Venus and Mercury) as being closer to the sun, and noticed that they revolved at a faster rate inside the Earth's orbit.

Before Copernicus, the sun was thought to be another planet. Placing the sun at the virtual center of the planetary system was the beginning of the Copernican revolution. By moving the Earth away from the center of the universe, where it was presumed to anchor all heavenly bodies, Copernicus was forced to address theories of gravity. Pre-Copernican gravitational explanations had posited a single center of gravity (the earth), but Copernicus theorized that each heavenly body might have its own gravitational qualities and asserted that heavy objects everywhere tended toward their own center. This insight would eventually lead to the theory of universal gravitation, but its impact was not immediate.

By 1543, Copernicus had become paralyzed on his right side and weakened both physically and mentally. The man who was clearly a perfectionist had no choice but to surrender control of his manuscript, *On the Revolutions,* in the last stages of printing. He entrusted his student George Rheticus with the manuscript, but when Rheticus was forced to leave Nuremberg, the manuscript fell into the hands of Lutheran theologian Andreas Osiander. Osiander, hoping to appease advocates of the geocentric theory, made several alterations without Copernicus's knowledge and consent. Osiander placed the word "hypothesis" on the title page, deleted important passages, and added his own sentences which diluted the impact and certainty of the work. Copernicus was said to have received a copy of the printed book in Frauenburg on his deathbed, unaware of Osiander's revisions. His ideas lingered in relative obscurity for nearly one hundred years, but the seventeenth century would see men like Galileo Galilei, Johannes Kepler and Isaac Newton build on his theories of a heliocentric universe, effectively obliterating Aristotelian ideas. Many have written about the unassuming Polish priest who would change the way people saw the universe, but the German writer and scientist Johann Wolfgang von Goethe may have been the most eloquent when he wrote of the contributions of Copernicus:

Of all discoveries and opinions, none may have exerted a greater effect on the human spirit than the doctrine of Copernicus . The world had scarcely become known as round and complete in itself when it was asked to waive the tremendous privilege of being the center of the universe. Never, perhaps, was a greater demand made on mankind—for by this admission so many things vanished in mist and smoke! What became of Eden, our world of innocence, piety and poetry; the testimony of the senses; the conviction of a poetic—religious faith? No wonder his contemporaries did not wish to let all this go and offered every possible resistance to a doctrine which in its converts authorized and demanded a freedom of view and greatness of thought so far unknown, indeed not even dreamed of.

—Johann Wolfgang von Goethe

INTRODUCTION

To the Reader Concerning the Hypotheses of this Work[1]

[1b][2] Since the newness of the hypotheses of this work—which sets the earth in motion and puts an immovable sun at the centre of the universe—has already received a great deal of publicity, I have no doubt that certain of the savants have taken grave offense and think it wrong to raise any disturbance among liberal disciplines which have had the right set-up for a long time now. If, however, they are willing to weigh the matter scrupulously, they will find that the author of this work has done nothing which merits blame. For it is the job of the astronomer to use painstaking and skilled observation in gathering together the history of the celestial movements, and then since he cannot by any line of reasoning reach the true causes of these movements—to think up or construct whatever causes or hypotheses he pleases such that, by the assumption of these causes, those same movements can be calculated from the principles of geometry for the past and for the future too. This artist is markedly outstanding in both of these respects: for it is not necessary that these hypotheses should be true, or even probably; but it is enough if they provide a calculus which fits the observations—unless by some chance there is anyone so ignorant of geometry and optics as to hold the epicycle of Venus as probable and to believe this to be a cause why Venus alternately precedes and follows the sun at an angular distance of up to 40° or more. For who does not see that it necessarily follows from this assumption that the diameter of the planet in its perigee should appear more than four times greater, and the body of the planet more than sixteen times greater, than in its apogee? Nevertheless the experience of all the ages is opposed to that.[3] There are also other things in this discipline which are just as absurd, but it is not necessary to examine them right now. For it is sufficiently clear that this art is absolutely and profoundly ignorant of the causes of the apparent irregular movements. And if it constructs and thinks up causes—and it has certainly thought up a good many—nevertheless it does not think them up in order to persuade anyone of their truth but only in order that they may provide a correct basis for calculation. But since for one and the same movement varying hypotheses are proposed

---

[1]This foreword, at first ascribed to Copernicus, is held to have been written by Andrew Osiander, a Lutheran theologian and friend of Copernicus, who saw the *De Revolutionibus* through the press.

[2]The numbers within the brackets refer to the pages of the first edition, published in 1543 at Nuremberg.

[3]Ptolemy makes Venus move on an epicycle the ratio of whose radius to the radius of the eccentric circle carrying the epicycle itself is nearly three to four. Hence the apparent magnitude of the planet would be expected to vary with the varying distance of the planet from the Earth, in the ratios stated by Osiander.

Moreover, it was found that, whenever the planet happened to be on the epicycle, the mean position of the sun appeared in line with *EPA*. And so, granted the ratios of epicycle and eccentric, Venus would never appear from the Earth to be at an angular distance of much more than 40° from the centre of her epicycle, that is to say, from the mean position of the sun, as it turned out by observation.

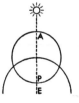

1

from time to time, as eccentricity or epicycle for the movement of the sun, the astronomer much prefers to take the one which is easiest to [ii^a] grasp. Maybe the philosopher demands probability instead; but neither of them will grasp anything certain or hand it on, unless it has been divinely revealed to him. Therefore let us permit these new hypotheses to make a public appearance among old ones which are themselves no more probable, especially since they are wonderful and easy and bring with them a vast storehouse of learned observations. And as far as hypotheses go, let no one expect anything in the way of certainty from astronomy, since astronomy can offer us nothing certain, lest, if anyone take as true that which has been constructed for another use, he go away from this discipline a bigger fool than when he came to it. Farewell.

Preface and Dedication to Pope Paul III

[ii^b] I can reckon easily enough, Most Holy Father, that as soon as certain people learn that in these books of mine which I have written about the revolutions of the spheres of the world I attribute certain motions to the terrestrial globe, they will immediately shout to have me and my opinion hooted off the stage. For my own works do not please me so much that I do not weigh what judgments others will pronounce concerning them. And although I realize that the conceptions of a philosopher are placed beyond the judgment of the crowd, because it is his loving duty to seek the truth in all things, in so far as God has granted that to human reason; nevertheless I think we should avoid opinions utterly foreign to rightness. And when I considered how absurd this "lecture" would be held by those who know that the opinion that the Earth rests immovable in the middle of the heavens as if their centre had been confirmed by the judgments of many ages—if I were to assert to the contrary that the Earth moves; for a long time I was in great difficulty as to whether I should bring to light my commentaries written to demonstrate the Earth's movement, or whether it would not be better to follow the example of the Pythagoreans and certain others who used to hand down the mysteries of their philosophy not in writing but by word of mouth and only to their relatives and friends—witness the letter of Lysis to Hipparchus. They however seem to me to have done that not, as some judge, out of a jealous unwillingness to communicate their doctrines but in order that things of very great beauty which have been investigated by the loving care of great men should not be scorned by those who find it a bother to expend any great energy on letters—except on the money-making variety—or who are provoked by the exhortations and examples of others to the liberal study of philosophy but on account of their natural [iii^a] stupidity hold the position among philosophers that drones hold among bees. Therefore, when I weighed these things in my mind, the scorn which I had to fear on account of the newness and absurdity of my opinion almost drove me to abandon a work already undertaken.

2

But my friends made me change my course in spite of my long-continued hesitation and even resistance. First among them was Nicholas Schonberg, Cardinal of Capua, a man distinguished in all branches of learning; next to him was my devoted friend Tiedeman Giese, Bishop of Culm, a man filled with the greatest zeal for the divine and liberal arts: for he in particular urged me frequently and even spurred me on by added reproaches into publishing this book and letting come to light a work which I had kept hidden among my things for not merely nine years, but for almost four times nine years. Not a few other learned and distinguished men demanded the same thing of me, urging me to refuse no longer—on account of the fear which I felt—to contribute my work to the common utility of those who are really interested in mathematics: they said that the absurder my teaching about the movement of the Earth now seems to very many persons, the more wonder and thanksgiving will it be the object of, when after the publication of my commentaries those same persons see the fog of absurdity dissipated by my luminous demonstrations. Accordingly I was led by such persuasion and by that hope finally to permit my friends to undertake the publication of a work which they had long sought from me.

But perhaps Your Holiness will not be so much surprised at my giving the results of my nocturnal study to the light—after having taken such care in working them out that I did not hesitate to put in writing my conceptions as to the movement of the Earth—as you will be eager to hear from me what came into my mind that in opposition to the general opinion of mathematicians and almost in opposition to common sense I should dare to imagine some movement of the Earth. And so I am unwilling to hide from Your Holiness that nothing except my knowledge that mathematicians have not agreed with one another in their researches moved me to think out a different scheme of drawing up the movements of the spheres of the world. For in the first place mathematicians are so uncertain about the movements of the sun and moon that they can neither demonstrate nor observe the unchanging magnitude of the [ii^b] revolving year. Then in setting up the solar and lunar movements and those of the other five wandering stars, they do not employ the same principles, assumptions, or demonstrations for the revolutions and apparent movements. For some make use of homocentric circles only, others of eccentric circles and epicycles, by means of which however they do not fully attain what they seek. For although those who have put their trust in homocentric circles have shown that various different movements can be composed of such circles, nevertheless they have not been able to establish anything for certain that would fully correspond to the phenomena. But even if those who have thought up eccentric circles seem to have been able for the most part to compute the apparent movements numerically by those means, they have in the meanwhile admitted a great deal which seems to contradict the first principles of regularity of movement. Moreover, they have not been able to discover or to infer the chief point of all, *i.e.*, the form of the world and the certain commensurability of its parts. But they are in exactly the same fix as someone taking from different places hands, feet, head, and the other limbs—shaped very beautifully but not with reference to one body and without correspondence to one another—so that such parts made up a monster rather than a man. And so, in the process of demonstration which they call "method," they are found either to have omitted something necessary or to have admitted something foreign which by no means pertains to the matter; and they would by no means have been in this fix, if they had followed sure principles. For if the hypotheses they assumed were not false, everything which followed from the hypotheses would have been verified without fail; and though what I am saying may be obscure right now, nevertheless it will become clearer in the proper place.

Accordingly, when I had meditated upon this lack of certitude in the traditional mathematics concerning the composition of movements of the spheres of the world, I began to be annoyed that the philosophers, who in other respects had made a very careful scrutiny of the least details of the world, had discovered no sure scheme for the movements of the machinery of the world, which has been built for us by the Best and Most Orderly Workman of all. Wherefore I took the trouble to reread all the books by philosophers which I could get hold of, to see if any of them even supposed that the movements of the spheres of the world [*iv*ᵃ] were different from those laid down by those who taught mathematics in the schools. And as a matter of fact, I found first in Cicero that Nicetas thought that the Earth moved. And afterwards I found in Plutarch that there were some others of the same opinion: I shall copy out his words here, so that they may be known to all:

> *Some think that the Earth is at rest; but Philolaus the Pythagorean says that it moves around the fire with an obliquely circular motion, like the sun and moon. Herakleides of Pontus and Ekphantus the Pythagorean do not give the Earth any movement of locomotion, but rather a limited movement of rising and setting around its centre, like a wheel.*[1]

Therefore I also, having found occasion, began to meditate upon the mobility of the Earth. And although the opinion seemed absurd, nevertheless because I knew that others before me had been granted the liberty of constructing whatever circles they pleased in order to demonstrate astral phenomena, I thought that I too would be readily permitted to test whether or not, by the laying down that the Earth had some movement, demonstrations less shaky than those of my predecessors could be found for the revolutions of the celestial spheres.

[1] *De placitis philosophorum*, III, 13.

And so, having laid down the movements which I attribute to the Earth farther on in the work, I finally discovered by the help of long and numerous observations that if the movements of the other wandering stars are correlated with the circular movement of the Earth, and if the movements are computed in accordance with the revolution of each planet, not only do all their phenomena follow from that but also this correlation binds together so closely the order and magnitudes of all the planets and of their spheres or orbital circles and the heavens themselves that nothing can be shifted around in any part of them without disrupting the remaining parts and the universe as a whole.

Accordingly, in composing my work I adopted the following order: in the first book I describe all the locations of the spheres or orbital circles together with the movements which I attribute to the earth, so that this book contains as it were the general set-up of the universe. But afterwards in the remaining books I correlate all the movements of the other planets and their spheres or orbital circles with the mobility of the Earth, so that it can be gathered from that how far the apparent movements of the remaining planets and their orbital circles can be saved by being correlated with the movements of the Earth. And I have no doubt that talented and learned mathematicians will agree with me, if—as philosophy [iv^b] demands in the first place—they are willing to give not superficial but profound thought and effort to what I bring forward in this work in demonstrating these things. And in order that the unlearned as well as the learned might see that I was not seeking to flee from the judgment of any man, I preferred to dedicate these results of my nocturnal study to Your Holiness rather than to anyone else; because, even in this remote corner of the earth where I live, you are held to be most eminent both in the dignity of your order and in your love of letters and even of mathematics; hence, by the authority of your judgment you can easily provide a guard against the bites of slanderers, despite the proverb that there is no medicine for the bite of a sycophant.

But if perchance there are certain "idle talkers" who take it upon themselves to pronounce judgment, although wholly ignorant of mathematics, and if by shamelessly distorting the sense of some passage in Holy Writ to suit their purpose, they dare to reprehend and to attack my work; they worry me so little that I shall even scorn their judgments as foolhardy. For it is not unknown that Lactantius, otherwise a distinguished writer but hardly a mathematician, speaks in an utterly childish fashion concerning the shape of the Earth, when he laughs at those who have affirmed that the Earth has the form of a globe. And so the studious need not be surprised if people like that laugh at us. Mathematics is written for mathematicians; and among them, if I am not mistaken, my labours will be seen to contribute something to the ecclesiastical commonwealth, the principate of which Your Holiness now holds. For not many years ago under Leo X when the Lateran Council was considering the question of reforming the Ecclesiastical Calendar, no decision was reached, for the sole reason that the magnitude of the year and the months and the movements of the sun and moon had not yet been measured with sufficient accuracy. From that time on I gave attention to making more exact observations of these things and was encouraged to do so by that most distinguished man, Paul, Bishop of Fossombrone, who had been present at those deliberations. But what have I accomplished in this matter I leave to the judgment of Your Holiness in particular and to that of all other learned mathematicians. And so as not to appear to Your Holiness to make more promises concerning the utility of this book than I can fulfill, I now pass on to the body of the work.

BOOK ONE[1]

Among the many and varied literary and artistic studies upon which the natural talents of man are nourished, I think that those above all should be embraced and pursued with the most loving care which have to do with things that are very beautiful and very worthy of knowledge. Such studies are those which deal with the godlike circular movements of the world and the course of the stars, their magnitudes, distances, risings and settings, and the causes of the other appearances in the heavens; and which finally explicate the whole form. For what could be more beautiful than the heavens which contain all beautiful things? Their very names make this clear: *Caelum* (heavens) by naming that which is beautifully carved; and *Mundus* (world), purity and elegance. Many philosophers have called the world a visible god on account of its extraordinary excellence. So if the worth of the arts were measured by the matter with which they deal, this art—which some call astronomy, others astrology, and many of the ancients the consummation of mathematics—would be by far the most outstanding. This art which is as it were the head of all the liberal arts and the one most worthy of a free man leans upon nearly all the other branches of mathematics. Arithmetic, geometry, optics, geodesy, mechanics, and whatever others, all offer themselves in its service. And since a property of all good arts is to draw the mind of man away from the vices and direct it to better things, these arts can do that more plentifully, over and above the unbelievable pleasure of mind (which they furnish). For who, after applying himself to things which he sees established in the best order and directed by divine ruling, would not through diligent contemplation of them and through a certain habituation be awakened to that which is best and would not wonder at the Artificer of all things, in Whom is all happiness and every good? For the divine Psalmist surely did not say gratuitously that he took pleasure in the workings of God and rejoiced in the works of His hands, unless by means of these things as by some sort of vehicle we are transported to the contemplation of the highest Good.

Now as regards the utility and ornament which they confer upon a commonwealth—to pass over the innumerable advantages they give to private citizens—Plato makes an extremely good point, for in the seventh book of the *Laws* he says that this study should be pursued in especial, that through it the orderly arrangement of days into months and years and the determination of the times for solemnities and sacrifices should keep the state alive and watchful; and he says that if anyone denies that this study is necessary for a man who is going to take up any of the highest branches of learning, then such a person is thinking foolishly; and he thinks that it is impossible for anyone to become godlike or be called so who has no necessary knowledge of the sun, moon, and the other stars.

[1]The three introductory paragraphs are found in the Thorn centenary and Warsaw editions

However, this more divine than human science, which inquires into the highest things, is not lacking in difficulties. And in particular we see that as regards its principles and assumptions, which the Greeks call "hypotheses," many of those who undertook to deal with them were not in accord and hence did not employ the same methods of calculation. In addition, the courses of the planets and the revolution of the stars cannot be determined by exact calculations and reduced to perfect knowledge unless, through the passage of time and with the help of many prior observations, they can, so to speak, be handed down to posterity. For even if Claud Ptolemy of Alexandria, who stands far in front of all the others on account of his wonderful care and industry, with the help of more than forty years of observations brought this art to such a high point that there seemed to be nothing left which he had not touched upon; nevertheless we see that very many things are not in accord with the movements which should follow from his doctrine but rather with movements which were discovered later and were unknown to him. Whence even Plutarch in speaking of the revolving solar year says, "So far the movement of the stars has overcome the ingenuity of the mathematicians." Now to take the year itself as my example, I believe it is well known how many different opinions there are about it, so that many people have given up hope of risking an exact determination of it. Similarly, in the case of the other planets I shall try—with the help of God, without Whom we can do nothing—to make a more detailed inquiry concerning them, since the greater the interval of time between us and the founders of this art—whose discoveries we can compare with the new ones made by us—the more means we have of supporting our own theory. Furthermore, I confess that I shall expound many things differently from my predecessors—although with their aid, for it was they who first opened the road of inquiry into these things.

1. THE WORLD IS SPHERICAL

[1ª] In the beginning we should remark that the world is globe-shaped; whether because this figure is the most perfect of all, as it is an integral whole and needs no joints; or because this figure is the one having the greatest volume and thus is especially suitable for that which is going to comprehend and conserve all things; or even because the separate parts of the world *i.e.*, the sun, moon, and stars are viewed under such a form; or because everything in the world tends to be delimited by this form, as is apparent in the case of drops of water and other liquid bodies, when they become delimited of themselves. And so no one would hesitate to say that this form belongs to the heavenly bodies.

2. THE EARTH IS SPHERICAL TOO

The Earth is globe-shaped too, since on every side it rests upon its centre. But it is not perceived straightway to be a perfect sphere, on account of the great height of its mountains and the lowness of its valleys, though they modify its universal roundness to only a very small extent.

That is made clear in this way. For when people journey northward from anywhere, the northern vertex of the axis of daily revolution gradually moves overhead, and the other moves downward to the same extent; and many stars situated to the north are seen not to set, and many to the south are seen not to rise any more. So Italy does not see Canopus, which is visible to Egypt. And Italy sees the last star of Fluvius, which is not visible to this region situated in a more frigid zone. Conversely, for people who travel southward, the second group of stars becomes higher in the sky; while those become lower which for us are high up.

Moreover, the inclinations of the poles have everywhere the same ratio with places at equal distances from the poles of the Earth and that [1^b] happens in no other figure except the spherical. Whence it is manifest that the Earth itself is contained between the vertices and is therefore a globe.

Add to this the fact that the inhabitants of the East do not perceive the evening eclipses of the sun and moon; nor the inhabitants of the West, the morning eclipses; while of those who live in the middle region—some see them earlier and some later.

Furthermore, voyagers perceive that the waters too are fixed within this figure; for example, when land is not visible from the deck of a ship, it may be seen from the top of the mast, and conversely, if something shining is attached to the top of the mast, it appears to those remaining on the shore to come down gradually, as the ship moves from the land, until finally it becomes hidden, as if setting.

Moreover, it is admitted that water, which by its nature flows, always seeks lower places—the same way as earth—and does not climb up the shore any farther than the convexity of the shore allows. That is why the land is so much higher where it rises up from the ocean.

3. HOW LAND AND WATER MAKE UP A SINGLE GLOBE

And so the ocean encircling the land pours forth its waters everywhere and fills up the deeper hollows with them. Accordingly it was necessary for there to be less water than land, so as not to have the whole earth soaked with water—since both of them tend toward the same centre on account of their weight—and so as to leave some portions of land—such as the islands discernible here and there—for the preservation of living creatures. For what is the continent itself and the *orbis terrarum* except an island which is larger than the rest? We should not listen to certain Peripatetics who maintain that there is ten times more water than land and who arrive at that conclusion because in the transmutation of the elements the liquefaction of one part of earth results in ten parts of water. And they say that land has emerged for a certain distance because, having hollow spaces inside, it does not balance everywhere with respect to weight and so the centre of gravity is different from the centre of magnitude. But they fall into error through ignorance of geometry; for they do not know that there cannot be seven times more water than land and some part of the land still remain dry, unless the land abandon its centre of gravity and give place to the waters as being heavier. For spheres are to one another as the cubes of their diameters. If therefore there were seven parts of water and one part of land, [2ª] the diameter of the land could not be greater than the radius of the globe of the waters. So it is even less possible that the water should be ten times greater. It can be gathered that there is no difference between the centres of magnitude and of gravity of the Earth from the fact that the convexity of the land spreading out from the ocean does not swell continuously, for in that case it would repulse the sea-waters as much as possible and would not in any way allow interior seas and huge gulfs to break through. Moreover, from the seashore outward the depth of the abyss would not stop increasing, and so no island or reef or any spot of land would be met with by people voyaging out very far. Now it is well known that there is not quite the distance of two miles—at practically the centre of the *orbis terrarum*—between the Egyptian and the Red Sea. And on the contrary, Ptolemy in his *Cosmography* extends inhabitable lands as far as the median circle, and he leaves that part of the Earth as unknown, where the moderns have added Cathay and other vast regions as far as 60° longitude, so that inhabited land extends in longitude farther than the rest of the ocean does. And if you add to these the islands discovered in our time under the princes of Spain and Portugal and especially America—named after the ship's captain who discovered her—which they consider a second *orbis terrarum* on account of her so far unmeasured magnitude—besides many other islands heretofore unknown, we would not be greatly surprised if there were antiphodes or antichthones. For reasons of geometry compel us to believe that America is situated diametrically opposite to the India of the Ganges.

And from all that I think it is manifest that the land and the water rest upon one centre of gravity; that this is the same as the centre of magnitude of the land, since land is the heavier; that parts of land which are as it were yawning are filled with water; and that accordingly there is little water in comparison with the land, even if more of the surface appears to be covered by water.

Now it is necessary that the land and the surrounding waters have the figure which the shadow of the Earth casts, for it eclipses the moon by projecting a perfect circle upon it. Therefore the Earth is not a plane, as Empedocles and Anaximenes opined; or a tympanoid, as Leucippus; or a scaphoid, as Heracleitus; or hollowed out in any other way, as Democritus; or again a cylinder, as Anaximander; and it is not infinite in its lower part, with the density increasing rootwards, as Xenophanes thought; but it is perfectly round, as the philosophers perceived.

4. THE MOVEMENT OF THE CELESTIAL BODIES IS REGULAR, CIRCULAR, AND EVERLASTING—OR ELSE COMPOUNDED OF CIRCULAR MOVEMENTS

[2ᵇ] After this we will recall that the movement of the celestial bodies is circular. For the motion of a sphere is to turn in a circle; by this very act expressing its form, in the most simple body, where beginning and end cannot be discovered or distinguished from one another, while it moves through the same parts in itself.

But there are many movements on account of the multitude of spheres or orbital circles.[1] The most obvious of all is the daily revolution—which the Greeks call νυχ-θήμερν; *i.e.*, having the temporal span of a day and a night. By means of this movement the whole world—with the exception of the Earth—is supposed to be borne from east to west. This movement is taken as the common measure of all movements, since we measure even time itself principally by the number of days.

Next, we see other as it were antagonistic revolutions; *i.e.*, from west to east, on the part of the sun, moon, and the wandering stars. In this way the sun gives us the year, the moon the months—the most common periods of time; and each of the other five planets follows its own cycle. Nevertheless these movements are manifoldly different from the first movement. First, in that they do not revolve around the same poles as the first movement but follow the oblique ecliptic; next, in that they do not seem to move in their circuit regularly. For the sun and moon are caught moving at times more slowly and at times more quickly. And we perceive the five wandering stars sometimes even to retrograde and to come to a stop between these two movements. And though the sun always proceeds straight ahead along its route, they wander in various ways, straying sometimes towards the south, and at other times towards the north—whence they are called "planets." Add to this the fact that sometimes they are nearer the Earth—and are then said to be at their perigee—and at other times are farther away—and are said to be at their apogee.

We must however confess that these movements are circular or are composed of many circular movements, in that they maintain these irregularities in accordance with a constant law and with fixed periodic returns: and that could not take place, if they were not circular. For it is only the circle which can bring back what is past and over with; and in this way, for example, the sun by a movement composed of circular movements brings back to us the inequality of days and nights and the four seasons of the

---

[1]The "orbital circle" (*orbis*) is the great circle whereon the planet moves in its sphere (*sphaera*). Copernicus; uses the word *orbis* which designates a circle primarily rather than a sphere because, while the sphere may be necessary for the mechanical explanation of the movement, only the circle is necessary for the mathematical.

year. [3ª] Many movements are recognized in that movement, since it is impossible that a simple heavenly body should be moved irregularly by a single sphere. For that would have to take place either on account of the inconstancy of the motor virtue—whether by reason of an extrinsic cause or its intrinsic nature—or on account of the inequality between it and the moved body. But since the mind shudders at either of these suppositions, and since it is quite unfitting to suppose that such a state of affairs exists among things which are established in the best system, it is agreed that their regular movements appear to us as irregular, whether on account of their circles having different poles or even because the earth is not at the centre of the circles in which they revolve. And so for us watching from the Earth, it happens that the transits of the planets, on account of being at unequal distances from the Earth, appear greater when they are nearer than when they are farther away, as has been shown in optics: thus in the case of equal arcs of an orbital circle which are seen at different distances there will appear to be unequal movements in equal times. For this reason I think it necessary above all that we should note carefully what the relation of the Earth to the heavens is, so as not—when we wish to scrutinize the highest things—to be ignorant of those which are nearest to us, and so as not—by the same error—to attribute to the celestial bodies what belongs to the Earth.

5. Does the Earth Have a Circular Movement? And of Its Place

Now that it has been shown that the Earth too has the form of a globe, I think we must see whether or not a movement follows upon its form and what the place of the Earth is in the universe. For without doing that it will not be possible to find a sure reason for the movements appearing in the heavens. Although there are so many authorities for saying that the Earth rests in the centre of the world that people think the contrary supposition inopinable and even ridiculous; if however we consider the thing attentively, we will see that the question has not yet been decided and accordingly is by no means to be scorned. For every apparent change in place occurs on account of the movement either of the thing seen or of the spectator, or on account of the necessarily unequal movement of both. For no movement is perceptible relatively to things moved equally in the same directions—I mean relatively to the thing seen and the spectator. Now it is from the Earth that the celestial circuit is beheld and presented to our sight. Therefore, if some movement should belong to the Earth [3ᵇ] it will appear, in the parts of the universe which are outside, as the same movement but in the opposite direction, as though the things outside were passing over. And the daily revolution in especial is such a movement. For the daily revolution appears to carry the whole universe along, with the exception of the Earth and the things around it. And if you admit that the heavens possess none of this movement but that the Earth turns from west to east, you will find—if you make a serious examination—that as regards the apparent rising and setting of the sun, moon, and stars the case is so. And since it is the heavens which contain and embrace all things as the place common to the universe, it will not be clear at once why movement should not be assigned to the contained rather than to the container, to the thing placed rather than to the thing providing the place.

As a matter of fact, the Pythagoreans Herakleides and Ekphantus were of this opinion and so was Hicetas the Syracusan in Cicero; they made the Earth to revolve at the centre of the world. For they believed that the stars set by reason of the interposition of the Earth and that with cessation of that they rose again. Now upon this assumption there follow other things, and a no smaller problem concerning the place of the Earth, though it is taken for granted and believed by nearly all that the Earth is the centre of the world. For if anyone denies that the Earth occupies the midpoint or centre of the world yet does not admit that the distance (between the two) is great enough to be compared with (the distance to) the sphere of the fixed stars but is considerable and quite apparent in relation to the orbital circles of the sun and the planets; and if for that reason he thought that their movements appeared irregular because they are organized around a different centre from the centre of the Earth, he might perhaps be able to bring forward a perfectly sound reason for movement which appears irregular. For the fact that the wandering stars are seen to be sometimes nearer the Earth and at other times farther away necessarily argues that the centre of the Earth is not the centre of their circles. It is not yet clear whether the Earth draws near to them and moves away or they draw near to the Earth and move away.

And so it would not be very surprising if someone attributed some other movement to the earth in addition to the daily revolution. As a matter of fact, Philolaus the Pythagorean—no ordinary mathematician, whom Plato's biographers say Plato went to Italy for the sake of seeing—is supposed to have held that the Earth moved in a circle and wandered in some other movements and was one of the planets.

Many however have believed that they could show by geometrical reasoning that the Earth is in the middle of the world; that it has the proportionality of a point in relation to the immensity of the heavens, occupies the central position, and for this reason is immovable, because, when the universe moves, the centre [4a] remains unmoved and the things which are closest to the centre are moved the most slowly.

6. On the Immensity of the Heavens in Relation to the Magnitude of the Earth

It can be understood that this great mass which is the Earth is not comparable with the magnitude of the heavens, from the fact that the boundary circles—for that is the translation of the Greek ὁρίζοντες—cut the whole celestial sphere into two halves; for that could not take place if the magnitude of the Earth in comparison with the heavens, or its distance from the centre of the world, were considerable. For the circle bisecting a sphere goes through the centre of the sphere, and is the greatest circle which it is possible to circumscribe.

Now let the horizon be the circle *ABCD*, and let the Earth, where our point of view is, be *E*, the centre of the horizon by which the visible stars are separated from those which are not visible. Now with a dioptra or horoscope or level placed at *E*, the beginning of Cancer is seen to rise at point *C*; and at the same moment the beginning of Capricorn appears to set at *A*. Therefore, since *AEC* is in a straight line with the dioptra, it is clear that this line is a diameter of the ecliptic, because the six signs bound a semicircle, whose centre *E* is the same as that of the horizon. But when a revolution has taken place and 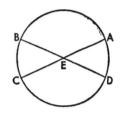 the beginning of Capricorn arises at *B*, then the setting of Cancer will be visible at *D*, and *BED* will be a straight line and a diameter of the ecliptic. But it has already been seen that the line *AEC* is a diameter of the same circle; therefore, at their common section, point *E* will be their centre. So in this way the horizon always bisects the ecliptic, which is a great circle of the sphere. But on a sphere, if a circle bisects one of the great circles, then the circle bisecting is a great circle. Therefore the horizon is a great circle; and its centre is the same as that of the ecliptic, as far as appearance goes; although nevertheless the line passing through the centre of the Earth and the line touching to the surface are necessarily different; but on account of their immensity in comparison with the Earth they are like parallel lines, which on account of the great distance between the termini appear to be one line, when the space contained between them [4ᵇ] is in no perceptible ratio to their length, as has been shown in optics.

From this argument it is certainly clear enough that the heavens are immense in comparison with the Earth and present the aspect of an infinite magnitude, and that in the judgment of sense-perception the Earth is to the heavens as a point to a body and as a finite to an infinite magnitude. But we see that nothing more than that has been shown, and it does not follow that the Earth must rest at the centre of the world. And we should be even more surprised if such a vast world should wheel completely around during the space of twenty-four hours rather than that its least part, the Earth, should. For saying that the centre is immovable and that those things which are closest to the centre are moved least does not argue that the Earth rests at the centre of the world. That is no different from saying that the heavens revolve but the poles are at rest and those things which are closest to the poles are moved least. In this way Cynosura (the pole star) is seen to move much more slowly than Aquila or Canicula because, being very near to the pole, it describes a smaller circle, since they are all on a single sphere, the movement of which stops at its axis and which does not allow any of its parts to have movements which are equal to one another. And nevertheless the revolution of the whole brings them round in equal times but not over equal spaces.

The argument which maintains that the Earth, as a part of the celestial sphere and as sharing in the same form and movement, moves very little because very near to its centre advances to the following position: therefore the Earth will move, as being a body and not a centre, and will describe in the same time arcs similar to, but smaller than, the arcs of the celestical circle. It is clearer than daylight how false that is; for there would necessarily always be noon at one place and midnight at another, and so the daily risings and settings could not take place, since the movement of the whole and the part would be one and inseparable.

But the ratio between things separated by diversity of nature is so entirely different that those which describe a smaller circle turn more quickly than those which describe a greater circle. In this way Saturn, the highest of the wandering stars, completes its revolution in thirty years, and the moon which is without doubt the closest to the Earth completes its circuit in a month, and finally the Earth itself will be considered to complete a circular movement in the space of a day and a night. So this same problem concerning the daily revolution comes up again. And also the question about the place of the Earth becomes even less certain on account of what was just said. For that demonstration proves nothing except that the heavens are of an indefinite magnitude with respect to the Earth. But it is not at all clear how far this immensity stretches out. On the contrary, since the minimal and indivisible corpuscles, which are called atoms, are not perceptible to sense, they do not, when taken in twos or in some small number, constitute a visible body; but they can be taken in such a large quantity that there will at last be enough to form a visible magnitude. So it is as regards the place of the earth; for although it is not at the centre of the world, nevertheless the distance is as nothing, particularly in comparison with the sphere of the fixed stars.

7. WHY THE ANCIENTS THOUGHT THE EARTH WAS AT REST AT THE MIDDLE OF THE WORLD AS ITS CENTRE

[5ª] Wherefore for other reasons the ancient philosophers have tried to affirm that the Earth is at rest at the middle of the world, and as principal cause they put forward heaviness and lightness. For Earth is the heaviest element; and all things of any weight are borne towards it and strive to move towards the very centre of it.

For since the Earth is a globe towards which from every direction heavy things by their own nature are borne at right angles to its surface, the heavy things would fall on one another at the centre if they were not held back at the surface; since a straight line making right angles with a plane surface where it touches a sphere leads to the centre. And those things which are borne toward the centre seem to follow along in order to be at rest at the centre. All the more then will the Earth be at rest at the centre; and, as being the receptacle for falling bodies, it will remain immovable because of its weight.

They strive similarly to prove this by reason of movement and its nature. For Aristotle says that the movement of a body which is one and simple is simple, and the simple movements are the rectilinear and the circular. And of rectilinear movements, one is upward, and the other is downward. As a consequence, every simple movement is either toward the centre, *i.e.*, downward, or away from the centre, *i.e.*, upward, or around the centre, *i.e.*, circular. Now it belongs to earth and water, which are considered heavy, to be borne downward, *i.e.*, to seek the centre: for air and fire, which are endowed with lightness, move upward, *i.e.*, away from the centre. It seems fitting to grant rectilinear movement to these four elements and to give the heavenly bodies a circular movement around the centre. So Aristotle. Therefore, said Ptolemy of Alexandria, if the Earth moved, even if only by its daily rotation, the contrary of what was said above would necessarily take place. For this movement which would traverse the total circuit of the Earth in twenty-four hours would necessarily be very headlong and of an unsurpassable velocity. Now things which are suddenly and violently whirled around are seen to be utterly unfitted for reuniting, and the more unified are seen to become dispersed, unless some constant force constrains them to stick together. And a long time ago, he says, the scattered Earth would have passed beyond the heavens, as is certainly ridiculous; [5b] and *a fortiori* so would all the living creatures and all the other separate masses which could by no means remain unshaken. Moreover, freely falling bodies would not arrive at the places appointed them, and certainly not along the perpendicular line which they assume so quickly. And we would see clouds and other things floating in the air always borne toward the west.

8. Answer to the Aforesaid Reasons and Their Inadequacy

For these and similar reasons they say that the Earth remains at rest at the middle of the world and that there is no doubt about this. But if someone opines that the Earth revolves, he will also say that the movement is natural and not violent. Now things which are according to nature produce effects contrary to those which are violent. For things to which force or violence is applied get broken up and are unable to subsist for a long time. But things which are caused by nature are in a right condition and are kept in their best organization. Therefore Ptolemy had no reason to fear that the Earth and all things on the Earth would be scattered in a revolution caused by the efficacy of nature, which is greatly different from that of art or from that which can result from the genius of man. But why didn't he feel anxiety about the world instead, whose movement must necessarily be of greater velocity, the greater the heavens are than the Earth? Or have the heavens become so immense, because an unspeakably vehement motion has pulled them away from the centre, and because the heavens would fall if they came to rest anywhere else?

Surely if this reasoning were tenable, the magnitude of the heavens would extend infinitely. For the farther the movement is borne upward by the vehement force, the faster will the movement be, on account of the ever-increasing circumference which must be traversed every twenty-four hours: and conversely, the immensity of the sky would increase with the increase in movement. In this way, the velocity would make the magnitude increase infinitely, and the magnitude the velocity. And in accordance with the axiom of physics that *that which is infinite cannot be traversed or moved in any way*, then the heavens will necessarily come to rest.

But they say that beyond the heavens there isn't any body or place or void or any-thing at all; and accordingly it is not possible for the heavens to move outward; in that case it is rather surprising that something can be held together by nothing. But if the heavens were infinite and were finite only with respect to a hollow space inside, then it will be said with more truth that there is nothing outside the heavens, since anything [6ᵃ] which occupied any space would be in them; but the heavens will remain immo-bile. For movement is the most powerful reason wherewith they try to conclude that the universe is finite.

But let us leave to the philosophers of nature the dispute as to whether the world is finite or infinite, and let us hold as certain that the Earth is held together between its two poles and terminates in a spherical surface. Why therefore should we hesitate any longer to grant to it the movement which accords naturally with its form, rather than put the whole world in a commotion—the world whose limits we do not and can-not know? And why not admit that the appearance of daily revolution belongs to the heavens but the reality belongs to the Earth? And things are as when Aeneas said in Virgil: "We sail out of the harbor, and the land and the cities move away." As a matter of fact, when a ship floats on over a tranquil sea, all the things outside seem to the voy-agers to be moving in a movement which is the image of their own, and they think on the contrary that they themselves and all the things with them are at rest. So it can eas-ily happen in the case of the movement of the Earth that the whole world should be believed to be moving in a circle. Then what would we say about the clouds and the other things floating in the air or falling or rising up, except that not only the Earth and the watery element with which it is conjoined are moved in this way but also no small part of the air and whatever other things have a similar kinship with the Earth? whether because the neighbouring air, which is mixed with earthly and watery matter, obeys the same nature as the Earth or because the movement of the air is an acquired one, in which it participates without resistance on account of the contiguity and perpetual rotation of the Earth. Conversely, it is no less astonishing for them to say that the highest region of the air follows the celestial movement, as is shown by those stars which appear suddenly—I mean those called "comets" or "bearded stars" by the Greeks. For that place is assigned for their generation; and like all the other stars they rise and set. We can say that that part of the air is deprived of terrestrial motion on account of its great distance from the Earth. Hence the air which is nearest to the Earth and the things floating in it will appear tranquil, unless they are driven to and fro by the wind or some other force, as happens. For how is the wind in the air different from a current in the sea?

But we must confess that in comparison with the world the movement of falling and of rising bodies is twofold and is in general compounded of the rectilinear and the circular. As regards things which move downward on account of their weight [6^b] because they have very much earth in them, doubtless their parts possess the same nature as the whole, and it is for the same reason that fiery bodies are drawn upward with force. For even this earthly fire feeds principally on earthly matter; and they define flame as glowing smoke. Now it is a property of fire to make that which it invades to expand; and it does this with such force that it can be stopped by no means or contrivance from breaking prison and completing its job. Now expanding movement moves away from the centre to the circumference; and so if some part of the Earth caught on fire, it would be borne away from the centre and upward. Accordingly, as they say, a simple body possesses a simple movement—this is first verified in the case of circular movement—as long as the simple body remain in its unity in its natural place. In this place, in fact, its movement is none other than the circular, which remains entirely in itself, as though at rest. Rectilinear movement, however, is added to those bodies which journey away from their natural place or are shoved out of it or are outside it somehow. But nothing is more repugnant to the order of the whole and to the form of the world than for anything to be outside of its place. Therefore rectilinear movement belongs only to bodies which are not in the right condition and are not perfectly conformed to their nature—when they are separated from their whole and abandon its unity. Furthermore, bodies which are moved upward or downward do not possess a simple, uniform, and regular movement—even without taking into account circular movement. For they cannot be in equilibrium with their lightness or their force of weight. And those which fall downward possess a slow movement at the beginning but increase their velocity as they fall. And conversely we note that this earthly fire—

and we have experience of no other—when carried high up immediately dies down, as if through the acknowledged agency of the violence of earthly matter.

Now circular movement always goes on regularly, for it has an unfailing cause; but (in rectilinear movement) the acceleration stops, because, when the bodies have reached their own place, they are no longer heavy or light, and so the movement ends. Therefore, since circular movement belongs to wholes and rectilinear to parts, we can say that the circular movement stands with the rectilinear, as does animal with sick. And the fact that Aristotle divided simple movement into three genera: away from the centre, toward the centre, and around the centre, will be considered merely as an act of reason, just as we distinguish between line, point, and surface, though none of them can subsist without the others or [7ª] without body.

In addition, there is the fact that the state of immobility is regarded as more noble and godlike than that of change and instability, which for that reason should belong to the Earth rather than to the world. I add that it seems rather absurd to ascribe movement to the container or to that which provides the place and not rather to that which is contained and has a place, *i.e.*, the Earth. And lastly, since it is clear that the wandering stars are sometimes nearer and sometimes farther away from the Earth, then the movement of one and the same body around the centre—and they mean the centre of the Earth—will be both away from the centre and toward the centre. Therefore it is necessary that movement around the centre should be taken more generally; and it should be enough if each movement is in accord with its own centre. You see therefore that for all these reasons it is more probably that the Earth moves than that it is at rest—especially in the case of the daily revolution, as it is the Earth's very own. And I think that is enough as regards the first part of the question.

9. Whether Many Movements Can Be Attributed to the Earth, and Concerning the Centre of the World

Therefore, since nothing hinders the mobility of the Earth, I think we should now see whether more than one movement belongs to it, so that it can be regarded as one of the wandering stars. For the apparent irregular movement of the planets and their variable distances from the Earth—which cannot be understood as occurring in circles homocentric with the Earth—make it clear that the Earth is not the centre of their circular movements. Therefore, since there are many centres, it is not foolhardy to doubt whether the centre of gravity of the Earth rather than some other is the centre of the world. I myself think that gravity or heaviness is nothing except a certain natural appetency implanted in the parts by the divine providence of the universal Artisan, in order that they should unite with one another in their oneness and wholeness and come together in the form of a globe. It is believable that this affect is present in the sun, moon, and the other bright planets and that through its efficacy they remain in the spherical figure in which they are visible, though they nevertheless accomplish their circular movements in many different ways. Therefore if the Earth too possesses movements different from the one around its centre, then they will necessarily be movements which similarly appear on the outside in the many bodies; and we find the yearly revolution among these movements. For if the annual revolution were changed from being solar to being terrestrial, and immobility were granted to the sun, [7b] the risings and settings of the signs and of the fixed stars—whereby they become morning or evening stars—will appear in the same way; and it will be seen that the stoppings, retrogressions, and progressions of the wandering stars are not their own, but are a movement of the Earth and that they borrow the appearances of this movement. Lastly, the sun will be regarded as occupying the centre of the world. And the ratio of order in which these bodies succeed one another and the harmony of the whole world teaches us their truth, if only—as they say—we would look at the thing with both eyes.

10. On the Order of the Celestial Orbital Circles

I know of no one who doubts that the heavens of the fixed stars is the highest up of all visible things. We see that the ancient philosophers wished to take the order of the planets according to the magnitude of their revolutions, for the reason that among things which are moved with equal speed those which are the more distant seem to be borne along more slowly, as Euclid proves in his *Optics*. And so they think that the moon traverses its circle in the shortest period of time, because being next to the Earth, it revolves in the smallest circle. But they think that Saturn, which completes the longest circuit in the longest period of time, is the highest. Beneath Saturn, Jupiter. After Jupiter, Mars.

There are different opinions about Venus and Mercury, in that they do not have the full range of angular elongations from the sun that the others do.[1] Wherefore some place them above the sun, as Timaeus does in Plato; some, beneath the sun, as Ptolemy and a good many moderns. Alpetragius makes Venus higher than the sun and Mercury lower. Accordingly, as the followers of Plato suppose that all the planets—which are otherwise dark bodies—shine with light received from the sun, they think that if the planets were below the sun, they would on account of their slight distance from the sun be viewed as only half—or at any rate as only partly—spherical. For the light which they receive is reflected by them upward for the most part, *i.e.*, towards the sun, as we see in the case of the new moon or the old. Moreover, they say that necessarily the sun would sometimes be obscured through their interposition and that its light would be eclipsed in proportion to their magnitude; and as that has never appeared to take place, they think that these planets cannot by any means be below the sun.[1]

[1]The greatest angular elongation of Venus from the sun is approximately 45°; that of Mercury, approximately 24°; while Saturn, Jupiter, and Mars have the full range of possible angular elongation, *i.e.*, up to 180°.

On the contrary, those who place Venus and Mercury below the sun claim as a reason the amplitude of the space which they find between the sun and the moon. [8a] For they find that the greatest distance between the Earth and the moon, *i.e.*, $64^1/_6$ units, whereof the radius of the Earth is one, is contained almost 18 times in the least distance between the sun and the Earth. This distance is 1160 such units, and therefore the distance between the sun and the moon is 1096 such units. And then, in order for such a vast space not to remain empty, they find that the intervals between the perigees and apogees—according to which they reason out the thickness of the spheres[2]—add up to approximately the same sum: in such fashion that the apogee of the moon may be succeeded by the perigee of Mercury, that the apogee of Mercury may be followed by the perigee of Venus, and that finally the apogee of Venus may nearly touch the perigee of the sun. In fact they calculate that the interval between the perigee and the apogee of Mercury contains approximately $177^1/_2$ of the aforesaid units and that the remaining space is nearly filled by the 910 units of the interval between the perigee and apogee of Venus.[3] Therefore they do not admit that these planets have a certain opacity, like that of the moon; but that they shine either by their own proper light or because their entire bodies are impregnated with sunlight, and that accordingly they do not obscure the sun, because it is an extremely rare occurrence for them to be interposed between our sight and the sun, as they usually withdraw (from the sun) latitudinally. In addition, there is the fact that they are small bodies in comparison with the sun, since Venus even though larger than Mercury can cover scarcely one one-hundredth part of the sun, as al-Battani the Harranite maintains, who holds that the diameter of the sun is ten times greater, and therefore it would not be easy to see such a little speck in the midst of such beaming light. Averroes, however, in his paraphrase of Ptolemy records having seen something blackish, when he observed the conjunction of the sun and Mercury which he had computed. And so they judge that these two planets move below the solar circle.

[1]The transit of Venus across the face of the sun was first observed—by means of a telescope—in 1639.
[2]That is to say, the thickness of the sphere would measured by the ratio of the diameter of the epicycle to the diameter of the sphere, or, in the accompanying diagram, by the distance between the inmost and the outmost of the three homocentric circles.
[3]The succession of the orbital circles according to their perigees and apogees may be represented in the following diagram, which has been drawn to scale.

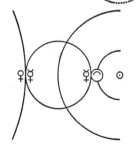

But how uncertain and shaky this reasoning is, is clear from the fact that though the shortest distance of the moon is 38 units whereof the radius of the Earth is one unit—according to Ptolemy, but more than 49 such units by a truer evaluation, as will be shown below—nevertheless we do not know that this great space contains anything except air, or if you prefer, what they call the fiery element.

Moreover, there is the fact that the diameter of the epicycle of Venus—by reason of which Venus has an angular digression of approximately 45° on either side of the sun—would have to be six times greater than the distance from the centre of the Earth to its perigee, as will be shown in the proper place.[1] Then what will they say is contained in all this space, which [8ᵇ] is so great as to take in the Earth, air, ether, moon and Mercury, and which moreover the vast epicycle of Venus would occupy if it revolved around an immobile Earth?

Furthermore, how unconvincing is Ptolemy's argument that the sun must occupy the middle position between those planets which have the full range of angular elongation from the sun and those which do not is clear from the fact that the moon's full range of angular elongation proves its falsity.

But what cause will those who place Venus below the sun, and Mercury next, or separate them in some other order—what cause will they allege why these planets do not also make longitudinal circuits separate and independent of the sun, like the other planets[2]—if indeed the ratio of speed or slowness does not falsify their order? Therefore it will be necessary either for the Earth not to be the centre to which the order of the planets and their orbital circles is referred, or for there to be no sure reason for their order and for it not to be apparent why the highest place is due to Saturn rather than to Jupiter or some other planet. Wherefore I judge that what Martianus Capella—who wrote the *Encyclopedia*—and some other Latins took to be the case is by no means to be despised. For they hold that Venus and Mercury circle around the sun as a centre; and they hold that for this reason Venus and Mercury do not have any farther elongation from the sun than the convexity of their orbital circles permits; for they do not make a circle around the earth as do the others, but have perigee and apogee interchangeable (in the sphere of the fixed stars). Now what do they mean except that the centre of their spheres is around the sun? Thus the orbital circle of Mercury will be enclosed within the orbital circle of Venus—which would

---

[1]According to Ptolemy, the ratio of the radius of Venus' epicycle to the radius of its eccentric is between 2 to 3 and 3 to 4, or approximately 43¹/₆ to 60. Now since at perigee the epicycle subtracts from the mean distance, or radius of the eccentric circle, that which at apogee it adds to the mean distance, the ratio of Venus' distance at perigee to its distance at apogee is approximately 1 to 6. That is to say, in the passage from apogee to perigee, the ratio of increase, in the apparent magnitude of the planet should be approximately 36 to 1, as the apparent magnitude varies inversely in the ratio of the square of the distance. But no such increase in the magnitude of the planet is apparent. This opposition between an appearance and the consequences of an hypothesis made to save another appearance is still present within Copernicus' own scheme.

[2]Ptolemy makes the centres of the epicycles of Venus and Mercury travel around the Earth longitudinally at the same rate as the mean sun, and in such fashion that the mean sun is always on the straight line extending from the centre of the Earth through the centres of their epicycles, while the centres of the epicycles of the upper planets may be at any angular distance from the mean sun.

have to be more than twice as large—and will find adequate room for itself within that amplitude.[1] Therefore if anyone should take this as an occasion to refer Saturn, Jupiter, and Mars also to this same centre, provided he understands the magnitude of those orbital circles to be such as to comprehend and encircle the Earth remaining within them, he would not be in error, as the table of ratios of their movements makes clear.[2] For it is manifest that the planets are always nearer the Earth at the time of their evening rising, *i.e.*, when they are opposite to the sun and the Earth is in the middle between them and the sun. But they are farthest away from the Earth at the time of their evening setting, *i.e.*, when they are occulted in the neighbourhood of the sun, namely, when we have the sun between them and the Earth. All that shows clearly enough that their centre is more directly related to the sun and is the

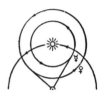

[1]As in the following diagram which has been drawn to scale.

[2]Take the case of Mars. In Ptolemy, the ratio of its epicycle to its eccentric is $39^1/_2$ to 60, or approximately 2 to 3. Mars has 37 cycles of anomaly, or movement on the epicycle, and 42 cycles of longitude, or movement of the epicycle on the eccentric, in 79 solar years; or for the sake of easiness let us say that the ratio of the sun's movement to either of the planets' two movements is 2 to 1. Copernicus is here suggesting that if the centre of the planet's movement is placed around the moving sun, then the Ptolemaic cycles of anomaly will represent the number of times the sun has overtaken the planet in longitude: thus the 37 cycles of anomaly plus the 42 cycles of longitude add up to the 79 solar revolutions. That is to say, the sun will now be traveling around the Earth on a circle which has the same relative magnitude as the Martian epicycle in Ptolemy and bears an epicycle having the same relative magnitude as Ptolemy's Martian eccentric circle, on which epicycle Mars travels in the opposite direction at half the speed of the sun. Under both hypotheses the appearances from the Earth will be the same, as can be seen in the following diagrams.

For according to the Ptolemaic hypothesis, let the Earth be at the center of the approximately homocentric circles of the sun, Maps, and the ecliptic. Let the radius of the planet's epicycle be to the radius of the planet's eccentric as 2 to 3. Now, first, let the sun be viewed at the beginning of Leo, and let the planet at the perigee of its epicycle be viewed at the beginning of Aquarius, in opposition to the sun. Next, let the sun move 240° eastwards, to the beginning of Aries; and during the same interval let the epicycle move 120° eastwards, to the beginning of Gemini, and the planet 120° eastwards on the epicycle. Now the planet will be found to appear in Taurus, about 36° west of the sun.

But if according to the semi-Copernican hypothesis, the sun is made to revolve around the Earth on a circle having the same relative magnitude as Mars' Ptolemaic epicycle, while Mars is placed on an epicycle which has the same relative magnitude as its Ptolemaic eccentric and has its centre at the sun; and if the apparent positions of Mars and the sun are first the same as before, and the sun moves 240° eastwards, bearing along the deferent of Mars, while Mars moves 120° westwards on its epicycle; then Mars will once more be found to appear in Taurus, approximately 36° west of the sun.

PTOLEMAIC HYPOTHESIS

Movement of Sun=240°
Movement of Eccentric=120°
Movement of Epicycle=120°

SEMI-COPERNICAN HYPOTHESIS

Movement of Sun=240°
Movement of Mars=120°

same as that to which Venus and Mercury refer their revolutions.[1] But as they all have one common centre, it is necessary that the space left between the convex orbital circle of Venus and the concave orbital circle of Mars should be viewed as an orbital circle [9ª] or sphere homocentric with them in respect to both surfaces, and that it should receive the Earth and its satellite the moon and whatever is contained beneath the lunar globe. For we can by no means separate the moon from the Earth, as the moon is incontestably very near to the Earth—especially since we find in this expanse a place for the moon which is proper enough and sufficiently large. Therefore we are not ashamed to maintain that this totality—which the moon embraces—and the centre of the Earth too traverse that great orbital circle among the other wandering stars in an annual revolution around the sun; and that the centre of the world is around the sun. I also say that the sun remains forever immobile and that whatever apparent movement belongs to it can be verified of the mobility of the Earth; that the magnitude of the world is such that, although the distance from the sun to the Earth in relation to whatsoever planetary sphere you please possesses magnitude which is sufficiently manifest in proportion to these dimensions, this distance, as compared with the sphere of the fixed stars, is imperceptible. I find it much more easy to grant that than to unhinge the understanding by an almost infinite multitude of spheres—as those who keep the earth at the centre of the world are forced to do. But we should rather follow the wisdom of nature, which, as it takes very great care not to have produced anything superfluous or useless, often prefers to endow one thing with many effects. And though all these things are difficult, almost inconceivable, and quite contrary to the opinion of the multitude, nevertheless in what follows we will with God's help make them clearer than day—at least for those who are not ignorant of the art of mathematics.

Therefore if the first law is still safe—for no one will bring forward a better one than that the magnitude of the orbital circles should be measured by the magnitude of time—then the order of the spheres will follow in this way—beginning with the highest: the first and highest of all is the sphere of the fixed stars, which comprehends itself and all things, and is accordingly immovable. In fact it is the place of the universe, *i.e.*, it is that to which the movement and position of all the other stars are referred. For in the deduction of terrestrial movement, we will however give the cause why there are appearances such as to make people believe that even the sphere of the fixed stars somehow moves. Saturn, the first of the wandering stars follows; it completes its circuit in 30 years. After it comes Jupiter moving in a 12-year period

Conjunction   Opposition
ACCORDING TO PTOLEMY

[1]Copernicus is asking what reason there is why the planets are always found to be at their apogees at the time of conjunction with the sun, and at their perigees at the time of opposition, since according to the Ptolemaic scheme the reverse is also possible—as is evident from the accompanying diagram.
But if the sun and not the Earth is the centre of the planet's movements, the reason is obvious.

Conjunction   Opposition
ACCORDING TO COPERNICUS

of revolution. Then Mars, which completes a revolution every 2 years. The place fourth in order is occupied by the annual revolution [9ᵇ] in which we said the Earth together with the orbital circle of the moon as an epicycle is comprehended. In the fifth place, Venus, which completes its revolution in 7¹/₂ months. The sixth and final place is occupied by Mercury, which completes its revolution in a period of 88 days.[1] In the center of all rests the sun. For who would place this lamp of a very beautiful temple in another or better place than this wherefrom it can illuminate everything at the same time? As a matter of fact, not unhappily do some call it the lantern; others, the mind and still others, the pilot of the world. Trismegistus calls it a "visible god"; Sophocles' Electra, "that

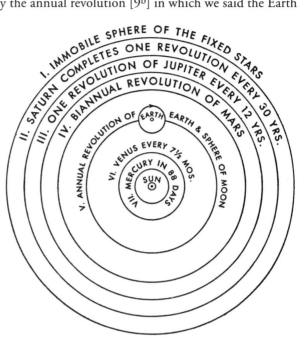

[1]In order to see how Copernicus, derived the length of his periods of revolution, consider the following Ptolemaic ratios for the lower planets:

	Cycles of anomaly	Cycles of longitude	Solar years
Mercury	145	46 +	46 +
Venus	5	8 —	8 —

It is noteworthy that the number of cycles of longitude in one year is equal to the number of solar cycles. Moreover, the two planets have a limited angular elongation from the sun. In order to explain these two peculiar appearances Copernicus sets the Earth in motion on the circumference of a circle which encloses the orbits of Venus and Mercury, with the sun at the centre of all three orbits. Thus the planet's cycles of anomaly in so many years become the number of times the planet has overtaken the Earth, as they revolve around the sun. That is to say, in so many solar years the planet will have traveled around the sun a number of times which is equal to the sum of its cycles of anomaly and its cycles in longitude. Thus, for example, Venus travels around the sun approximately 13 times in 8 solar years; hence its period of revolution is approximately 7¹/₂ months; and similarly, that of Mercury is approximately 88 days—although for some obscure reason Copernicus actually writes down 9 months for Venus (*nono mense reducitur*) and 80 days for Mercury (*octaginta dierum spatio circumcurrens*).

The reader may intuit from the following diagrams the equipollence, with respect to the appearances, of the Ptolemaic and the Copernican explanations of the movement of Venus.

Now, on Ptolemy's hypothesis, let the Earth be placed at the centre of the ecliptic, the solar circle, and the orbital circle of Venus, which carries the planetary epicycle. The radius of the epicycle is to that of the orbital circle approximately as 3 is to 4. First let the sun be situated at the middle of Scorpio, and let Venus be in conjunction with the sun and at the perigee of its epicycle. Next let the sun move 180° eastwards to the middle of Taurus, and similarly the centre of the epicycle; during this same interval the planet will move 112¹/₂° eastwards on its epicycle and will be found to appear in the middle of Aries approximately, or 30° west of the sun.

But according to the Copernican hypothesis, let us place the sun at the centre of the orbital circles of Venus and the Earth, which preserve the relative magnitudes of the Ptolemaic epicycle and orbital circle of Venus, but let us keep the Earth at the centre of the ecliptic, as far as appearances go, since the distance between the Earth and the sun is imperceptible in comparison with the magnitude of the sphere of the fixed stars. Now if the Earth is placed in the middle of Taurus, as viewed from the sun, and the planet at its perigee between the Earth and the sun, in such fashion that Venus and the sun would appear in the middle of Scorpio, while Venus moves eastwards 292¹/₂°, then the sun will be found to appear in the middle of Taurus, and the planet itself in middle of Aries or 30° west of the sun.

But let us turn to the three upper planets.

	Cycles of anomaly	Cycles of longitude	Solar years
Mars	37	42 +	79
Jupiter	65	6 —	71 —
Saturn	57	2 +	59 —

It is here noteworthy that according to the Ptolemaic hypothesis the sum of the revolutions of the eccentric circle and the revolutions in anomaly is equal to the number of solar cycles; and also that, the conjunctions with the sun take place at the planet's apogee, and the oppositions at its perigee.

(continued p. 26)

which gazes upon all things." And so the sun, as if resting on a kingly throne, governs the family of stars which wheel around. Moreover, the Earth is by no means cheated of the services of the moon; but, as Aristotle says in the *De Animalibus*, the earth has the closest kinship with the moon. The Earth moreover is fertilized by the sun and conceives offspring every year.

Therefore in this ordering we find [10ª] that the world has a wonderful commensurability and that there is a sure bond of harmony for the movement and magnitude of the orbital circles such as cannot be found in any other way.[1] For now the careful observer can note why progression and retrogradation appear greater in Jupiter than in

But according to Copernicus the Ptolemaic cycles of anomaly will now represent the number of times the Earth has overtaken the planet; and the period of revolution in longitude will stay the alone. Thus, for example, Saturn will have two revolutions in longitude in 59 years, or one revolution around the sun in about 30 years. The planet will be revolving directly on its eccentric circle instead of on its Ptolemaic epicycle, and the Earth will now be revolving on an inner circle which has the same relative magnitude as the former epicycle. The two hypotheses, of course, are equipollent here too, with respect to appearances.

In other words, in constructing a theory to account for four coincidences which were left unexplained by Ptolemy, namely, (1) the equality between the number of cycles in longitude and the solar cycles, in the two lower planets; (2) the equality between the solar cycles and the sum of the cycles of anomaly and longitude, in the upper planets; (3) the limited angular digressions of Mercury and Venus away from the sun; and (4) the apogeal conjunctions and perigeal oppositions of Saturn, Jupiter, and Mars; Copernicus has telescoped the eccentric circle of Venus and that of Mercury into one circle carrying the Earth; and he has furthermore collapsed the three epicycles of Saturn, Jupiter, and Mars into this same one circle. That is to say, one circle is now doing the work of five.

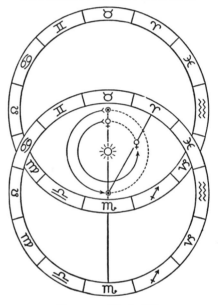

COPERNICAN HYPOTHESIS

Movement of Earth=180°
Movement of Venus=292½°

[1]Let us recall the Ptolemaic ratios between the radius of the epicycle and that of the eccentric circle, and also the eccentricity.

	Epicycle	Eccentric	Eccentricity
Mercury	22^1/$_2$	60	3
Venus	43^1/$_6$	60	1^1/$_4$
Mars	39^1/$_2$	60	6
Jupiter	11^1/$_2$	60	2^2/$_5$
Saturn	6^1/$_2$	60	3^1/$_4$

By the Ptolemaic scheme it is impossible to compute the magnitudes of the eccentric circles themselves relative to one another, as there is no common measure. But now that the eccentric circles of Mercury and Venus and the epicycles of Mars, Jupiter, and Saturn have all been reduced to the orbital circle of the Earth, it is easy to calculate the relative magnitudes of the orbital circles—heretofore the epicycles of the lower planets and the eccentric circles of the upper—since, by reason of the necessary commensurability between epicycle and eccentric, they are all commensurable with the orbital circle of the Earth. Thus, for example, if we take the distance from the Earth to the sun as 1, the planets will observe the following approximate distances from the sun.

Mercury	^{1}/$_3$	Earth	1	Jupiter	5
Venus	^{3}/$_4$	Mars	1^1/$_2$	Saturn	9

Saturn and smaller than in Mars; and in turn greater in Venus than in Mercury.[1] And why these reciprocal events appear more often in Saturn than in Jupiter, and even less often in Mars and Venus than in Mercury.[2] In addition, why when Saturn, Jupiter, and Mars are in opposition (to the mean position of the sun) they are nearer to the Earth than at the time of their occultation and their reappearance. And especially why at the times when Mars is in opposition to the sun, it seems to equal Jupiter in magnitude and to be distinguished from Jupiter only by a reddish color, but when discovered through careful observation by means of a sextant is found with difficulty among the stars of second magnitude?[3] All these things proceed from the same cause, which resides in the movement of the Earth.

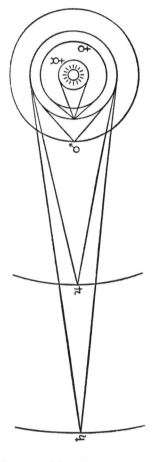

But that there are no such appearances among the fixed stars argues that they are at an immense height away, which makes the circle of annual movement or its image disappear from before our eyes since every visible thing has a certain distance beyond which it is no longer seen, as is shown in optics. For the brilliance of their lights shows that there is a very great distance between Saturn the highest of the planets and the sphere of the fixed stars. It is by this mark in particular that they are distinguished from the planets, as it is proper to have the greatest difference between the moved and the unmoved. How exceedingly fine is the godlike work of the Best and Greatest Artist!

11. A DEMONSTRATION OF THE THREEFOLD MOVEMENT OF THE EARTH

Therefore since so much and such great testimony on the part of the planets is consonant with the mobility of the Earth, we shall now give a summary of its movement,

[1] In the three upper planets, the angles which measure the apparent progression and retrogradation have as their vertex the centre of the planet and as their sides the tangents drawn to the orbital circle of the Earth. In the two lower planets, however, the vertex of the angle is at the centre of the Earth and the sides are the tangents drawn to the orbital circle of the planet. It is easy to see that, on account of the relative magnitudes of the orbital circles, the arcs of progression and retrogradation will appear smaller in Saturn than in Jupiter, and smaller in Jupiter than in Mars, and greater in Venus than in Mercury.

[2] The interchanges of progression and retrogradation are proportional to the number of times the Earth overtakes the outer planets and the inner planets overtake the Earth. Now the Earth overtakes Saturn more often than Jupiter, Jupiter more often than Mars, Mars more often than overtaken by Venus, and overtaken less often by Venus than by Mercury. Hence the frequency of progression and retrogradation is in that order.

[3] According to the Ptolemaic scheme, it can be inferred only from the changes in magnitude of the planet Mars what its relative distances from the Earth are at perigee and apogee. But according to the Copernican scheme, it follows from the relative distances of the planet at perigee and at apogee—which are as 1 to 5—that the apparent diameter of the planet should vary inversely in that ratio—assuming that the planet could be seen when in conjunction with the sun.

insofar as the appearances can be shown forth by its movement as by an hypothesis. We must allow a threefold movement altogether.

The first—which we said the Greeks called νυχθημἐρινος—is the proper circuit of day and night, which goes around the axis of the earth from west to east—as the world is held to move in the opposite direction—and describes the equator or the equinoctial circle—which some, imitating the Greek expression [10[b]] ἰσηἐρινος call the equidial.

The second is the annual movement of the centre, which describes the circle of the (zodiacal) signs around the sun similarly from west to east, *i.e.*, towards the signs which follow (from Aries to Taurus) and moves along between Venus and Mars, as we said, together with the bodies accompanying it. So it happens that the sun itself seems to traverse the ecliptic with a similar movement. In this way, for example, when the centre of the Earth is traversing Capricorn, the sun seems to be crossing Cancer; and when Aquarius, Leo, and so on, as we were saying.

It has to be understood that the equator and the axis of the Earth have a variable inclination with the circle and the plane of the ecliptic. For if they remained fixed and only followed the movement of the centre simply, no inequality of days and nights would be apparent, but it would always be the summer solstice or the winter solstice or the equinox, or summer or winter, or some other season of the year always remaining the same. There follows then the third movement, which is the declination: it is also an annual revolution but one towards the signs which precede (from Aries to Pisces), or westwards, *i.e.*, turning back counter to the movement of the centre; and as a consequence of these two movements which are nearly equal to one another but in opposite directions, it follows that the axis of the Earth and the greatest of the parallel circles on it, the equator, always look towards approximately the same quarter of the world, just as if they remained immobile. The sun in the meanwhile is seen to move along the oblique ecliptic with that movement with which the centre of the earth moves, just as if the centre of the earth were the centre of the world—provided you

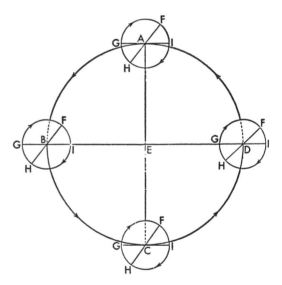

remember that the distance between the sun and the earth in comparison with the sphere of the fixed stars is imperceptible to us.

Since these things are such that they need to be presented to sight rather than merely to be talked about, let us draw the circle *ABCD*, which will represent the annual circuit of the centre of the earth in the plane of the ecliptic, and let *E* be the sun around its centre. I will cut this circle into four equal parts by means of the diameters *AEC* and *BED*. Let the point *A* be the beginning of Cancer; *B* of Libra; *E* of Capricorn; and *D* of Aries. Now let us put the centre of the earth first at *A*, around which we shall describe the terrestrial equator *FGHI*, but not in the same plane (as the ecliptic) except that the diameter *GAI* is the common section of the circles, *i.e.*, of the equator and the ecliptic. Also let the diameter *FAH* be drawn at right angles to *GAI*; and let *F* be the limit of the greatest southward declination (of the equator), and *H* of the northward declination. With this set-up, the Earth-dweller will see the sun—which is at the centre *E*—at the point of the winter solstice in Capricorn—[11ª] which is caused by the greatest northward declination at *H* being turned toward the sun; since the inclination of the equator with respect to line *AE* describes by means of the daily revolution the winter tropic, which is parallel to the equator at the distance comprehended by the angle of inclination *EAH*. Now let the centre of the Earth proceed from west to east; and let *F*, the limit of greatest declination, have just as great a movement from east to west, until at *B* both of them have traversed quadrants of circles. Meanwhile, on account of the equality of the revolutions, angle *EAI* will always remain equal to angle *AEB*; the diameters will always stay parallel to one another—*FAH* to *FBH* and *GAI* to *GBI*; and the equator will remain parallel to the equator. And by reason of the cause spoken of many times already, these lines will appear in the immensity of the sky as the same. Therefore from the point *B* the beginning of Libra, *E* will appear to be in Aries, and the common section of the two circles (of the ecliptic and the equator) will fall upon line *GBIE*, in respect to which the daily revolution has no declination; but every declination will be on one side or the other of this line. And so the sun will be soon in the spring equinox. Let the centre of the Earth advance under the same conditions; and when it has completed [11ᵇ] a semicircle at *C*, the sun will appear to be entering Cancer. But since *F* the southward declination of the equator is now turned toward the sun, the result is that the sun is seen in the north, traversing the summer tropic in accordance with angle of inclination *ECF*. Again, when *F* moves on through the third quadrant of the circle, the common section *GI* will fall on line *ED*, whence the sun, seen in Libra, will appear to have reached the autumn equinox. But then as, in the same progressive movement, *HF* gradually turns in the direction of the sun, it will make the situation at the beginning return, which was our point of departure.

In another way: Again in the underlying plane let *AEC* be both the diameter (of the ecliptic) and its common section with the circle perpendicular to its plane. In this circle let *DGFI*, the meridian passing through the poles of the Earth be described around *A* and *C*, in turn, *i.e.*, in Cancer and in Capricorn. And let the axis of the Earth be *DF*, the north pole *D*, the south pole *F*, and *GI* the diameter of the equator. Therefore when *F* is turned in the direction of the sun, which is at E, and the incli-

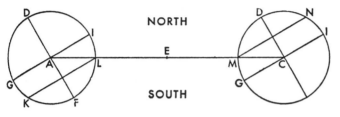

nation of the equator is northward in proportion to angle *IAE*, then the movement around the axis will describe—with the diameter *KL* and at the distance *LI*—parallel to the equator the southern circle, which appears with respect to the sun as the tropic of Capricorn. Or—to speak more correctly—this movement around the axis describes, in the direction of *AE*, a conic surface, which has the centre of the earth as its vertex and a circle parallel to the equator as its base.[1] Moreover in the opposite sign, *C*, the same things take place but conversely. Therefore it is clear how the two mutually opposing movements, *i.e.*, that of the centre and that of the inclination, force the axis of the Earth to remain balanced in the same way and to keep a similar position, and how they make all things appear as if they were movements of the sun.

Now we said that the yearly revolutions of the centre and of the declination were approximately equal, because if they were exactly so, then the points of equinox and solstice and the obliquity of the ecliptic in relation to the sphere of the fixed stars could not change at all. But as the difference is very slight, [12ª] it is not revealed except as it increases with time: as a matter of fact, from the time of Ptolemy to ours there has been a precession of the equinoxes and solstices of about 21°. For that reason some have believed that the sphere of the fixed stars was moving, and so they choose a ninth higher sphere. And when that was not enough, the moderns added a tenth, but without attaining the end which we hope we shall attain by means of the movement of the Earth. We shall use this movement as a principle and a hypothesis in demonstrating other things.

12. ON THE STRAIGHT LINES IN A CIRCLE

Because the proofs which we shall use in almost the entire work deal with straight lines and arcs, with plane and spherical triangles, and because Euclid's *Elements*, although they clear up much of this, do not have what is here most required, namely, how to find the sides from the angles and the angles from the sides,

---

[1]Or, in other words, the axis of the terrestrial equator describes around the axis of the terrestrial ecliptic a double conic surface having its vertices at the centre of the Earth, in a period of revolution equal approximately to that of the Earth's centre.

since the angle does not measure the subtending straight line—just as the line does not measure the angle—but the arc does, there has accordingly been found a method whereby the lines subtending any arc may become known. By means of these lines, or chords, it is possible to determine the arc corresponding to the angle: and conversely by means of the arc to determine the straight line, or chord, which subtends the angle. So it does not seem irrelevant, if we treat of these lines, and also of the sides and angles of plane and spherical triangles—which Ptolemy discussed a few at a time here and there—in order that these questions may be answered here once and for all and that what we are going to teach may become clearer. Now, by the common agreement of mathematicians, we divide the circle into 360 degrees. Now the ancients employed a diameter of 120 parts. But in order to avoid the complication of minutes and seconds in the multiplication and division of the numbers attached to the lines, as the lines are usually incommensurable in length, and often in square too; some of their successors established a rational diameter of 1,200,000 parts or of 2,000,000 parts, or of some other rational quantity—from the time when Arabic numerals came into general use. This mathematical notation surpasses any other— Greek or Latin—[12ᵇ] in a certain singular ease of employment and readily accommodates itself to every class of computation. For that reason we too have taken a division of the diameter into 200,000 parts as sufficient to exclude any very noticeable error. For as regards things which are not related as number to number, it is enough to attain a close approximation. But we will unfold this in six theorems and a problem—following Ptolemy fairly closely.

FIRST THEOREM

The diameter of a circle being given, the sides of the triangle, tetragon, hexagon, and decagon, which the same circle circumscribes, are also given.

Half the diameter, or the radius, is equal to the side of the hexagon, (Euclid, IV, 15); the square on the side of the triangle is three times the square on the side of the

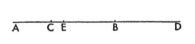

hexagon, (Euclid, XIII, 12); and the square on the side of the tetragon is twice the square on the side of the hexagon, Euclid as is shown in Euclid's *Elements* (IV, 9 and I, 47). Therefore the side of a hexagon is given in length as 100,000 parts, that of the tetragon as 141,422 parts, and that of the triangle as 173,205 parts.

Now, let *AB* be the side of the hexagon; and by Euclid, II, 11, or VI, 30, let it be cut in mean and extreme ratio at point *C*; and let *CB* be the greater segment to which its equal *BD* is added. Therefore the whole *ABD* will have been cut in extreme and mean ratio, and the lesser segment *BD* will be the side of the decagon inscribed in the circle, and *AB* will be the side of the inscribed hexagon, as is made clear by Euclid, XIII, 5 and 9.

But *BD* will be given in this way: let *AB* be bisected at *E*, and it will be clear from Euclid, XIII, 3 that

$$\text{sq. } EBD = 5 \text{ sq. } EB.$$

But

$$EB = 50,000.$$

Whence

$$5 \text{ sq. } EB \text{ is given.}$$

Hence

$$EBD = 111,803.$$

And

$$BD = EBD - EB = 111,803 - 50,000 = 61,803,$$

which is the side of the decagon sought.

Moreover the side of the pentagon, the square on which is equal to the sum of the squares on the side of the hexagon and on the side of the decagon (*Elements*, XIII, 10), is given as 117,557 parts.

Therefore the diameter of the circle being given, the sides of the triangle, tetragon, pentagon, hexagon and decagon, which may be inscribed in the same circle, have been given—as was to be shown.

PORISM

Furthermore, it is clear that when the chord subtending an arc has been given, that chord too can be found which subtends the rest [13ª] *of the semicircle.*

Since the angle in a semicircle is right, and in right triangles the square on the chord subtending the right angle, *i.e.*, the square on the diameter, is equal to the sum of the squares on the sides comprehending the right angle; therefore—since the side of the decagon, which subtends 36° of the circumference, has been shown to have 61,803 parts whereof the diameter has 200,000 parts—the chord which subtends the remaining 144° of the semicircle has 190,211 parts.

And in the case of the side of the pentagon, which is equal to 117,557 parts of the diameter and subtends an arc of 72°, a straight line of 161,803 parts is given, and it subtends remaining 108° of the circle.

SECOND THEOREM

If a quadrilateral is inscribed in a circle, the rectangle comprehended by the diagonals is equal to the two rectangles which are comprehended by the two pairs of opposite sides.

For let the quadrilateral *ABCD* be inscribed in a circle; I say that the rectangle comprehended by the diagonals *AC* and *DB* is equal to those comprehended by *AB*, *CD* and by *AD*, *BC*.

For let us make

$$\text{angle } ABE = \text{angle } CBD.$$

Therefore by addition

$$\text{angle } ABD = \text{angle } EBC,$$

taking angle *EBD* as common to both. Moreover

$$\text{angle } ACB = \text{angle } BDA$$

because they stand on the same segment of the circle; and accordingly the two similar triangles *BCE* and *BDA* will have their sides proportional. Hence

$$BC : BD = EC : AD.$$

And

$$\text{rect. } EC,\ BD = \text{rect. } BC,\ AD.$$

But also the triangles *ABE* and *CBD* are similar, because

$$\text{angle } ABE = \text{angle } CBD.$$

And

$$\text{angle } BAC = \text{angle } BDC,$$

because they intercept the same arc of the circle.

So again,

$$AB : BD = AE : CD$$

And

$$\text{rect. } AB,\ DC = \text{rect. } AE,\ BD.$$

But it has already been made clear that

$$\text{rect. } AD,\ BC = \text{rect. } BD,\ EC.$$

Accordingly, taken as a whole,

$$\text{rect. } BD,\ AC = \text{rect. } AD,\ BC + \text{rect. } AB,\ CD,$$

as it was opportune to have shown.

THIRD THEOREM

Hence if straight lines subtending unequal arcs in a semicircle are given, the chord subtending the arc whereby the greater arc exceeds the smaller is also given.

[13ᵇ] In the semicircle *ABCD* with diameter *AD*, let the straight lines *AB* and *AC*

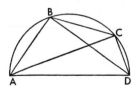

subtending unequal arcs be given. To us, who wish to discover the chord subtending *BC*, there are given by means of the aforesaid the chords *BD* and *CD* subtending the remaining arcs of the semicircle, and these chords bound the quadrilateral *ABCD* in the semicircle. The diagonals *AC* and *BD* have been given together with the three sides *AB*, *AD*, and *CD*. And, as has already been shown,

$$\text{rect. } AC, BD = \text{rect. } AB, CD + \text{rect. } AD, BC.$$

Therefore,

$$\text{rect. } AD, BC = \text{rect. } AC, BD - \text{rect. } AB, CD.$$

Accordingly, in so far as the division may be carried out,

$$(AC\text{-}BD - AB\text{-}CD) \div AD = BC,$$

which was sought.

Further when, for example, the sides of the pentagon and hexagon are given from the above, by this computation a line is given subtending 12°—which is the difference between the arcs—and it is equal to 20,905 parts of the diameter.

FOURTH THEOREM

Given a chord subtending any arc, the chord subtending half of the arc is also given.

Let us describe the circle *ABC*, whose diameter is *AC*, and let the arc *BC* be given together with the chord subtending it, and let the line *EF* from the centre *E* cut *BC* at right angles. Accordingly by Euclid, III, 3, it will bisect chord *BC* at *F*, and the arc at *D*. Let the chords subtending arcs *AB* and *BD* be drawn. Since the triangles *ABC* and *EFC* are right and also similar—for they have angle *ECF* in common; therefore, as

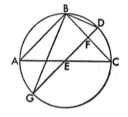

$$CF = {}^1/_2 \; BFC,$$

so

$$EF = {}^1/_2 \; AB.$$

But chord *AB* is given, for it subtends the remaining arc of the semicircle. Therefore *EF* is given; and so is line *DF* the remainder of the radius. Let the diameter *DEG* be completed, and let *BG* be joined. Therefore in triangle *BDG* line *BF* falls from the right angle at *B* perpendicular to the base. Accordingly,

$$\text{rect. } GD, DF = \text{sq. } BD.$$

Therefore *BD* is given in length, and it subtends half of the arc *BDC*.

And since a chord subtending 12° has already been given, the chord subtending 6° is given as 10,467 parts; that subtending 3°, as 5235 parts; that subtending $1^1/_2$°, as 2618 parts; and that subtending 45', as 1309 parts.

[14ᵃ] FIFTH THEOREM

Again, when chords are given subtending two arcs, the chord subtending the whole arc made up of them is also given.

Let there be given in the circle the two chords subtending the arcs *AB* and *BC*; I say that the chord subtending the whole arc *ABC* is also given.

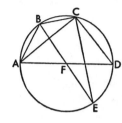

For let the diameters *AFD* and *BFE* be drawn, and also the chords *BD* and *CE*, which are given by means of the foregoing, on account of chords *AB* and *BC* being given; and

chord *DE* = chord *AB*.

The joining of *CD* completes the quadrilateral *BCDE*, whose diagonals *BD* and *CE* are given together with the three sides *BC*, *DE*, and *BE*; and the remaining side *CD* will be given by the second theorem; accordingly chord *CA* which subtends the remaining part of the semicircle will be given, and it subtends the whole arc *ABC* and is what was sought.

Furthermore, since so far there have been discovered chords which subtend 3°, $1^1/_2$°, and $^3/_4$°; by means of these intervals a table can be constructed with the most exact ratios. Nevertheless if we ascend through the degrees and add one arc to another arc either by halves or by some other mode, there is not unjustified doubt concerning the chords subtending those arcs, as the graphical ratios by which they can be shown are lacking to us. Nothing, however, prevents us from going on with that by some mode which is this side of error perceptible to sense and which is least unconsonant with the assumed number. This was what Ptolemy too sought as regards the chords subtending arcs of 1° or of $^1/_2$°; and he admonished us in the first place.

SIXTH THEOREM

The ratio of the arcs is greater than the ratio of the greater to the smaller of the chords.

Let there be in a circle two unequal successive arcs *AB* and *BC*, and let *BC* be the greater. I say that

arc *BC* : arc *AB* > chord *BC* : chord *AB*.

These chords comprehend angle *B*, and let that be bisected by line *BD*. And let *AC* be joined, which cuts *BD* at point *E*. Similarly let *AD* and *CD* be joined; then

AD = *CD*,

because they subtend equal arcs.

Accordingly, since in triangle *ABC*, the line which bisects the angle also cuts *AC* [14^b] at *E*, then

EC, segment of base : *AE* = *BC* : *AB* (Euclid, VI, 3)

and since

BC > *AB*,

then

EC > *EA*.

Let *DF* be erected perpendicular to *AC*; it will bisect *AC* at point *F*. And *F* must necessarily be found in the greater segment *EC*. And since in every triangle the greater angle is subtended by the greater side, in the triangle *DEF*

$$\text{side } DE > \text{side } DF,$$

and further,

$$AD > DE,$$

wheretofore the circumference described with *D* as center and *DE* as radius will cut *AD* and pass beyond *DF*. Therefore let it cut *AD* at *H*, and let it be extended in the straight line *DFI*. Since

$$\text{sect. } EDI > \text{trgl. } EDF,$$

while

$$\text{trgl. } DEA > \text{sect. } DEH,$$

therefore

$$\text{trgl. } DEF : \text{trgl. } DEA < \text{sect. } DEI : \text{sect. } DEH.$$

But sectors are proportional to their arcs or to the angles at the centre; while triangles under the same vertex are proportional to their bases. Accordingly

$$\text{angle } EDF : \text{angle } ADE > \text{base } EF : \text{base } AE.$$

Therefore, *componendo*,

$$\text{angle } FDA : \text{angle } ADE > \text{base } AF : \text{base } AE.$$

And, in the same way,

$$\text{angle } CDA : \text{angle } ADE > \text{base } AC : \text{base } AE.$$

But, *separando*,

$$\text{angle } CDE : \text{angle } EDA > \text{base } CE : \text{base } EA.$$

But

$$\text{angle } CDE : \text{angle } EDA = \text{arc } CB : \text{arc } AB.$$

And

$$\text{base } CE : \text{base } AE = \text{chord } CB : \text{chord } AB.$$

Therefore

$$\text{arc } CB : \text{arc } AB > \text{chord } BC : \text{chord } AB,$$

as was to be shown.

PROBLEM

But since the arc is always greater than the straight line subtending it—as the straight line is the shortest of those lines which have the same termini—nevertheless in going from greater to lesser sections of the circle, the inequality approaches equality, so that finally the circular line and the straight line go out of existence simultaneously at the point of tangency on the circle. Therefore it is necessary that just before that moment they differ from one another by no discernible difference.

For example, let arc AB be 3° and arc AC $1^1/_2$°. It has been shown that
$$\text{ch. } AB = 5235,$$
where diameter = 200,000,

and that
$$\text{ch. } AC = 2618.$$

And though
$$\text{arc } AB = 2 \text{ [15}^a\text{] arc } AC,$$
yet
$$\text{ch. } AB < 2 \text{ ch. } AC$$
and
$$\text{ch. } AC - 2617 = 1.$$

But if we make
$$\text{arc } AB = 1^1/_2°$$
and
$$\text{arc } AC = {}^3/_4°,$$
then
$$\text{ch. } AB = 2618$$
and
$$\text{ch. } AC = 1309,$$

and even though chord AC ought to be greater than half of chord AD, it is seen to be no different from the half. And the ratios of the arcs and the straight lines are now apparently the same. Therefore, since we see that we have come so far that the difference between the straight and the circular line evades sense-perception as completely as if there were only one line, we do not hesitate to take 1309 as subtending $^3/_4$° and in the same ratio to fit the chord to the degree and to the remaining parts (of the degree); and so with the addition of $^1/_4$° to the $^3/_4$° we establish 1° as subtended by 1745, $^1/_2$° by $872^1/_2$, and $^1/_3$° by approximately 582.

Nevertheless I think it will be enough if in the table we give only the halves of the chords subtending twice the arc, whereby we may concisely comprehend in the quadrant what it used to be necessary to spread out over the semicircle; and especially because the halves come more frequently into use in demonstration and calculation than the whole chords do. Now we have set forth a table increasing by $^1/_6$°'s and having three columns. In the first column are the degrees and sixth parts of a degree. The second contains the numerical length of half the chord subtending twice the arc. The third contains the difference between the numerical lengths of each half chord, and by means of these differences we can make proportional additions in taking half-chords of a particular number of minutes. The table follows:

Table of the Chords in a Circle

Arcs		Halves of the chords subtending twice the arcs	Differences between each half-chord	Arcs		Halves of the chords subtending twice the arcs	Differences between each half-chord	Arcs		Halves of the chords subtending twice the arcs	Differences between each half-chord
Deg.	Min.			Deg.	Min.			Deg.	Min.		
0	10	291	291	7	40	13341	288	15	10	26163	280
0	20	582	291	7	50	13629	288	15	20	26443	281
0	30	873	290	8	0	13917	288	15	30	26724	280
0	40	1163	291	8	10	14205	288	15	40	27004	280
0	50	1454	291	8	20	14493	288	15	50	27284	280
1	0	1745	291	8	30	14781	288	16	0	27564	279
1	10	2036	291	8	40	15069	287	16	10	27843	279
1	20	2327	290	8	50	15356	287	16	20	28122	279
1	30	2617	291	9	0	15643	288	16	30	28401	279
1	40	2908	291	9	10	15931	287	16	40	28680	279
1	50	3199	291	9	20	16218	287	16	50	28959	278
2	0	3490	291	9	30	16505	287	17	0	29237	278
2	10	3781	290	9	40	16792	286	17	10	29515	278
2	20	4071	291	9	50	17078	287	17	20	29793	278
2	30	4362	291	10	0	17365	286	17	30	30071	277
2	40	4653	290	10	10	17651	286	17	40	30348	277
2	50	4943	291	10	20	17937	286	17	50	30625	277
3	0	5234	290	10	30	18223	286	18	0	30902	276
3	10	5524	290	10	40	18509	286	18	10	31178	276
3	20	5814	291	10	50	18795	286	18	20	31454	276
3	30	6105	290	11	0	19081	285	18	30	31730	276
3	40	6395	290	11	10	19366	286	18	40	32006	276
3	50	6685	290	11	20	19652	285	18	50	32282	275
4	0	6975	290	11	30	19937	285	19	0	32557	275
4	10	7265	290	11	40	20222	285	19	10	32832	274
4	20	7555	290	11	50	20507	284	19	20	33106	275
4	30	7845	290	12	0	20791	285	19	30	33381	274
4	40	8135	290	12	10	21076	284	19	40	33655	274
4	50	8425	290	12	20	21360	284	19	50	33929	273
5	0	8715	290	12	30	21644	284	20	0	34202	273
5	10	9005	290	12	40	21928	284	20	10	34475	273
5	20	9295	290	12	50	22212	283	20	20	34748	273
5	30	9585	289	13	0	22495	283	20	30	35021	272
5	40	9874	290	13	10	22778	284	20	40	35293	272
5	50	10164	289	13	20	23062	282	20	50	35565	272
6	0	10453	289	13	30	23344	283	21	0	35837	271
6	10	10742	289	13	40	23627	283	21	10	36108	271
6	20	11031	289	13	50	23910	282	21	20	36379	271
6	30	11320	289	14	0	24192	282	21	30	36650	270
6	40	11609	289	14	10	24474	282	21	40	36920	270
6	50	11898	289	14	20	24756	282	21	50	37190	270
7	0	12187	289	14	30	25038	281	22	0	37460	270
7	10	12476	288	14	40	25319	282	22	10	37730	269
7	20	12764	289	14	50	25601	281	22	20	37999	269
7	30	13053	288	15	0	25882	281	22	30	38268	269

TABLE OF THE CHORDS IN A CIRCLE

Arcs		Halves of the chords subtending twice the arcs	Differences between each half-chord	Arcs		Halves of the chords subtending twice the arcs	Differences between each half-chord	Arcs		Halves of the chords subtending twice the arcs	Differences between each half-chord
Deg.	Min.			Deg.	Min.			Deg.	Min.		
22	40	38587	268	30	10	50252	251	37	40	61107	230
22	50	38805	268	30	20	50503	251	37	50	61337	229
23	0	39073	268	30	30	50754	250	38	0	61566	229
23	10	39341	267	30	40	51004	250	38	10	61795	229
23	20	39608	267	30	50	51254	250	38	20	62024	227
23	30	39875	266	31	0	51504	249	38	30	62251	228
23	40	40141	267	31	10	51753	249	38	40	62479	227
23	50	40408	266	31	20	52002	248	38	50	62706	226
24	0	40674	265	31	30	52250	248	39	0	62932	226
24	10	40939	265	31	40	52498	247	39	10	63158	225
24	20	41204	265	31	50	52745	247	39	20	63383	225
24	30	41469	265	32	0	52992	246	39	30	63608	224
24	40	41734	264	32	10	53238	246	39	40	63832	224
24	50	41998	264	32	20	53484	246	39	50	64056	223
25	0	42262	263	32	30	53730	245	40	0	64279	222
25	10	42525	263	32	40	53975	245	40	10	64501	222
25	20	42788	263	32	50	54220	244	40	20	64723	222
25	30	43051	262	33	0	54464	244	40	30	64945	221
25	40	43313	262	33	10	54708	243	40	40	65166	220
25	50	43575	262	33	20	54951	243	40	50	65386	220
26	0	43837	261	33	30	55194	242	41	0	65606	219
26	10	44098	261	33	40	55436	242	41	10	65825	219
26	20	44359	261	33	50	55678	241	41	20	66044	218
26	30	44620	260	34	0	55919	241	41	30	66262	218
26	40	44880	260	34	10	56160	240	41	40	66480	217
26	50	45140	259	34	20	56400	241	41	50	66697	216
27	0	45399	259	34	30	56641	239	42	0	66913	216
27	10	45658	259	34	40	56880	239	42	10	67129	215
27	20	45917	258	34	50	57119	239	42	20	67344	215
27	30	46175	258	35	0	57358	238	42	30	67559	214
27	40	46433	257	35	10	57596	237	42	40	67773	214
27	50	46690	257	35	20	57833	237	42	50	67987	213
28	0	46947	257	35	30	58070	237	43	0	68200	212
28	10	47204	256	35	40	58307	236	43	10	68412	212
28	20	47460	256	35	50	58543	236	43	20	68624	211
28	30	47716	255	36	0	58779	235	43	30	68835	211
28	40	47971	255	36	10	59014	234	43	40	69046	210
28	50	48226	255	36	20	59248	234	43	50	69256	210
29	0	48481	254	36	30	59482	234	44	0	69466	209
29	10	48735	254	36	40	59716	233	44	10	69675	208
29	20	48989	253	36	50	59949	232	44	20	69883	208
29	30	49242	253	37	0	60181	232	44	30	70091	207
29	40	49495	253	37	10	60413	232	44	40	70298	207
29	50	49748	252	37	20	60645	231	44	50	70505	206
30	0	50000	252	37	30	60876	231	45	0	70711	205

TABLE OF THE CHORDS IN A CIRCLE

Arcs		Halves of the chords subtending twice the arcs	Differences between each half-chord	Arcs		Halves of the chords subtending twice the arcs	Differences between each half-chord	Arcs		Halves of the chords subtending twice the arcs	Differences between each half-chord
Deg.	Min.			Deg.	Min.			Deg.	Min.		
45	10	70916	205	52	40	79512	176	60	10	86747	145
45	20	71121	204	52	50	79688	176	60	20	86892	144
45	30	71325	204	53	0	79864	174	60	30	87036	142
45	40	71529	203	53	10	80038	174	60	40	87178	142
45	50	71732	202	53	20	80212	174	60	50	87320	142
46	0	71934	202	53	30	80386	172	61	0	87462	141
46	10	72136	201	53	40	80558	172	61	10	87603	140
46	20	72337	200	53	50	80730	172	61	20	87743	139
46	30	72537	200	54	0	80902	170	61	30	87882	138
46	40	72737	199	54	10	81072	170	61	40	88020	138
46	50	72936	199	54	20	81242	169	61	50	88158	137
47	0	73135	198	54	30	81411	169	62	0	88295	136
47	10	73333	198	54	40	81580	168	62	10	88431	135
47	20	73531	197	54	50	81748	167	62	20	88566	135
47	30	73728	196	55	0	81915	167	62	30	88701	134
47	40	73924	195	55	10	82082	166	62	40	88835	133
47	50	74119	195	55	20	82248	165	62	50	88968	133
48	0	74314	194	55	30	82413	164	63	0	89101	131
48	10	74508	194	55	40	82577	164	63	10	89232	131
48	20	74702	194	55	50	82741	163	63	20	89363	130
48	30	74896	194	56	0	82904	162	63	30	89493	129
48	40	75088	192	56	10	83066	162	63	40	89622	129
48	50	75280	191	56	20	83228	161	63	50	89751	128
49	0	75471	190	56	30	83389	160	64	0	89879	127
49	10	75661	190	56	40	83549	159	64	10	90006	127
49	20	75851	189	56	50	83708	159	64	20	90133	125
49	30	76040	189	57	0	83867	158	64	30	90258	125
49	40	76299	188	57	10	84025	157	64	40	90383	124
49	50	76417	187	57	20	84182	157	64	50	90507	124
50	0	76604	187	57	30	84339	156	65	0	90631	122
50	10	76791	186	57	40	84495	155	65	10	90753	122
50	20	76977	185	57	50	84650	155	65	20	90875	121
50	30	77162	185	58	0	84805	154	65	30	90996	120
50	40	77347	184	58	10	84959	153	65	40	91116	119
50	50	77531	184	58	20	85112	152	65	50	91235	119
51	0	77715	182	58	30	85264	151	66	0	91354	118
51	10	77897	182	58	40	85415	151	66	10	91472	118
51	20	78079	182	58	50	85566	151	66	20	91590	116
51	30	78261	181	59	0	85717	149	66	30	91706	116
51	40	78442	180	59	10	85866	149	66	40	91822	114
51	50	78622	179	59	20	86015	148	66	50	91936	114
52	0	78801	179	59	30	86163	147	67	0	92050	114
52	10	78980	178	59	40	86310	147	67	10	92164	112
52	20	79158	177	59	50	86457	145	67	20	92276	112
52	30	79335	177	60	0	86602	145	67	30	92388	111

TABLE OF THE CHORDS IN A CIRCLE

Arcs		Halves of the chords subtending twice the arcs	Differences between each half-chord	Arcs		Halves of the chords subtending twice the arcs	Differences between each half-chord	Arcs		Halves of the chords subtending twice the arcs	Differences between each half-chord
Deg.	Min.			Deg.	Min.			Deg.	Min.		
67	40	92499	110	75	10	96667	75	82	40	99182	37
67	50	92609	109	75	20	96742	73	82	50	99219	36
68	0	92718	109	75	30	96815	72	83	0	99255	35
68	10	92827	108	75	40	96887	72	83	10	99290	34
68	20	92935	107	75	50	96959	71	83	20	99324	33
68	30	93042	106	76	0	97030	69	83	30	99357	32
68	40	93148	105	76	10	97099	70	83	40	99389	32
68	50	93253	105	76	20	97169	68	83	50	99421	31
69	0	93358	104	76	30	97237	67	84	0	99452	30
69	10	93462	103	76	40	97304	67	84	10	99482	29
69	20	93565	102	76	50	97371	66	84	20	99511	28
69	30	93667	102	77	0	97437	65	84	30	99539	28
69	40	93769	101	77	10	97502	64	84	40	99567	27
69	50	93870	99	77	20	97566	64	84	50	99594	26
70	0	93969	99	77	30	97630	62	85	0	99620	24
70	10	94068	99	77	40	97692	62	85	10	99644	24
70	20	94167	97	77	50	97754	61	85	20	99668	24
70	30	94264	97	78	0	97815	60	85	30	99692	22
70	40	94361	96	78	10	97875	59	85	40	99714	22
70	50	94457	95	78	20	97934	58	85	50	99736	20
71	0	94552	94	78	30	97992	58	86	0	99756	20
71	10	94646	93	78	40	98050	57	86	10	99776	19
71	20	94739	93	78	50	98107	56	86	20	99795	18
71	30	94832	92	79	0	98163	55	86	30	99813	17
71	40	94924	91	79	10	98218	54	86	40	99830	17
71	50	95015	90	79	20	98272	53	86	50	99847	16
72	0	95105	90	79	30	98325	53	87	0	99863	15
72	10	95195	89	79	40	98378	52	87	10	99878	14
72	20	95284	88	79	50	98430	51	87	20	99892	13
72	30	95372	87	80	0	98481	50	87	30	99905	12
72	40	95459	86	80	10	98531	49	87	40	99917	11
72	50	95545	85	80	20	98580	49	87	50	99928	11
73	0	95630	85	80	30	98629	47	88	0	99939	10
73	10	95715	84	80	40	98676	47	88	10	99949	9
73	20	95799	83	80	50	98723	46	88	20	99958	8
73	30	95882	82	81	0	98769	45	88	30	99966	7
73	40	95964	81	81	10	98814	44	88	40	99973	6
73	50	96045	81	81	20	98858	44	88	50	99979	6
74	0	96126	80	81	30	98902	42	89	0	99985	4
74	10	96206	79	81	40	98944	42	89	10	99989	4
74	20	96285	78	81	50	98986	41	89	20	99993	3
74	30	96363	77	82	0	99027	40	89	30	99996	2
74	40	96440	77	82	10	99067	39	89	40	99998	1
74	50	96517	75	82	20	99106	38	89	50	99999	1
75	0	96592	75	82	30	99144	38	90	0	100000	0

13. ON THE SIDES AND ANGLES OF PLANE
RECTILINEAR TRIANGLES

I

[19b] *The sides of a triangle whose angles are given are given.*

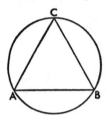

I say let there be the triangle *ABC,* around which a circle is circumscribed, by Euclid, IV, 5. Therefore arcs *AB, BC,* and *CA* will be given in degrees whereof 360° are equal to two right angles. Now given the arcs, the subtending sides of the triangle inscribed in the circle are also given by the table drawn up, where the diameter is assumed to have 200,000 parts.

II

But if two sides of the triangle are given together with one of the angles, the remaining side and the remaining angles may become known.

For the given sides are either equal or unequal. But the given angle is either right or acute or obtuse. Again, the given sides either comprehend the angle or they do not comprehend it.

Therefore in triangle *ABC* first let the two given sides *AB* and *AC* be equal, and let them comprehend the given angle *A.*

Therefore the remaining angles at base *BC* are also given—since they are equal—as half of the remainder, when *A* is subtracted from two right angles. And if the angle given first was at the base, then its equal is soon given, and from the two of them the remaining angle that goes to make up two right angles. But given the angles of a triangle, the sides are given; and moreover the base *BC* is given from the table in the parts whereof *AB* or *AC* as radius has 100,000 parts or whereof the diameter has 200,000 parts.

III

But if the angle BAC comprehended by the given sides is right, the same thing will result.

Since it is obvious that

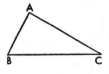

[20a] sq. *AB* + sq. *AC* = sq. *BC;*

therefore *BC* is given in length and the sides in their ratio to one another. But the segment of a circle which comprehends a right triangle is a semicircle, and base *BC* is the diameter. Therefore *AB* and *AC* as subtending the remaining angles *C* and *B* will be given in the parts whereof *BC* has 200,000 parts. And the ratio of the table will reveal the angles in the degrees whereof 180° are equal to two right angles.

The same thing will result if *BC* is given together with one of the sides comprehending the right angle, as I judge has been clearly established.

IV

But now let the given angle ABC be acute, and also let it be comprehended by the given sides AB and BC.

And from point *A* drop a perpendicular to *BC* extended, if necessary, according as it falls inside or outside the triangle, and let it be *AD*. By this perpendicular the two right triangles *ABD* and *ADC* are distinguished, and since the angles in *ABD* are given—for *D* is a right angle, and *B* is given by hypothesis; therefore *AD* and *BD* are given by the table as subtending angles *A* and *B* in the parts whereof *AB*, the diameter of the circle, has 200,000 parts. And in the same ratio wherein *AB* was given in length, *AD* and *BD* are given similarly; and *CD*, which is the difference between *BC* and *BD*, is given also.

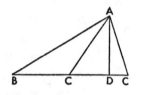

Therefore in the right triangle *ADC*, the sides *AD* and *CD* being given, *AC* the side sought and angle *ACD* are given according to what has been shown above.

V

And it will not turn out differently, if angle *B* is obtuse.

For the perpendicular *AD* dropped from point *A* to straight line *BC* extended makes the triangle *ABD* have its angles given. For angle *ABD*, which is exterior to angle *ABC*, is given; and

angle *D* = 90°.

Therefore sides *BD* and *AD* are given in the parts whereof *AB* has 200,000. And since *BA* and *BC* have a given ratio to one another, therefore *AB* too is given in the same parts, wherein *BD* and the whole *CBD* are given.

Accordingly in the right triangle *ADC*, since the two sides *AD* and *CD* are given, side *AC* and angles *BAC* and *ACB*, which were sought for, are also given.

VI

But let either of the given sides, *AC* or *AB*, be the one subtending the given angle *B*. [20b] Therefore *AC* is given by the table in parts whereof the diameter of the circle circumscribing the triangle *ABC* has 200,000 and according to the given ratio of *AC* to *AB*. *AB* is given in similar parts, and by the table the angle *ACB* is given together with the remaining angle *BAC*, by which chord *CB* is also given. And by this ratio they are given in any magnitude.

VII

Given all the sides of the triangle, the angles are given.

It is too well known to be worth mentioning that each angle of an equilateral triangle is one third of two right angles.

It is also clear in the case of an isosceles triangle. For each of the equal sides is to the third side as half of the diameter is to the side subtending the arc by which the angle comprehended by the equal sides is given according to the table, wherein the 360° around the centre are equal to four right angles.[1] Then the two angles at the base are given as half of the supplementary angle.

Therefore it now remains to show this in the case of scalene triangles, which we divide in the same way into right triangles. Therefore let there be the scalene triangle *ABC* of which the sides are given, and upon the side which is the longest, namely *BC*, drop the perpendicular *AD*. Now Euclid, II, 13 tells us that if *AB* subtends the acute angle, then

(sq. *AC* + sq. *BC*) − sq. *AB* = 2 rect. *BC, CD*.

Now it is necessary for angle *C* to be acute; for otherwise *AB* would be the longest side contrary to the hypotheses, according to Euclid, I, 17 – 19. Therefore *BD* and *DC* are given, and there will be the right triangles *ABC* and *ADC* with their sides and angles given—as has so often happened before—and so the angles of triangle *ABC* which were sought become established.

Another way. Similarly Euclid, III, 36 will perhaps give us an easy method, if with *BC* the shorter side as radius and with point *C* as centre, we describe a circle which will cut either one or both of the remaining sides.

First, let it cut both: *AB* at point *E* and *AC* at *D*; and let line *ADC* extended to point *F* in order to complete the diameter *DCF*. And with this construction it is clear from that proposition of Euclid that

[21ª] rect. *FA, AD* = rect. *BA, AE*,

since each is equal to the square on the tangent to the circle from *A*. But the whole *AF* is given, as all its segments are given, since

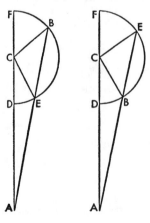

[1]As in the subjoined figure:

44

radius CF = radius CD = BC,

and

$AD = CA - CD$.

Wherefore, as the rectangle BA, AE is given, AE also is given in length; and so is the remainder BE subtending arc BE. By joining EC we shall have the isosceles triangle BCE with all its sides given. Therefore the angle EBC is given. Hence in the triangle ABC the remaining angles at C and at A may become known by means of what has been shown above.

However, let the circle not cut AB as in the other figure, where AB falls upon the concave circumference; nevertheless BE will be given, and in the isosceles triangle BCE angle CBE will be given and also the exterior angle ABC. And by the same method as before the remaining angles are given.

And we have said enough concerning rectilinear triangles, in which a great part of geodesy consists. Now let us turn to spherical triangles.

14. ON SPHERICAL TRIANGLES

In this place we take that triangle as spherical which is comprehended by three arcs of great circles on a spherical surface. But we take the difference and magnitude of the angles from the arc of a great circle, *i.e.*, a great circle described with the point of section as a pole; and this arc is the arc intercepted by the quadrants of the circles comprehending the angle. For as the arc thus intercepted is to the whole circumference, so is the angle of section to four right angles—which we have said contain 360 equal degrees.

I

[21^b] *If there are three arcs of the great circles of a sphere, and if any two of them joined together are longer than the third; it is clear that a spherical triangle can be constructed from them.*

For Euclid, XI, 23 shows in the case of angles what is here proposed in the case of arcs. Since there is the same ratio between angles as between arcs, and since the great circles are those circles which pass through the centre of the sphere; it is manifest that those three sectors of circles, *i.e.*, the sectors to which the three arcs belong, form a solid angle at the centre of the sphere. Therefore what was proposed has been established.

II

Any arc of a (spherical) triangle must be less than a semicircle.

For the semicircle makes no angle at the centre but falls upon it in a straight line.

45

But the remaining two angles which intercept the arcs cannot complete a solid angle at the centre, and so they cannot complete a spherical triangle.

And I think this is the reason why Ptolemy in his exposition of triangles of this genus, especially as regards the figure of the spherical sector, argues that none of the arcs taken together must be greater than a semicircle.

III

In spherical triangles having a right angle, the chord subtending twice the side opposite the right angle is to a chord subtending twice one of the sides comprehending the right angle as the diameter of the sphere is to the chord which subtends the angle comprehended in the great circle of the sphere by the first side and by the remaining side.

For let there be the spherical triangle *ABC*, of which the angle at *C* is right. I say that

ch. 2 *AB* : ch. 2 *BC* = dmt. sph. : ch. 2 *BAC* gr. circ. sph.

With *A* as a pole draw *DE* the arc of a great circle, and let *ABD* and *ACE* the quadrants of the circles be completed. And from the centre *F* of the sphere draw the common sections of the circles: *FA* the common section of circles *ABD* and *ACE*, [22ª] *FE* of circles *ACE* and *DE*, and *FD* of circles *ABD* and *DE*; and moreover, *FC* of the circles *AC* and *BC*. Then draw *BG* at right angles to *FA*, *BI* at right angles to *FC*, and *DK* at right angles to *FE*; and let *GI* be joined.

Since if a circle cuts a circle described through its poles, it cuts it at right angles; therefore the angle *AED* will be right; and angle *ACB* is right by hypothesis; and each of the planes *EDF* and *BCF* is perpendicular to plane *AEF*. Wherefore if a line be erected in the underlying plane of *AFE* at right angles 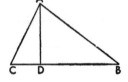 to point *K* in the common section, this line and *KD* will comprehend a right angle, by the definition of planes which are perpendicular to one another. Wherefore, by Euclid, XI, 4, line *KD* is perpendicular to circle *AEF*. But *BI* was erected in the same relation to the same plane; and so by Euclid, XI, 6, *DK* is parallel to *BI* and *FD* is parallel to *GB*, because

angle *FGB* = angle *GFD* = 90°.

And by Euclid, XI, 10,

angle *FDK* = angle *GBI*.

But

angle *FKD* = 90°,

and by definition

GI is perpendicular to *IB*.

Accordingly the sides of similar triangles are proportional; and

$$DF : BG = DK : BI.$$

But
$$BI = {}^1/_2 \text{ ch. } 2 \ CB,$$
since it is at right angles to the radius from center F; and for the same reason,
$$BG = {}^1/_2 \text{ ch. } 2 \ BA,$$
$$DK = {}^1/_2 \text{ ch. } 2 \ DE, \text{ or } {}^1/_2 \text{ ch. } 2 \ DAE,$$
and
$$DF = {}^1/_2 \text{ dmt. sph.,}$$
Therefore it is clear that
$$\text{ch. } 2 \ AB : \text{ch. } 2 \ BC = \text{dmt.} : \text{ch. } 2 \ DAE \text{ (or ch. } 2 \ DE),$$
as it was time to show.

IV

In any triangle having a right angle, if another angle and any side are given, the remaining angle and the remaining sides will be given.

For let there be the triangle ABC having the angle A right and having one of the other two angles, namely B, given.

Let us take three cases of the given side. For it is either adjacent to both the given angles, as AB, or only to the right angle, as AC, or is opposite the right angle, as BC.

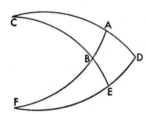

Therefore let AB be the side given first; and with C as a pole let arc [22^b] DE of the great circle be described. Let the quadrants CAD and CBE be completed; and let AB and DE be extended, until they cut one another at point F. Therefore conversely the pole of CAD will be at F, because

$$\text{angle } A = \text{angle } D = 90°.$$

And since, if in a sphere the great circles cut one another at right angles, they will bisect one another and pass through the poles of one another; therefore ABF and DEF are quadrants of circles. And since AB is given, BF the remainder of the quadrant is also given; and the vertical angle EBF is equal to the given angle ABC. But by what has been shown above

$$\text{ch. } 2 \ BF : \text{ch. } 2 \ EF = \text{dmt. sph.} : \text{ch. } 2 \ EBF.$$

But three of the chords have been given:
$$\text{dmt. sph.,}$$
$$\text{ch. } 2 \ BF,$$
$$\text{ch. } 2 \ EBF,$$
or the half-chords; and therefore by Euclid, VI, 15, there is also given

$^1/_2$ ch. 2 *EF*;

and by the table the arc *EF* itself and *DE* the remainder of the quadrant, or the angle at *C*, which was sought. Similarly and alternately,

ch. 2 *DE* : ch. 2 *AB* = ch. 2 *EBC* : ch. 2 *CB*.

But *DE*, *AB*, and *CE* on the quadrants of the circle have already been given; and therefore the fourth chord, subtending twice arc *CB*, will be given, and also the side *CB*, which was sought.

And since

ch. 2 *CB* : ch. 2 *CA* = ch. 2 *BF* : ch. 2 *EF*,

because they both have the ratio of

dmt. sph. : ch. 2 *CBA*,

and because things which have the same ratio to one and the same thing have the same ratio to one another; therefore with the three chords *BF*, *EF*, and *CB* given, the fourth chord *CA* is also given; and arc *CA* is the third side of the triangle *ABC*.

But now let *AC* be the side assumed as given, and let our problem be to find the sides *AB* and *BC* together with the remaining angle *C*. Again similarly and by inversion,

ch. 2 *CA* : ch. 2 *CB* = ch. 2 *ABC* : dmt.

Hence the side *CB* is given, and also *AD* and *BE* the remainders of the quadrants of the circles. And so again,

ch. 2 *AD* : ch. 2 *BE* = ch. 2 *ABF*, *i.e.*, dmt., : ch. 2 *BF*.

Therefore arc *BF* is given, and the side *AB*, which is the remainder.

And similarly,

ch. 2 *BC* : ch. 2 *AB* = 2 ch. *CBE* : ch. 2 *DE*.

Hence arc *DE*, or twice the remaining angle at *C*, will be given.

Furthermore, if it was *BC* which was assumed, again as before, *AC* and the remainders *AD* and *BE* will be given. Hence arc *BF* and the remaining side *AB* are given by means of the diameter and the chords [23ª] subtending them, as has often been said. And as in the preceding theorem, by means of arcs *BC*, *AB*, and *CBE* being given, the arc *ED*, *i.e.*, the remaining angle at *C*, which we were seeking, is discovered.

And so again in the triangle *ABC* with two angles *A* and *B* given, of which *A* is right, and with one of the three sides given, the third angle and the remaining sides are given, as was to be shown.

V

The sides of a right triangle, of which the angles are given, are also given.

Let the preceding diagram be kept. On account of the angle *C* being given, the arc *DE* and *EF* the remainder of the quadrant are given. And since *BEF* is a right angle, because *BE* was let fall from the pole of arc *DEF*; and since angle *EBF* is equal to its vertical angle, which was given; therefore the triangle *BEF*, having the right angle *E* and the angle at *B* given together with the side *EF*, has its sides and angles given by the preceding theorem. Therefore *BF* is given, and so is *AB* the remainder of the quadrant. And similarly in the triangle *ABC* the remaining sides *AC* and *BC* are shown as above.

VI

If in the same sphere two triangles have right angles and another angle equal to another angle and one side equal to one side—whether the sides be adjacent to the equal angles or lie opposite one of the equal angles—they will have the remaining sides equal to the remaining sides and the remaining angle equal to the remaining angle.

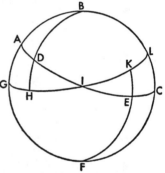

Let there be the hemisphere *ABC*, in which the two triangles *ABD* and *CEF* are taken. Let the angles at *A* and *C* be right; and furthermore let the angle *ADB* be equal to *CEF*, and one side to one side. And first let the equal sides be adjacent to the equal angles, *i.e.*, let *AD* be equal to *CE*. I say moreover that side *AB* is equal to side *CF*, side *BD* to *EF*, and the remaining angle *ABD* to the remaining angle *CFE*.

For with *B* and *F* as poles, draw *GHI* and *IKL* the quadrants of the great circles. And let quadrants *ADI* and *CEI* be completed. They necessarily cut one another at the pole of the hemisphere, point *I*, [23^b] because

$$\text{angle } A = \text{angle } C = 90°$$

and quadrants *GHI* and *CEI* have been drawn through the poles of the circle *ABC*.

Therefore, since it has been assumed that

$$\text{side } AD = \text{side } CE$$

then by subtraction

$$\text{arc } DI = \text{arc } EI.$$

And

$$\text{angle } IDH = \text{angle } IEK;$$

for they are placed at the vertices of the angles assumed as equal; and

$$\text{angle } H = \text{angle } K = 90°.$$

As things which have the same ratio to the same are in the same ratio; and since by Theorem III in this chapter,

$$\text{ch. 2 } ID : \text{ch. 2 } HI = \text{dmt. sph.} : \text{ch. 2 } IDH,$$

and

$$\text{ch. } EI : \text{ch. 2 } KI = \text{dmt. sph.} : \text{ch. 2 } IEK;$$

therefore

$$\text{ch. 2 } ID : \text{ch. 2 } HI = \text{ch. 2 } EI : \text{ch. 2 } IK.$$

And by Euclid's *Elements*, V, 14, since

$$\text{ch. 2 } DI = \text{ch. 2 } IE$$

therefore

$$\text{ch. 2 } HI = \text{ch. 2 } IK.$$

And as in equal circles equal chords cut off equal arcs, and as the parts of multiples are in the same ratio (as the multiples); therefore the plain arcs *IH* and *IK* will be equal; and so will *GH* and *KL* the remainders of the quadrants. Whence it is clear that

$$\text{angle } B = \text{angle } F,$$

and since, by the inverse of the third theorem,

$$\text{ch. 2 } AD : \text{ch. 2 } BD = \text{ch. 2 } HG : \text{ch 2 } BDH, \text{ or dmt.},$$

and

$$\text{ch. 2 } EC : \text{ch. 2 } EF = \text{ch. 2 } KL : \text{ch. 2 } FEK, \text{ or dmt.},$$

wherefore

$$\text{ch. 2 } AD : \text{ch. 2 } BD = \text{ch. 2 } EC : \text{ch. 2 } EF$$

and

$$AD = CE.$$

Therefore, by Euclid's *Elements*, V, 14,

$$\text{arc } BD = \text{arc } EF,$$

on account of the chords subtending twice the area being equal.

In the same way with *BD* and *EF* equal, we will show that the remaining sides and angles are equal.

And in turn, if sides *AB* and *CF* are assumed to be equal, the results will follow the same identity of ratio.

VII

Now also even if there is no right angle, but provided that the sides which are adjacent to the equal angles are equal to one another, the same thing will be shown.

In this way if in the two triangles *ABD* and *CEF*

$$\text{angle } B = \text{angle } F$$

and

$$\text{angle } D = \text{angle } E,$$

and if side *BD* is adjacent to the equal [24ª], angles
and

side *BD* = side *EF,*

I say that again the triangles are equilateral and
equiangular.

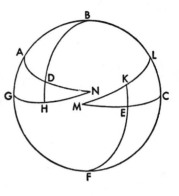

For once more with *B* and *F* as poles, describe
GH and *KL,* the arcs of the great circles. And let *AD*
and *GH* extended intersect at *N*; and let *EC* and *LK*
similarly extended intersect at *M.*

Therefore since in the two triangles *HDN* and
EKM

angle *HDN* = angle *KEM,*

because they are placed at the vertex of the angles assumed equal; and since

angle *H* = angle *K* = 90°

on account of the intersection of circles described through the poles of one another;
and

side *DH* = side *EK*;

therefore the triangles are equiangular and equilateral by the preceding proof. And
again because

arc *GH* = arc *KL*

on account of its being assumed that

angle *B* = angle *F*;

therefore by addition

arc *GHN* = arc *MKL,*

by the axiom concerning the addition of equals. And therefore there are these two triangles *AGN* and *MCL* where

side *GN* = side *ML,*

angle *ANG* = angle *CML,*

and

angle *G* = angle *L* = 90°.

So the triangles will have their sides and angles equal. Therefore when equals have been
subtracted from equals, the remainders will be equal:

arc *AD* = arc *CE,*

arc *AB* = arc *CF,*

and

angle *BAD* = angle *ECF,*

as was to be shown.

VIII

Now further, if two triangles have two sides equal to two sides and an angle equal to an angle, whether the angle which the equal sides comprehend, or an angle at the base, they will also have base equal to base and the remaining angles equal to the remaining angles.

As in the preceding diagram, let

$$\text{side } AB = \text{side } CF$$

and

$$\text{side } AD = \text{side } CE.$$

And first let

$$\text{angle } A = \text{angle } C,$$

which is comprehended by the equal sides. I say also that

$$\text{base } BD = \text{base } EF,$$
$$\text{angle } B = \text{angle } F,$$

and

$$\text{angle } BDA = \text{angle } CEF.$$

For we shall have the two triangles *AGN* and *CLM*, where

$$\text{angle } G = \text{angle } L = 90°.$$

And since

$$\text{angle } GAN = 180° - \text{angle } BAD,$$

and

$$\text{angle } MCL = 180° - \text{angle } ECF,$$

then

$$\text{angle, } GAN = \text{angle } MCL.$$

Therefore the triangles are equiangular and equilateral.
Wherefore since

$$\text{arc } AN = \text{arc } CM$$

and

$$\text{arc } AD = \text{arc } CE,$$

then by subtraction

$$\text{arc } DN = \text{arc } ME.$$

But it has already been made clear that

$$\text{angle } DNH = \text{angle } EMK,$$

and

$$\text{angle } H = \text{angle } K = 90°.$$

Therefore the two triangles *DHN* and *EMK* will also be equiangular and equilateral. [24b] Hence (by the subtraction of equals)

$$\text{arc } BD = \text{arc } EF$$

and

$$\text{arc } GH = \text{arc } KL.$$

Hence

$$\text{angle } B = \text{angle } F,$$

and

$$\text{angle } ADB = \text{angle } FEC.$$

But if instead of sides AD and CE it be assumed that

$$\text{base } BD = \text{base } EF,$$

which are opposite the equal angles; and if the rest stays the same; then the proof will be similar. For since

$$\text{exterior angle } GAN = \text{exterior angle } MCL,$$
$$\text{angle } G = \text{angle } L = 90°,$$

and

$$\text{side } AG = \text{side } CL;$$

in the same way as before we shall have the two triangles AGN and MCL as equiangular and equilateral. And moreover, as parts of them,

$$\text{trgl. } DNH = \text{trgl. } MEK,$$

because

$$\text{angle } H = \text{angle } K = 90°,$$
$$\text{angle } DNH = \text{angle } KME,$$

and by subtraction from the quadrant

$$\text{side } DH = \text{side } EK.$$

Whence the same things follow as before.

IX

Moreover, in isosceles spherical triangles the angles at the base are equal to one another.

Let there be triangle ABC, where

$$\text{side } AB = \text{side } AC.$$

I say that on the base angle ABC = angle ACB.

From the vertex A drop a great circle which will cut the base at right angles, *i.e.*, a circle through the poles of the base; and let this circle be AD. Therefore, since in the two triangles ABD and ADC

$$\text{side } BA = \text{side } AC,$$

and

$$\text{side } AD = \text{side } AD,$$

and

$$\text{angle } BDA = \text{angle } CDA = 90°,$$

it is clear from what was shown above that

$$\text{angle } ABC = \text{angle } ACB,$$

as was to be shown.

PORISM

Hence it follows that the arc from the vertex of an isosceles triangle which falls at right angles upon the base will at the same time bisect the base and the angle comprehended by the equal sides, and vice versa. And that is clear from what has been shown above.

X

If two triangles in the same sphere have the sides of the one severally equal to the sides of the other, they will have the angles of the one severally equal to the angles of the other.

For in each triangle the three segments of great circles form pyramids which have as their apexes the centre of the sphere and as their bases the plane triangles which are comprehended by the straight lines subtending the arcs of the convex triangles. And those pyramids are similar and [25ª] equal by the definition of similar and equal solid figures (Euclid, XI, Def. 10); now the ratio of similarity is that the angles taken in any order will be severally equal to one another. Therefore the triangles will have their angles equal to one another.

In particular, those who define similarity of figures more generally say that similar figures are those which have similar declinations, and have corresponding angles equal to one another. Whence I think it is manifest that in a sphere the triangles which are equilateral are similar, just as in the case of plane triangles.

XI

Every triangle which has two sides and an angle given will have the remaining sides and angles given.

For if the two sides are given as equal, the angles at the base will be equal, and by drawing an arc from the vertex at right angles to the base, what is sought will easily be found by means of the Porism to the ninth theorem.

But if however the sides given are unequal, as in triangle *ABC*, where angle *A* is
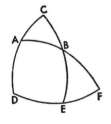
given together with two sides, the sides either comprehend the given angle or do not comprehend it: First, let the given sides *AB* and *AC* comprehend it. And with *C* as a pole draw arc *DEF* of a great circle; and let the quadrants *CAD* and *CBE* be completed; and let *AB* extended cut *DE* at point *F.* So also in the triangle *ADF,*

$$\text{side } AD = 90° - \text{arc } AC;$$

and

$$\text{angle } BAD = 180° - \text{angle } CAB.$$

54

For the ratios and dimensions of these angles are the same as those of angles occurring at the intersection of straight lines and planes. And

$$\text{angle } D = 90°.$$

Therefore by the fourth theorem of this chapter, triangle *ADF* will have its sides and angles given. And again in triangle *BEF* angle *F* has been found, and

$$\text{angle } E = 90°$$

on account of the intersection of circles through the poles of one another; and

$$\text{side } BF = \text{arc } ABF - \text{arc } AB.$$

Therefore by the same theorem triangle *BEF* also will have its angles and sides given. Whence *BC* the side sought is given, as

$$BC = 90° - BE,$$

and *BC* is the side sought. And

$$\text{arc } DE = \text{arc } DEF - \text{arc } EF.$$

And so angle *C* is given. Any by means of angle *EBF*, the vertical angle *ABC*, which was sought, is given.

But if in place of side *AB*, side *CB* which is opposite to the given angle is assumed, the same thing will result. For *AD* and *BE* the remainders of quadrants are given; and by the same argument the two triangles *ADF* and *BEF* will have their sides and angles given, as before.

Whence, as was intended, *ABC* the triangle set before us will have its sides and angles given.

[25ᵇ] XII

Furthermore, if any two angles are given together with one side, there will be the same result.

For let the construction in the previous figure stay; and in triangle *ABC* let the two angles *ACB* and *BAC* be given together with side *AC*, which is adjacent to both angles. Now if one of the angles given were right, then everything else would follow from the ratios by the preceding fourth theorem. But we wish to keep the theorems different and to have neither of the angles right. Therefore

$$AD = 90° - AC.$$

And

$$\text{angle } BAD = 180° - \text{angle } BAC.$$

And

$$\text{angle } D = 90°.$$

Therefore by the fourth theorem in this chapter, triangle *AFD* will have its angles and sides given. But through angle *C* being given, the arc *DE* is given, and so is the remainder

$$\text{arc } EF = 90° - \text{arc } DE.$$

And

$$\text{angle } BEF = 90°;$$

and

$$\text{angle } F = \text{angle } F.$$

In the same way by the fourth theorem BE and BF are given; and through them we can discover sides AB and BC, which were sought.

Moreover, if one of the given angles is opposite the given side, namely if angle ABC is given in place of angle ACB, and if the rest stayed the same, then it can be shown in similar fashion that the whole triangle ADF will be established as having its sides and angles given; and similarly the part of it which is triangle BEF; since on account of angle F being common to both, angle EBF being at the vertex of the given angle, and angle E being right, it is shown as above that all the sides are given. And from that there follows what I said. For all these things are always tied together by a mutual and perpetual bond, as befits the form of a globe.

XIII

Finally, all the sides of a triangle being given, the angles are given.

Let all the sides of triangle ABC be given: I say that all the angles too are found.

For the triangle either will have equal sides or it will not. First therefore let AB and AC be equal. It is clear that the halves of chords subtending twice those sides will be equal. And let these halves be BE and CE, which on account of being at an equal distance from the centre of the sphere will cut one another at point E in DE the common section of the circles, as is clear from Euclid, III, Def. 4, [26a] and its converse.

But by Euclid, III, 3, in plane ABD

$$\text{angle } DEB = 90°;$$

and in plane ACD similarly

$$\text{angle } DEC = 90°.$$

Therefore by Euclid, XI, Def. 3, BEC is the angle of inclination of the planes; and we shall find it as follows; for since there is a straight line subtending BC, we shall have a rectilinear triangle BEC with its sides given on account of their arcs being given; and then since the angles may be found, we shall have the angle BEC, which was sought, *i.e.*, we shall have the spherical angle BAC; and we shall have the others as above.

But if the triangle is scalene, as in the second figure, it is clear that the halves of the chords subtending twice the sides will by no means touch one another. For if

$$\text{arc } AC > \text{arc } AB,$$

then, as

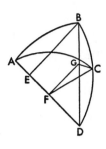

$$CF = {}^1/_2 \text{ ch. } 2\ AC,$$

CF will fall lower down. But if

$$\text{arc } AC < \text{arc } AB,$$

then *CF* will fall higher up, according as such lines become nearer and farther away from the centre, by Euclid, III, 15. Now however let *FG* be drawn parallel to *BE*; and at point *G* let it cut *BD* the common section of the two circles (*AB* and *BC*). And let *GC* be joined. Therefore it is clear that

$$\text{angle } EFG = \text{angle } AEB = 90°$$

And too

$$\text{angle } EFC = 90°;$$

for

$$CF = {}^1/_2 \text{ ch. } 2\ AC.$$

Therefore angle *CFG* will be the angle of section of circles *AB* and *AC*; and we shall find this angle too. For

$$DF : FG = DE : EB,$$

since triangles *DFG* and *DEB* are similar. Therefore *FG* is given in the parts wherein *FC* is also given; and

$$DG : DB = DE : EB.$$

Hence *DG* will be given in the same parts whereof *DC* has 100,000. But as the angle *GDC* is given through the arc *BC*, therefore by the second theorem on plane triangles the side *GC* is given in the same parts wherein the remaining sides of the plane triangle *GFC* are given. Therefore by the last theorem on plane triangles we shall have the angle *GFC*, *i.e.*, the spherical angle *BAC*, which was sought; and then we shall find the remaining angles by the eleventh theorem on spherical triangles.

XIV

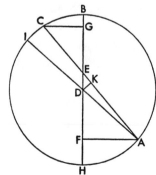

If a given arc of a circle is cut anywhere so that both of the segments together are less than a semicircle, and if the ratio of half of the chord subtending twice one segment to the half of the chord subtending twice the other segment is given, [26ᵇ] the arcs of those segments will also be given.

For let arc *ABC* be given, around centre *D*; and let *ABC* be cut at point *B* anywhere, but in such a way that the segments are less than a semicircle; and let

$$^1/_2 \text{ ch. } 2\ AB : {}^1/_2 \text{ ch. } 2\ BC$$

be somehow given in length: I say that the axes *AB* and *BC* are also given.

For let the straight line *AC* be drawn, which the diameter cuts at point *E*; and from the termini *A* and *C* let the perpendiculars *AF* and *CG* fall upon the diameter. And of necessity

$$AF = \frac{1}{2} \text{ ch. } 2\ AB$$

and

$$CG = \frac{1}{2} \text{ ch. } 2\ BC.$$

Therefore in the right triangles *AEF* and *CEG*

$$\text{angle } AEF = \text{angle } CEG,$$

because they are vertical angles. And the triangles which are therefore equiangular and similar have the sides opposite the equal angles proportional:

$$AF : CG = AE : EC.$$

Therefore we shall have *AE* and *EC* in the parts wherein *AF* or *GC* has been given. But the chord subtending arc *ABC* is given in the parts wherein the radius *DEB*, *AK* the half of chord *AC*, and the remainder *EK* are given. Let *DA* and *DK* be joined, and they will be given in the parts wherein *BD* is given: *DK* will be given as half of the chord subtending the remaining segment which is supplementary to arc *ABC* and is comprehended by angle *DAK*. And therefore angle *ADK* is given, which comprehends half of arc *ABC*. But in the triangle *EDK* having two sides given and angle *EKD* right, angle *EDK* will also be given. Hence the whole angle *EDA* comprehending the arc *AB* will be given. Thereby also the remainder *CB* will be manifest. And it was this that we were trying to show.

XV

If all the angles of a triangle are given, even though now is a right angle, all the sides are given.

Let there be the triangle *ABC*, all the angles of which are given but none of which is right. I say that all the sides are given too.

For from some one of the angles, say *A*, drop the arc *AD* through the poles of *CB*. *AD* will cut *BC* at right angles, and it will fall within the triangle, unless one of the angles at the base—angle *B* or angle *C*—is obtuse and the other acute. If that were the case, the arc would have to be drawn from the obtuse angle to the base. So with the quadrants *BAF*, *CAG*, and *DAE* completed and with *B* and *C* as poles, let the arcs *EF* and *EG* [27ª] be drawn.

Therefore

$$\text{angle } F = \text{angle } G = 90°.$$

Therefore in the right triangle *EAF*

$$\frac{1}{2} \text{ ch. } 2\ AE : \frac{1}{2} \text{ ch. } 2\ EF = \frac{1}{2} \text{ dmt. sph. } : \frac{1}{2} \text{ ch. } 2\ EAF$$

Similarly in right triangle *AEG*

$^1/_2$ ch. 2 *AE* : $^1/_2$ ch. 2 *EG* = $^1/_2$ dmt. sph. : $^1/_2$ ch. 2 *EAG*.

Therefore, *ex aequali*,

$^1/_2$ ch. 2 *EF* : $^1/_2$ ch. 2 *EG* = $^1/_2$ ch. 2 *EAF* : $^1/_2$ ch. 2 *EAG*.

And because arcs *FE* and *EG* arc given, since

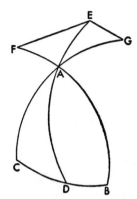

arc *FE* = 90° − angle *B*

and

arc *EG* = 90° − angle *C*;

thence we shall have the ratio between angles *EAF* and *EAG* given, *i.e.*, the ratio between *BAD* and *CAD*, which are their vertical angles. Now the whole angle *BAC* has been given; therefore by the foregoing theorem, angles *BAD* and *CAD* will also be given. Then by the fifth theorem we shall determine sides *AB*, *BD*, *AC*, *CD*, and the whole of arc *BC*.

This much said enroute concerning triangles, according as they are necessary for our undertaking, will be sufficient. For if they had to be treated in greater detail, the work would be of unusual size.

[27ᵇ] BOOK TWO

Since we have expounded briefly the three terrestrial movements, by means of which we promised to demonstrate all the planetary appearances, now we shall fulfill our promise by proceeding from the whole to the parts and examining and investigating particular questions to the extent of our powers. Now we shall begin with the best-known movement of all, the revolution of day and night—which we said the Greeks called νυχθήμερος and which we have taken as belonging wholly and immediately to the terrestrial globe, since from this movement arise the months, years, and other variously named periods of time, as number from unity. Therefore we shall say only a few words about the inequality of days and nights, the rising and setting of the sun and of the parts of the ecliptic and the signs, and the consequences of this type of revolution; for many people have written about these subjects copiously enough and what they say is in harmony and agreement with our conceptions. It is of no importance if we take up in an opposite fashion what others have demonstrated by means of a motionless earth and a giddy world and race with them toward the same goal, since things related reciprocally happen to be inversely in harmony with one another. Nevertheless we shall omit nothing necessary. But no one should be surprised if we still speak of the rising and setting of the sun and stars, *et cetera*; but he should realize that we are speaking in the usual manner of speech which can be recognized by all and that we are nevertheless always keeping in mind that: "To us who are being carried by the Earth, the sun and the moon seem to pass over; and the stars return to their former positions and again move away."

1. ON THE CIRCLES AND THEIR NAMES

We have said that the equator is the greatest of the parallel circles on the terrestrial globe described around the axis of its daily revolution and that the ecliptic is the circle through the middle [28ᵃ] of the signs under which the centre of the Earth moves in a circle in its annual revolution.

But since the ecliptic crosses the equator obliquely; in proportion to the inclination of the axis of the Earth to it, it describes in the course of the daily revolution two circles which touch it on either side of the equator, as if the farthest limits to its obliquity. These circles are called the tropics. For on them the sun appears to make its "tropes," *i.e.*, its winter and summer changes of direction. Whence the northern circle used to be called the tropic of the summer solstice and the other the tropic of the shortest day, as was set forth in our summary exposition of the circular movements of the Earth.

Next follows the so-called horizon, which the Latins call the boundary circle; for it is the boundary between that part of the world which is visible to us and that part which lies hidden. All stars which set are seen to have their rising on it; and it has its centre on the surface of the Earth and its pole at the point directly overhead. But since it is impossible to compare the Earth with the immensity of the heavens—for according to our hypothesis even the total distance between the sun and the moon is indiscernible beside the magnitude of the heavens—the circle of the horizon appears to bisect the heavens, as if it went through the centre of the world, as we demonstrated in the beginning.

But when the horizon is oblique to the equator, it too touches on either side of the equator twin parallel circles, *i.e.*, the northern circle of the always visible stars and the southern circle of the always hidden stars. The first circle was called the arctic, and the second the antarctic by Proclus and the Greeks; and they become greater or smaller in proportion to the obliquity of the horizon or the elevation of the pole of the equator.[1]

There remains the meridian circle which passes through the poles of the horizon and through the poles of the equator too and hence is perpendicular to both circles. The sun's reaching it gives us midday and midnight.

But these two circles which have their centres on the surface of the Earth, *i.e.*, the horizon and the meridian, are wholly consequent upon the movement of the Earth and upon our sight at some particular place. For the eye everywhere becomes as it were the centre of the sphere of all things which are visible to it on all sides.

Furthermore all the circles assumed on the Earth produce circles in the heavens as their likenesses and images, as will be shown more clearly in cosmography and in connection with the dimensions of the Earth. And these circles at any rate are the ones having proper names, though there are infinite ways of designating and naming others.

2. ON THE OBLIQUITY OF THE ECLIPTIC AND THE DISTANCE OF THE TROPICS AND HOW THEY ARE DETERMINED

[28ᵇ] Since the ecliptic lies between the tropics and crosses the equator obliquely, I therefore think that we should now try to observe what the distance between the tropics is and hence what the angle of section between the equator and the ecliptic is. For in order to perceive this by sense with the help of artificial instruments, by means of which the job can be done best, it is necessary to have a wooden square prepared, or preferably a square made from some other more solid material, from stone or metal; for the wood might not stay in the same condition on account of some alteration in the atmosphere and might mislead the observer. Now one surface of it should be very

---

[1] That is to say, the magnitude of the circle of the always visible stars varies inversely with the obliquity of the horizon and directly with the elevation of the poles of the equator.

carefully planed, and it should be of sufficient area to admit being divided into sections, that is, a side should be about 5 or 6 feet long. Now with one of the comers (of the square) as centre and with a side as radius, let a quadrant of a circle be drawn and divided into 90 equal degrees; and let each of the degrees be subdivided into 60 minutes, or whatever number can be taken. Next let a cylindrical pointer which has been well turned on a lathe be set up at the centre (of the quadrant) and fixed in such a way as to be perpendicular to the surface and to extend out from it a little, say perhaps a finger's width or less.

When the instrument has been prepared in this way, the next thing to do is to exhibit the line of the meridian on a piece of flooring which lies in the plane of the horizon and which has been made even as carefully as is possible by means of a hydroscope or ground-level, so as not to have a slope in any part of it. The piece of flooring should have a circle drawn on it and a cylinder erected at the center of the circle: we shall take observations and mark the point where at some time before midday the extremity of the shadow of the cylinder touches the circumference of the circle, We shall do the same thing in the afternoon, and then shall bisect the arc of the circle lying between the two points we have already marked. In this way a straight line drawn from the centre through the point of section will indicate infallibly for us the south and the north.

Accordingly the plane surface of the instrument should be set up on this piece of flooring as a base and fixed perpendicular to it with the centre (of the quadrant) to the south, so that a plumb-line from the centre would fall exactly at right angles to the meridian line. For it comes about in this way that the surface of the instrument exhibits the meridian circle. Hence on the days of summer and winter solstice the shadows of the sun at noon [29a] are to be observed according as they are cast by the pointer, or cylinder, from the centre (of the quadrant); and some mark is to be made on the arc of the quadrant, so that the place of the shadow may be kept more surely. And we shall note down the centre of the shadow in degrees and minutes as accurately as is possible. For if we do this, the arc between the marked shadows—the summer—and winter—solstitial shadows—will be found and will show us the distance between the tropics and also the total obliquity of the ecliptic.[1] By taking half of the arc, we shall have the distance of the tropics from the equator, and it will be clear what the angle of inclination is between the equator and the ecliptic.

Now Ptolemy took the interval between the aforesaid limits—the northern and the southern—as 47°42'40", whereof the circle has 360°, as he found had been observed by Hipparchus and Eratosthenes before his time; and there are 11P whereof the whole circle has 83P. Hence half the arc—and half the arc has 23°51'20", whereof

---

[1] Since the distance between the sun and the Earth is imperceptible in relation to the radius of the sphere of the fixed stars, the centre of the quadrant may be taken as the centre of the sphere of the fixed stars.

the circle has 360°—showed the distance of the tropics from the equator and what the angle of section with the ecliptic was. Accordingly Ptolemy believed that these things were invariably such and would always remain so. But these distances have been found to have decreased continually from that time down to ours. For it has already been discovered by us and some of our contemporaries that the distance between the tropics is not more than 46°58' approximately and that the angle of section is 23°29'. Hence it is clear enough that the obliquity of the ecliptic is not fixed. More on this below, where we shall show by a probable enough conclusion that it was never greater than 23°52' and will not ever be less than 23°28'.

3. ON THE ARCS AND ANGLES OF THE INTERSECTIONS OF THE EQUATOR, ECLIPTIC, AND MERIDIAN, BY MEANS OF WHICH DECLINATIONS AND RIGHT ASCENSIONS ARE DETERMINED, AND ON THE COMPUTATION OF THESE ARCS AND ANGLES

Accordingly as we were saying in the case of the horizon that the parts of the world have their risings and settings on it, we say that the meridian circle [29b] halves the heavens. During the space of twenty-four hours this circle is crossed by both the ecliptic and the equator and divides both of their circumferences by cutting them at the spring and at the autumnal intersection and in turn has its circumference divided by the arc intercepted by those two circles. Since they are all great circles, they form a spherical right triangle; for the angle is right where the meridian circle by definition cuts the equator described through its poles. Now the arc of the meridian circle, or any arc of a circle passing through the poles (of the equator) and intercepted in this way is called the declination of a segment of the ecliptic; and the corresponding arc on the equator is called the right ascension occurring at the same time as the similar arc on the ecliptic.

All this is easily demonstrated in a convex triangle. For let the circle *ABCD* be a circle passing simultaneously through the poles of the ecliptic and of the equator—most people call this circle the "colure"—let the semi-circle of the ecliptic be *AEC*, the semicircle of the equator *BED*; let the spring equinox be at point *E*, the summer solstice at *A*, and the winter solstice at *C*. Now let *F* be taken as the pole of daily revolution, and on the ecliptic let

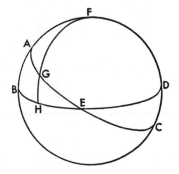

arc *EG* = 30°,

for example, and let it be cut off by *FGH* the quadrant of a circle.

Then it is clear that in triangle *EGH*

side *EG* = 30°,

angle *GEH* is given,

since at its least, in conformity with the greatest declination *AB*,

angle *GFH* = 23°28'

where 4 rt. angles = 360°;

and

angle *GHE* = 90°.

Therefore by the fourth theorem on sphericals, triangle *EHG* will have its sides and angles given. For it was shown that

ch. 2 *EG* : ch. 2 *GH* = ch. 2 *AGE*, or dmt. sph. : ch. 2 *AB*

and their halves are in the same ratio. And since

$^1/_2$ ch. 2 *AGE* = radius = 100,000,

$^1/_2$ ch. 2 *AB* = 39,822,

and

$^1/_2$ ch. 2 *EG* = 50,000;

and since, if four numbers are proportional, the product of the means is equal to the product of the extremes; therefore

$^1/_2$ ch. 2 *GH* = 19,911,

and hence, by the table,

arc *GH* = 11°29',

which is the declination of segment *EG*,

side *FG* = 78°31',

side *AG* = 60°,

since they are the remainders of the quadrants, and

angle *FAG* = 90°.

In the same way

[30ª] $^1/_2$ ch. 2 *FG* : $^1/_2$ ch. 2 *AG* = $^1/_2$ ch. 2 *FGH* : $^1/_2$ ch. 2 *BH*.

Now since three of these chords are given, the fourth will also be given, that is to say,

arc *BH* = 62°6',

which is the right ascension from the summer solstice, and

HE = 27°54'

from the spring equinox. Similarly, since

side *FG* = 78°31',

side *AF* = 64°30',

and

AGE = 90°;

then, since angles *AGF* and *HGE* are vertical angles,

angle *AGF* = angle *HGE* = 63°29¹/₂'.

In the rest we shall do as in this example. But we should not be ignorant of the fact that the meridian circle cuts the ecliptic at right angles in the signs where the ecliptic touches the tropics, for then the meridian circle cuts it through its poles, as we said. But at the equinoctial points the meridian makes an angle less than a right angle by the angle of inclination of the ecliptic, so that in conformity with the least inclination of the ecliptic it makes an angle of 66°32'.

Moreover we should note that equal sides and equal angles of the triangles follow upon equal arcs of the ecliptic being taken from the points of solstice or equinox. In this way if we

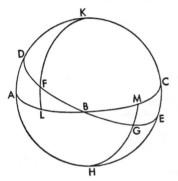

draw the equatorial arc *ABC* and the ecliptic *BDE* as intersecting at point *B*, where the equinox is, and if we take as equal the arcs *FB* and *BG* and also arcs *KFL* and *HGM* two quadrants of circle described through the poles of daily revolution; there will be the two triangles *FLB* and *BMG*, wherein

side *BF* = side *BG*

angle *FLB* = angle *GBM*

and

angle *FLB* = angle *GMB* = 90°.

Therefore by the sixth theorem on spherical triangles the sides and angles are equal. Hence

declination *FL* = declination *GM*

rt. ascension *LB* = rt. ascension *BM*

and

angle *LFB* = angle *MGB*.

This same fact will be manifest upon the assumption of equal arcs described from a point of solstice, for example, when *AB* and *BC* on different sides of their point of contact *B*, the solstice, are equally distant from it. For when the arcs *DA* and *DB* (and *DC*) have been drawn from the pole of the equator, there will similarly be the two triangles *ABD* and *DBC*.

Then

base *AB* = base *BC*

side *BD* is common

and

angle *ABD* = angle *CBD* = 90°.

Accordingly by the eighth theorem on sphericals the triangles will be shown to have equal sides and angles. It is clear from this that such angles and arcs laid out in one quadrant of the ecliptic [30ᵇ] are in accord with the remaining quadrants of the full circle.

We shall subjoin an example of these things in the tables. In the first column are placed the degrees of the ecliptic; in the following column, the declinations answering to those degrees; and in the third column the minutes, which are the differences between the particular declinations and the declinations which occur at the time of greatest obliquity of the ecliptic: the greatest of these differences is 24'.

We shall do the same thing in the table of ascensions and the table of meridian angles. For it is necessary for all things which are consequences of the obliquity of the ecliptic to be changed with a change in it. Furthermore, in the right ascensions an extremely slight difference is found, one which does not exceed $1/10$ "time" and which in the space of an hour makes only $1/150$ "time."—The ancients give the name of "time" to the parts of the equator which arise together with the parts of the ecliptic. Each of these circles, as we have often said, has 360 parts; but in order to distinguish between them, most of the ancients called the parts of ecliptic "degrees" and those of the equator "times"; and we will copy

TABLE OF DECLINATIONS OF THE DEGREES OF THE ECLIPTIC

Eclip-tic	Declination		Differ-ence	Eclip-tic	Declination		Differ-ence	Eclip-tic	Declination		Differ-ence
Deg.	Deg.	Min.	Min.	Deg.	Deg.	Min.	Min.	Deg.	Deg.	Min.	Min.
1	0	24	0	31	11	50	11	61	20	23	20
2	0	48	1	32	12	11	12	62	20	35	21
3	1	12	1	33	12	32	12	63	20	47	21
4	1	36	2	34	12	52	13	64	20	58	21
5	2	0	2	35	13	12	13	65	21	9	21
6	2	23	2	36	13	32	14	66	21	20	22
7	2	47	3	37	13	52	14	67	21	30	22
8	3	11	3	38	14	12	14	68	21	40	22
9	3	35	4	39	14	31	14	69	21	49	22
10	3	58	4	40	14	50	14	70	21	58	22
11	4	22	4	41	15	9	15	71	22	7	22
12	4	45	4	42	15	27	15	72	22	15	23
13	5	9	5	43	15	46	16	73	22	23	23
14	5	32	5	44	16	4	16	74	22	30	23
15	5	55	5	45	16	22	16	75	22	37	23
16	6	19	6	46	16	39	17	76	22	44	23
17	6	41	6	47	16	56	17	77	22	50	23
18	7	4	7	48	17	13	17	78	22	55	23
19	7	27	7	49	17	30	18	79	23	1	24
20	7	49	8	50	17	46	18	80	23	5	24
21	8	12	8	51	18	1	18	81	23	10	24
22	8	34	8	52	18	17	18	82	23	13	24
23	8	57	9	53	18	32	19	83	23	17	24
24	9	19	9	54	18	47	19	84	23	20	24
25	9	41	9	55	19	2	19	85	23	22	24
26	10	3	10	56	19	16	19	86	23	24	24
27	10	25	10	57	19	30	20	87	23	26	24
28	10	46	10	58	19	44	20	88	23	27	24
29	11	8	10	59	19	57	20	89	23	28	24
30	11	29	11	60	20	10	20	90	23	28	24

them for the remainder of the work.—Therefore since the difference is so small that it can be properly neglected, we are not peeved at having to place it in a separate column.

Hence these tables can be made to apply to any other obliquity of the ecliptic, if in conformity with the ratio of difference between the least and greatest obliquity of the ecliptic we make the proper corrections. For example, if with an obliquity of 23°34' we wish to know how great a declination follows from taking it distance of 30° from the equator along the ecliptic, we find that in the table there are 11°29' in the column of declinations and 11' in the column of differences. These 11' would be all added in the case of the greatest obliquity of the ecliptic, which is, as we said, an obliquity of 23°52'. But it has already been laid down that the obliquity is 23°34' and is accordingly greater than the least obliquity by 6', which are one quarter of 24', which is the excess of the greatest obliquity over the least. Now

$$3' : 11 \doteqdot 6' : 24'.$$

TABLE OF RIGHT ASCENSIONS

Ecliptic Deg.	Equator Deg.	Equator Min.	Difference Min.	Ecliptic Deg.	Equator Deg.	Equator Min.	Difference Min.	Ecliptic Deg.	Equator Deg.	Equator Min.	Difference Min.
1	0	55	0	31	28	54	4	61	58	51	4
2	1	50	0	32	29	51	4	62	59	54	4
3	2	45	0	33	30	50	4	63	60	57	4
4	3	40	0	34	31	46	4	64	62	0	4
5	4	35	0	35	32	45	4	65	63	3	4
6	5	30	0	36	33	43	5	66	64	6	3
7	6	25	1	37	34	41	5	67	65	9	3
8	7	20	1	38	35	40	5	68	66	13	3
9	8	15	1	39	36	38	5	69	67	17	3
10	9	11	1	40	37	37	5	70	68	21	3
11	10	6	1	41	38	36	5	71	69	25	3
12	11	0	2	42	39	35	5	72	70	29	3
13	11	57	2	43	40	34	5	73	71	33	3
14	12	52	2	44	41	33	6	74	72	38	2
15	13	48	2	45	42	32	6	75	73	43	2
16	14	43	2	46	43	31	6	76	74	47	2
17	15	39	2	47	44	32	5	77	75	52	2
18	16	34	3	48	45	32	5	78	76	57	2
19	17	31	3	49	46	32	5	79	78	2	2
20	18	27	3	50	47	33	5	80	79	7	2
21	19	23	3	51	48	34	5	81	80	12	1
22	20	19	3	52	49	35	5	82	81	17	1
23	21	15	3	53	50	36	5	83	82	22	1
24	22	10	4	54	51	37	5	84	83	27	1
25	23	9	4	55	52	38	4	85	84	33	1
26	24	6	4	56	53	41	4	86	85	38	0
27	25	3	4	57	54	43	4	87	86	43	0
28	26	0	4	58	55	45	4	88	87	48	0
29	26	57	4	59	56	46	4	89	88	54	0
30	27	54	4	60	57	48	4	90	90	0	0

When I add 3' to the 11°29', I shall have 11°32', which will then measure the declination of the arc of the ecliptic 30° from the equator.

The same thing can be done in the table of meridian angles and right ascensions, except that we must always add the differences in the case of right ascensions but subtract them in the case of the meridian angles, so that everything may proceed correctly in conformity with the time.

TABLE OF THE MERIDIAN ANGLES

Eclip-tic	Angle		Differ-ence	Eclip-tic	Angle		Differ-ence	Eclip-tic	Angle		Differ-ence
Deg.	Deg.	Min.	Min.	Deg.	Deg.	Min.	Min.	Deg.	Deg.	Min.	Min.
1	66	32	24	31	69	35	21	61	78	7	12
2	66	33	24	32	69	48	21	62	78	29	12
3	66	34	24	33	70	0	20	63	78	51	11
4	66	35	24	34	70	13	20	64	79	14	11
5	66	37	24	35	70	26	20	65	79	36	11
6	66	39	24	36	70	39	20	66	79	59	10
7	66	42	24	37	70	53	20	67	80	22	10
8	66	44	24	38	71	7	19	68	80	45	10
9	66	47	24	39	71	22	19	69	81	9	9
10	66	51	24	40	71	36	19	70	81	33	9
11	66	55	24	41	71	52	19	71	81	58	8
12	66	59	24	42	72	8	18	72	82	22	8
13	67	4	23	43	72	24	18	73	82	46	7
14	67	10	23	44	72	39	18	74	83	11	7
15	67	15	23	45	72	55	17	75	83	35	6
16	67	21	23	46	73	11	17	76	84	0	6
17	67	27	23	47	73	28	17	77	84	25	6
18	67	34	23	48	73	47	17	78	84	50	5
19	67	41	23	49	74	6	16	79	85	15	5
20	67	49	23	50	74	24	16	80	85	40	4
21	67	56	23	51	74	42	16	81	86	5	4
22	68	4	22	52	75	1	15	82	86	30	3
23	68	13	22	53	75	21	15	83	86	55	3
24	68	22	22	54	75	40	15	84	87	19	3
25	68	32	22	55	76	1	14	85	87	53	2
26	68	41	22	56	76	21	14	86	88	17	2
27	68	51	22	57	76	42	14	87	88	41	1
28	69	2	21	58	77	3	13	88	89	6	1
29	69	13	21	59	77	24	13	89	89	33	0
30	69	24	21	60	77	45	13	90	90	0	0

4. How to Determine the Declination and Right Ascension of any Star Which Is Placed Outside the Ecliptic But Whose Longitude and Latitude Have Been Established; and with What Degree of the Ecliptic It Halves the Heavens

[32ᵇ] These things have been set down concerning the ecliptic and the equator and their intersections. But as regards the daily revolution, it is of interest not only to know what parts of the ecliptic appear, by means of which the causes of the sun's appearing where it does are discovered, but also to know that there is a similar demonstration of the declination from the equator and of the right ascension in the case of those fixed or wandering stars which are outside the ecliptic but whose longitude and latitude have been given.

Therefore let the circle *ABCD* be described through the poles of the equator and of the ecliptic; let *AEC* be the semicircle of the equa-
tor above pole *F*; let *BED* be the semicircle of the ecliptic about pole *G*; and let its intersection with the equator be at point *E*. Now from the pole *G* let the arc *GHKL* be drawn through a star, and let the position of the star be given as point *H*, and let *FHMN* a quadrant of a circle fall through *H* from the pole of daily movement.

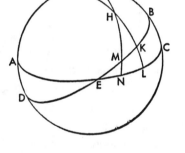

Then it is clear that the star which is at *H* falls upon the meridian at the same time as points *M* and *N* do, and that arc *HMN* is the declination of the star from the equator, and *EN* is its ascension in the right sphere, and those are what we are seeking. Accordingly since in triangle *KEL*

side *KE* is given,

angle *KEL* is given,

and

angle *EKL* = 90°,

therefore by the fourth theorem on spherical triangles,

side *KL* is given,

side *EL* is given,

and

angle *KLE* is given.

Therefore by addition

arc *HKL* is given.

And on that account, in triangle *HLN*,

angle *HLN* is given,

angle *LNH* = 90°,

and

side *HL* is given.

Therefore by the same fourth theorem on sphericals there are also given the remaining sides: *HN* the declination of the star, and *LN*, and the remaining distance *NE*, the right ascension, which measures the distance the sphere turns from the equinox to the star.

—Or in another way. If in the foregoing you take *KE* the arc of the ecliptic as the right ascension of *LE*, conversely *LE* will be given by the table of right ascensions; and so will *LK*, as the declination corresponding to *LE*; [33ª] and the angle *KLE* will be given by the table of meridian angles; and hence the remaining sides and angles, as we have showed, may be learned.—

Then by means of the right ascension *EN*, the number of degrees of *EM* the arc of the ecliptic are given. And in conformity with these things the star together with point *M* halves the heavens.

5. On the Sections of the Horizon

Now the horizon of a right sphere is different from the horizon of an oblique sphere. For the horizon to which the equator is perpendicular, or which passes through the poles of the equator, is called a right horizon.

We call the horizon which has some inclination with the equator the horizon of an oblique sphere.

Therefore on a right horizon all the stars rise and set, and the days are always equal to the nights. For this horizon bisects all the parallel circles described by the diurnal movement, and passes through their poles; and there occurs there what we have already explained in the case of the meridian circle. Here, however, we are taking the day as extending from sunrise to sunset, and not from light to darkness, as the crowds understand it, *i.e.*, from early morning twilight to the first street lights; but we shall say more on this subject in connection with the rising and setting of the signs.

On the contrary, where the axis of the Earth is perpendicular to the horizon there are no risings or settings, but all the stars turn in a gyre and are always visible or hidden, unless they are affected by some other motion, such as the annual movement around the sun. Consequently, there day lasts perpetually for the space of half a year and night for the rest of the time; and there is nothing else to differentiate summer and winter, since there the horizon coincides with the equator.

Furthermore, in an oblique sphere certain stars rise and set; and certain others are always visible or always hidden; and meanwhile the days and nights are unequal there where an oblique horizon touches two parallel circles in proportion to its inclination. And of these circles, the one nearer the visible pole is the boundary of the always visible stars, and conversely the circle nearer the hidden pole is the boundary for the always hidden stars. Therefore the horizon, as falling completely between these boundaries, cuts all the parallel circles in the middle into unequal arcs, except the equator, which is the greatest of the parallels; and great circles bisect one another. Therefore an oblique horizon in the upper hemisphere cuts off axes of parallels in the direction of the visible pole which are greater than the arcs which are toward the southern and hidden [33ᵇ] pole; and the converse is the case in the hidden hemisphere. The sun becomes visible in these horizons by reason of the diurnal movement and causes the inequality of days and nights.

6. WHAT THE DIFFERENCES BETWEEN THE MIDDAY SHADOWS ARE

There are differences between the midday shadows on account of which some people are called periscian, others amphiscian, and still others heteroscian. The periscian are those whom we might call "circumumbratile," that is to say, "throwing the shadow of the sun on every side." And they live where the distance between the vertex, or pole, of the horizon and the pole of the Earth is less or no greater than that between the tropic and the equator. For there the parallels which the horizon touches as the boundaries of the always apparent or always hidden stars are greater than, or equal to, the tropics. And so the summer sun high up among the always apparent stars at that time throws the shadow of a pointer in every direction. But where the horizon touches the tropics, the tropics become the boundaries of the always apparent and the always hidden stars. Wherefore instead of there being midnight the sun at its (winter) solstice seems to graze the Earth, at which time the whole circle of the ecliptic coincides with the horizon; and straightway six signs rise at the same time, and on the opposite side six signs set at the same time, and the pole of the ecliptic coincides with the pole of the horizon.

The amphiscian, who cast midday shadows on both sides, are those who live between the tropics. This is the space which the ancients called the middle zone. And since throughout that whole tract the circle of the ecliptic passes directly over head twice, as is shown in the second theorem of the *Phaenomena* of Euclid, the shadows of pointers are cast in two directions there: for as the sun moves back and forth, the pointers throw their shadows sometimes to the south and sometimes to the north.

The rest of us who inhabit the region between the two others are heteroscian, because we cast our midday shadows in only one direction, *i.e.*, towards the north.

Now the ancient mathematicians were accustomed to divide the world into seven climates, through Meröe, Siona, Alexandria, Rhodes, the Hellespont, the middle of the Pontus, Boristhenes, Byzantium, and so on with the single parallel circles taken according to the differences between the longest days and according to the lengths of the shadows, which they observed by means of pointers at noon on the days of equinoxes and solstices, and [34ª] according to the elevation of the pole or the latitude of some segment. Since these things have partly changed through time, they are not exactly the same as they once were, on account of the variable obliquity of the ecliptic, as we said, of which the ancients were ignorant; or, to speak more correctly, on account of the variable inclination of the equator to the plane of the ecliptic, upon which these things depend. But the elevations of the pole or the latitudes of the places and the equinoctial shadows agree with those which antiquity discovered and made note of. That would necessarily take place, since the equator depends upon the pole of the terrestrial globe. Wherefore those segments are not accurately enough designated and defined by shadows falling on special days, but more correctly by their distances from the equator, which remain perpetually. But although this variability of the tropics, because very slight, admits but slight diversity of days and of shadows in southern places, it becomes more apparent to those who are moving northward. Therefore as regards the shadows of pointers, it is clear that for any given altitude of the sun the length of the shadow is derivable and vice versa.

In this way if there is the pointer *AB* which casts a shadow *BC*; since the pointer is perpendicular to the plane of the horizon, angle *ABC* must always be right, by the definition of lines perpendicular to a plane. Wherefore if *AC* be joined, we shall have a right triangle *ABC*; and for a given altitude of the sun we shall have angle *ACB* given. And by the first theorem on plane triangles the ratio of the pointer *AB* to its shadow *BC* will be given, and *BC* will be given in length. Conversely, moreover, when *AB* and *BC* are given, it will be clear from the third theorem on plane triangles what angle *ACB* is and what the elevation of the sun making that shadow at that time is. In this way the ancients in describing the regions of the terrestrial globe gave the lengths of the midday shadows sometimes at the equinoxes and sometimes at the solstices.

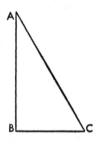

7. HOW THE LONGEST DAY, THE DISTANCE OF RISING, AND THE INCLINATION OF THE SPHERE ARE DERIVED FROM ONE ANOTHER, AND ON THE DIFFERENCES BETWEEN DAYS

[34ᵇ] In this way too for any obliquity of the sphere or inclination of the horizon we will demonstrate simultaneously the longest and the shortest day together with the distance of rising (of the sun) and the difference of the remaining days. Now the distance of rising is the arc of the horizon intercepted between the summer solstitial and the winter solstitial sunrises, or the sum of the distances of the solstitial from the equinoctial sunrise.

Therefore let *ABCD* be the meridian circle, and let *BED* be the semicircle of the horizon in the eastern hemisphere, and let *AEC* be the similar semicircle of the equa-

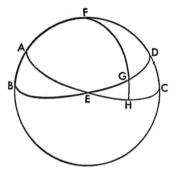

tor with *F* as its northern pole. Let point *G* be taken as the rising of the sun at the summer solstice, and let *FGH* the arc of a great circle be drawn. Therefore since the motion of the terrestrial sphere takes place around pole *F* of the equator, then necessarily points *G* an *H* will fit onto the meridian *ABCD* at the same time, since parallel circles are around the same poles through which pass the great circles which intercept similar arcs on the parallel circles.

Wherefore the selfsame time from the rising at *G* to midday measures also the arc *AEH*; and the time from midnight to sunrise measures *CH* the remaining and subterranean arc of the semicircle. Now *AEC* is a semicircle; and *AE* and *EC* are quadrants of circles, since they are drawn through the pole of *ABCD*. On that account, *EH* will be half the difference between the longest day and the equinox; and *EG* will be the distance between the equinoctial and the solstitial sunrise. Therefore since in triangle *EHG* angle *GEH*, the obliquity of the sphere, is established by means of arc *AB*, and angle *GHE* is right, and side *GH* is given as the distance from the summer tropic to the equator; the remaining sides are also given by the fourth theorem on sphericals: side *EH* as half the difference between the longest day and the equinox, and side *GE* as the distance of sunrise. Moreover, if together with side *GH* side *EH*, (half) the difference between the longest day and the equinox, or else *EG*, is given; angle *E* of the inclination of the sphere is given, and hence *FD* the elevation of the pole above the horizon.

But even if it is not the tropic but some other point *G* in the ecliptic which is taken, nevertheless arcs *EG* and *EH* will become manifest: since by the table of declinations set out above *GH* the arc of declination for that degree of the ecliptic becomes known, and the rest can be demonstrated in the same way.

Hence it also follows that the degrees of the ecliptic which are equally distant from the tropic cut off equal arcs of the horizon [35ᵃ] between the equinoctial sunrise and the same degrees and make the lengths of days and nights inversely equal. And that is so because the parallels which pass through each of those degrees of the ecliptic are equal, since each of the degrees has the same declination.

But when equal arcs are taken between the equinoctial intersection and the two degrees (on the ecliptic), again the distances of rising are equal but in different directions; and the duration of days and nights are inversely equal, because on each side of the equinox the durations describe equal arcs of parallels, according as the signs themselves which are equally distant from the equinox have equal declinations from the equator.

For in the same figure let GM and KN the arcs of parallels be described cutting the horizon BED at points G and K, and let LKO a quadrant of a great circle be drawn from the south pole L. Therefore since

$$\text{declination } HG = \text{declination } KO,$$

there will be two triangles DFG and BLK, wherein two sides of the one are equal to two sides of the other:

$$FG = LK$$

and the elevations of the poles are equal,

$$FD = LB,$$

and

$$\text{angle } D = \text{angle } B = 90°.$$

Therefore

$$\text{base } DG = \text{base } BK;$$

and hence, as the distances of sunrise are the remainders of the quadrants

$$GE = EK.$$

Wherefore since here too,

$$\text{side } EG = \text{side } EK,$$
$$\text{side } GH = \text{side } KO,$$

and

$$\text{vertical angle } KEO = \text{vertical angle } GEH;$$
$$\text{side } EH = \text{side } EO.$$

And

$$EH + 90° = OE + 90°.$$

Hence

$$\text{arc } AEH = \text{arc } OEC.$$

But since great circles described through the poles of parallel circles cut off similar arcs, GM and KN will be similar and equal, as was to be shown.

But all this can be shown differently. In the same way let the meridian circle $ABCD$ be described with centre E. Let the diameter of the equator and the common section of the two circles be AEC; let BED be the diameter of the horizon and the meridian line, let LEM be the axis of the sphere; and let L be the apparent pole and M the hidden. Let AF be taken as the distance of the summer solstice or as some other declination; and to AF let GF be drawn as the diameter of a parallel and its common section with the meridian; FG will cut the axis at K and the meridian line at N.

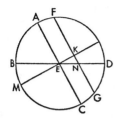

Therefore, [35^b] since by the definition of Posidonius those lines are parallel which neither move toward nor move away from one another but which everywhere make the perpendicular lines between them equal,

$$KE = {}^1/_2 \text{ ch. } 2 \ AF.$$

Similarly *KN* will be half of the chord subtending twice the arc of the parallel circle whose radius is *FK*. And twice this arc is the difference between the equinoctial day and the other day. And this is true because all the semicircles of which these lines are the common sections and diameters—namely, *BED* of the oblique horizon, *LEM* of the right horizon, *AEC* of the equator, and *FKG* of the parallel— are perpendicular to the plane of circle *ABCD*. And by Euclid's *Elements*, XI, 19, the common sections which they make with one another are perpendicular to the same plane at points *E*, *K*, and *N*: and by XI, 6, these common sections are parallel to one another.

And *K* is the centre of the parallel circle; and *E* is the centre of the sphere. Wherefore *EN* is half the chord subtending twice the arc of the horizon which is the difference between sunrise on the parallel and the equinoctial sunrise. Therefore, since the declination *AF* was given together with *FL* the remainder of the quadrant, *KE* half of the chord subtending twice arc *AF* and *FK* half the chord subtending twice arc *FL* will be established in terms of the parts whereof *AE* has 100,000. But in the right triangle *EKN* angle *KEN* is given by *DL* the elevation of the pole, and the remaining angle *KNE* is equal to *AEB*, because in the oblique sphere the parallels are equally inclined to the horizon; and the sides are given in the same parts whereof the radius has 100,000. Therefore *KN* will be given in the parts whereof *FK* the radius of the parallel has 100,000; for *KN* is equal to half the chord subtending the arc which measures the distance between the equinoctial day and a day on the parallel; and this arc is similarly given in the degrees whereof the parallel circle has 360°.

From this it is clear that

$$FK : KN = {}^1/_2 \text{ ch. } 2 \ FL : {}^1/_2 \text{ ch. } 2 \ AF \text{ comp. } {}^1/_2 \text{ ch. } 2 \ AB : {}^1/_2 \text{ ch. } 2 \ DL$$

and

$${}^1/_2 \text{ ch. } 2 \ FL : {}^1/_2 \text{ ch. } 2 \ AF \text{ comp. } {}^1/_2 \text{ ch. } 2 \ AB : {}^1/_2 \text{ ch. } 2 \ DL = FK : KE \text{ comp. } EK : KN.$$

That is to say, *EK* is taken as a mean between *FK* and *KN*. Similarly too

$$BE : EN = BE : EK \text{ comp. } KE : EN,$$

as Ptolemy shows in greater detail by means of spherical segments. So I think that not only the inequality of days and nights can be determined; but also that in the case of the moon and the stars whose declination on the parallels described through them by the daily movement has been given, the segments (of the parallels) which are above the horizon can be distinguished from those which are below; and hence the risings and settings (of the moon or stars) can be easily understood.

TABLE OF DIFFERENCE OF THE ASCENSIONS IN AN OBLIQUE SPHERE

Declination	Elevation of the Pole										
	31°		32°		33°		34°		35°		36°	
	Times	Min.	Times	Min.	Times	Min.	Times	Min.	Times	Min.	Times	Min.
1	0	36	0	37	0	39	0	40	0	42	0	44
2	1	12	1	15	1	18	1	21	1	24	1	27
3	1	48	1	53	1	57	2	2	2	6	2	11
4	2	24	2	30	2	36	2	42	2	48	2	55
5	3	1	3	8	3	15	3	23	3	31	3	39
6	3	37	3	46	3	55	4	4	4	13	4	23
7	4	14	4	24	4	34	4	45	4	56	5	7
8	4	51	5	2	5	14	5	26	5	39	5	52
9	5	28	5	41	5	54	6	8	6	22	6	36
10	6	5	6	20	6	35	6	50	7	6	7	22
11	6	42	6	59	7	15	7	32	7	49	8	7
12	7	20	7	38	7	56	8	15	8	34	8	53
13	7	58	8	18	8	37	8	58	9	18	9	39
14	8	37	8	58	9	19	9	41	10	3	10	26
15	9	16	9	38	10	1	10	25	10	49	11	14
16	9	55	10	19	10	44	11	9	11	25	12	2
17	10	35	11	1	11	27	11	54	12	22	12	50
18	11	16	11	43	12	11	12	40	13	9	13	39
19	11	56	12	25	12	55	13	26	13	57	14	29
20	12	38	13	9	13	40	14	13	14	46	15	20
21	13	20	13	53	14	26	15	0	15	36	16	12
22	14	3	14	37	15	13	15	49	16	27	17	5
23	14	47	15	23	16	0	16	38	17	17	17	58
24	15	31	16	9	16	48	17	29	18	10	18	52
25	16	16	16	56	17	38	18	20	19	3	19	48
26	17	2	17	45	18	28	19	12	19	58	20	45
27	17	50	18	34	19	19	20	6	20	54	21	44
28	18	38	19	24	20	12	21	1	21	51	22	43
29	19	27	20	16	21	6	21	57	22	50	23	45
30	20	18	21	9	22	1	22	55	23	51	24	48
31	21	10	22	3	22	58	23	55	24	53	25	53
32	22	3	22	59	23	56	24	56	25	57	27	0
33	22	57	23	54	24	19	25	59	27	3	28	9
34	23	55	24	56	25	59	27	4	28	10	29	21
35	24	53	25	57	27	3	28	10	29	21	30	35
36	25	53	27	0	28	9	29	21	30	35	31	52

TABLE OF DIFFERENCE OF THE ASCENSIONS IN AN OBLIQUE SPHERE

Declin ation	Elevation of the Pole										
	37°		38°		39°		40°		41°		42°	
	Times	Min.	Times	Min.	Times	Min.	Times	Min.	Times	Min.	Times	Min.
1	0	45	0	47	0	49	0	50	0	52	0	54
2	1	31	1	34	1	37	1	41	1	44	1	48
3	2	16	2	21	2	26	2	31	2	37	2	42
4	3	1	3	8	3	15	3	22	3	29	3	37
5	3	47	3	55	4	4	4	13	4	22	4	31
6	4	33	4	43	4	53	5	4	5	15	5	26
7	5	19	5	30	5	42	5	55	6	8	6	21
8	6	5	6	18	6	32	6	46	7	1	7	16
9	6	51	7	6	7	22	7	38	7	55	8	12
10	7	38	7	55	8	13	8	30	8	49	9	8
11	8	25	8	44	9	3	9	23	9	44	10	5
12	9	13	9	34	9	55	10	16	10	39	11	2
13	10	1	10	24	10	46	11	10	11	35	12	0
14	10	50	11	14	11	39	12	5	12	31	12	58
15	11	39	12	5	12	32	13	0	13	28	13	58
16	12	29	12	57	13	26	13	55	14	26	14	58
17	13	19	13	49	14	20	14	52	15	25	15	59
18	14	10	14	42	15	15	15	49	16	24	17	1
19	15	2	15	36	16	11	16	48	17	25	18	4
20	15	55	16	31	17	8	17	47	18	27	19	8
21	16	49	17	27	18	7	18	47	19	30	20	13
22	17	44	18	24	19	6	19	49	20	34	21	20
23	18	39	19	22	20	6	20	52	21	39	22	28
24	19	36	20	21	21	8	21	56	22	46	23	38
25	20	34	21	21	22	11	23	2	23	55	24	50
26	21	34	22	24	23	16	24	10	25	5	26	3
27	22	35	23	28	24	22	25	19	26	17	27	18
28	23	37	24	33	25	30	26	30	27	31	28	36
29	24	41	25	40	26	40	27	43	28	48	29	57
30	25	47	26	49	27	52	28	59	30	7	31	19
31	26	55	28	0	29	7	30	17	31	29	32	45
32	28	5	29	13	30	54	31	31	32	54	34	14
33	29	18	30	29	31	44	33	1	34	22	35	47
34	30	32	31	48	33	6	34	27	35	54	37	24
35	31	51	33	10	34	33	35	59	37	30	39	5
36	33	12	34	35	36	2	37	34	39	10	40	51

TABLE OF DIFFERENCE OF THE ASCENSIONS IN AN OBLIQUE SPHERE

Declin ation	Elevation of the Pole										
	43°		44°		45°		46°		47°		48°	
	Times	Min.	Times	Min.	Times	Min.	Times	Min.	Times	Min.	Times	Min.
1	0	56	0	58	1	0	1	2	1	4	1	7
2	1	52	1	56	2	0	2	4	2	9	2	13
3	2	48	2	54	3	0	3	7	3	13	3	20
4	3	44	3	52	4	1	4	9	4	18	4	27
5	4	41	4	51	5	1	5	12	5	23	5	35
6	5	37	5	50	6	2	6	15	6	28	6	42
7	6	34	6	49	7	3	7	18	7	34	7	50
8	7	32	7	48	8	5	8	22	8	40	8	59
9	8	30	8	48	9	7	9	26	9	47	10	8
10	9	28	9	48	10	9	10	31	10	54	11	18
11	10	27	10	49	11	13	11	37	12	2	12	28
12	11	26	11	51	12	16	12	43	13	11	13	39
13	12	26	12	53	13	21	13	50	14	20	14	51
14	13	27	13	56	14	26	14	58	15	30	16	5
15	14	28	15	0	15	32	16	7	16	42	17	19
16	15	31	16	5	16	40	17	16	17	54	18	34
17	16	34	17	10	17	48	18	27	19	8	19	51
18	17	38	18	17	18	58	19	40	20	23	21	9
19	18	44	19	25	20	9	20	53	21	40	22	29
20	19	50	20	35	21	21	22	8	22	58	23	51
21	20	59	21	46	22	34	23	25	24	18	25	14
22	22	8	22	58	23	50	24	44	25	40	26	40
23	23	19	24	12	25	7	26	5	27	5	28	8
24	24	32	25	28	26	26	27	27	28	31	29	38
25	25	47	26	46	27	48	28	52	30	0	31	12
26	27	3	28	6	29	11	30	20	31	32	32	48
27	28	22	29	29	30	38	31	51	33	7	34	28
28	29	44	30	54	32	7	33	25	34	46	36	12
29	31	8	32	22	33	40	35	2	36	28	38	0
30	32	35	33	53	35	16	36	43	38	15	39	53
31	34	5	35	28	36	56	38	29	40	7	41	52
32	35	38	37	7	38	40	40	19	42	4	43	57
33	37	16	38	50	40	30	42	15	44	8	46	9
34	38	58	40	39	42	25	44	18	46	20	48	31
35	40	46	42	33	44	27	46	23	48	36	51	3
36	42	39	44	33	46	36	48	47	51	11	53	47

TABLE OF DIFFERENCE OF THE ASCENSIONS IN AN OBLIQUE SPHERE

Declin ation	Elevation of the Pole										
	49°		50°		51°		52°		53°		54°
	Times	Min.	Times	Min.	Times	Min.	Times	Min.	Times	Min.	Times	Min.
1	1	9	1	12	1	14	1	17	1	20	1	23
2	2	18	2	23	2	28	2	34	2	39	2	45
3	3	27	3	35	3	43	3	51	3	59	4	8
4	4	37	4	47	4	57	5	8	5	19	5	31
5	5	47	5	50	6	12	6	26	6	40	6	55
6	6	57	7	12	7	27	7	44	8	1	8	19
7	8	7	8	25	8	43	9	2	9	23	9	44
8	9	18	9	38	10	0	10	22	10	45	11	9
9	10	30	10	53	11	17	11	42	12	8	12	35
10	11	42	12	8	12	35	13	3	13	32	14	3
11	12	55	13	24	13	53	14	24	14	57	15	31
12	14	9	14	40	15	13	15	47	16	23	17	0
13	15	24	15	58	16	34	17	11	17	50	18	32
14	16	40	17	17	17	56	18	37	19	19	20	4
15	17	57	18	39	19	19	20	4	20	50	21	38
16	19	16	19	59	20	44	21	32	22	22	23	15
17	20	36	21	22	22	11	23	2	23	56	24	53
18	21	57	22	47	23	39	24	34	25	33	26	34
19	23	20	24	14	25	10	26	9	27	11	28	17
20	24	45	25	42	26	43	27	46	28	53	30	4
21	26	12	27	14	28	18	29	26	30	37	31	54
22	27	42	28	47	29	56	31	8	32	25	33	47
23	29	14	30	23	31	37	32	54	34	17	35	45
24	31	4	32	3	33	21	34	44	36	13	37	48
25	32	26	33	46	35	10	36	39	38	14	39	59
26	34	8	35	32	37	2	38	38	40	20	42	10
27	35	53	37	23	39	0	40	42	42	33	44	32
28	37	43	39	19	41	2	42	53	44	53	47	2
29	39	37	41	21	43	12	45	12	47	21	49	44
30	41	37	43	29	45	29	47	39	50	1	52	37
31	43	44	45	44	47	54	50	16	52	53	55	48
32	45	57	48	8	50	30	53	7	56	1	59	19
33	48	19	50	44	53	20	56	13	59	28	63	21
34	50	54	53	30	56	20	59	42	63	31	68	11
35	53	40	56	34	59	58	63	40	68	18	74	32
36	56	42	59	59	63	47	68	26	74	36	90	0

TABLE OF DIFFERENCE OF THE ASCENSIONS IN AN OBLIQUE SPHERE

Declination	Elevation of the Pole										
	55°		56°		57°		58°		59°		60°	
	Times	Min.	Times	Min.	Times	Min.	Times	Min.	Times	Min.	Times	Min.
1	1	26	1	29	1	32	1	36	1	40	1	44
2	2	52	2	58	3	5	3	12	3	20	3	28
3	4	17	4	27	4	38	4	49	5	0	5	12
4	5	44	5	57	6	11	6	25	6	41	6	57
5	7	11	7	27	7	44	8	3	8	22	8	43
6	8	38	8	58	9	19	9	41	10	4	10	29
7	10	6	10	29	10	54	11	20	11	47	12	17
8	11	35	12	1	12	30	13	0	13	32	14	5
9	13	4	13	35	14	7	14	41	15	17	15	55
10	14	35	15	9	15	45	16	23	17	4	17	47
11	16	7	16	45	17	25	18	8	18	53	19	41
12	17	40	18	22	19	6	19	53	20	43	21	36
13	19	15	20	1	20	50	21	41	22	36	23	34
14	20	52	21	42	22	35	23	31	24	31	25	35
15	22	30	23	24	24	22	25	23	26	29	27	39
16	24	10	25	9	26	12	27	19	28	30	29	47
17	25	53	26	57	28	5	29	18	30	35	31	59
18	27	39	28	48	30	1	31	20	32	44	34	19
19	29	27	30	41	32	1	33	26	34	58	36	37
20	31	19	32	39	34	5	35	37	37	17	39	5
21	33	15	34	41	36	14	37	54	39	42	41	40
22	35	14	36	48	38	28	40	17	42	15	44	25
23	37	19	39	0	40	49	42	47	44	57	47	20
24	39	29	41	18	43	17	45	26	47	49	50	27
25	41	45	43	44	45	54	48	16	50	54	53	52
26	44	9	46	18	48	41	51	19	54	16	57	39
27	46	41	49	4	51	41	54	38	58	0	61	57
28	49	24	52	1	54	58	58	19	62	14	67	4
29	52	20	55	16	58	36	62	31	67	18	73	46
30	55	32	58	52	62	45	67	31	73	55	90	0
31	59	6	62	58	67	42	74	4	90	0		
32	63	10	67	53	74	12	90	0				
33	68	1	74	19	90	0			The vacant spaces go to			
34	74	33	90	0					stars which neither rise			
35	90	0							nor set			
36												

8. ON THE HOURS AND PARTS OF THE DAY AND NIGHT

[38ᵇ] Accordingly it is clear from this that if from the table we take the difference of days which correspond to the declination of the sun and is found under the given elevation of the pole and add it to a quadrant of a circle in the case of a northern declination and subtract it in the case of a southern declination, and then double the result, we shall have the length of that day and the span of night, which is the remainder of the circle.

Any of these segments divided by 15 "times" will show how many equal hours there are (in that day). But by taking a twelfth part of the segment we shall have the duration of one seasonal hour. Now the hours get their name from their day, whereof each hour is always the twelfth part. Hence the hours are found to have been called summer-solstitial, equinoctial, and winter-solstitial by the ancients.

But there were not any others in use at first except the twelve hours from sunrise to sunset; and they divided the night into four vigils or watches. This set-up of the hours lasted a long time by the tacit consent of mankind. And for its sake were water-clocks invented: by the addition and subtraction of dripping water people adjusted the hours to the different lengths of days, so as not to have distinctions in time obscured by a cloud. But afterwards when equal hours common to day and night came into general use, as making it easier to tell the time, then the seasonal hours became obsolete, so that if you asked any ordinary person whether it was the first, third or sixth, ninth, or eleventh hour of the day, he would not have any answer to make or would make one which had nothing to do with the matter. Furthermore, at present some measure the number of equal hours from noon, some from sunset, some from midnight, and others from sunrise, according as it is instituted by the state.

9. ON THE OBLIQUE ASCENSION OF THE PARTS OF THE ECLIPTIC AND HOW THE DEGREE WHICH IS IN THE MIDDLE OF THE HEAVENS IS DETERMINED WITH RESPECT TO THE DEGREE WHICH IS RISING

[39ᵃ] Now that the lengths and differences of days and nights have been expounded, there follows in proper order an exposition of oblique ascensions, that is to say, together with what "times" (of the equator) the dodekatemoria, *i.e.*, the twelve parts of the ecliptic, or some other arcs of it, cross the horizon. For the differences between right and oblique ascensions are the same as the differences between the equinox and a different day, as we set forth. Furthermore, the ancients borrowed the names of animals for twelve constellations of unmoving stars, and, beginning at the spring equinox, called them Aries, Taurus, Gemini, Cancer, and so on in order.

Therefore for the sake of greater clearness let the meridian circle *ABCD* be repeated; and the equatorial semicircle *AEC* and the horizon *BED*, which cut one another at point *E*. Now let point *H* be taken as the equinox. Let the ecliptic *FHI* pass through this point and cut the horizon at *L*; and through this intersection let *KLM* the quadrant of a great circle fall from *K* the pole of the equator. Thus it is perfectly clear that arc *HL* of the ecliptic and arc *HE* of the equator cross the horizon together, but that in the right sphere (arc *HL*) was rising together with arc *HEM*. Arc *EM* is the difference between these ascensions; and we have already shown that it is half the difference between the equinox and the different day. But in a northern declination what was there added (to the quadrant of a circle) is here subtracted (from the right ascension); but in a southern declination it is added to the right ascension, so that the ascension may become oblique. And hence the extent that the whole sign or some other arc of the ecliptic has emerged may become manifest by means of the numbered ascensions from beginning to end.

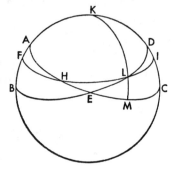

From this it follows that when some degree of the ecliptic is given, the rising of which has been measured from the equinox, the degree which is in the middle of the heavens is also given. For when the declination of a degree rising at *L* has been given as corresponding to arc *HL* the distance from the equinox, and arc *HEM* is the right ascension, and the whole *AHEM* is the arc of half a day: then the remainder *AH* is given. And *AH* is the right ascension of arc *FH*, which is also given by the table, or because angle *AHF* the angle of section is given together with side *AH*, and angle *FAH* is right. And so *FHL* the whole arc between the degree rising and the degree in the middle of the heavens is given.

Conversely, if the degree which is in the middle of the heavens, namely arc *FH*, is given first, we shall also know the sign which [39ᵇ] is rising. For arc *AF* the declination will be known; and by means of the angle of obliquity of the sphere, arc *AFB* and arc *FB* the remainder will become known. Now in triangle *BFL* angle *BFL* and side *FB* are given by the above, and angle *FBL* is right. Therefore side *FHL*, which was sought, is given; or by a different method, as below.

10. On the Angle of Section of the Ecliptic with the Horizon

Moreover, as the ecliptic is oblique to the axis of the sphere, it makes various angles with the horizon. For we have already said in the case of the differences of the shadows that two opposite degrees of the ecliptic pass through the axis of the horizon of those who live between the tropics. But I think it will be sufficient for our purpose if we demonstrate the angles which we the heteroscian inhabitants find. By means of these angles the universal ratio of them may easily be understood. Accordingly I think it is clear enough that in the oblique sphere when the equinox or the beginning of Aries is rising, the more the greatest southward declination increases—and this declination is measured from the beginning of Capricorn which is in the middle of the heavens at this time—the more the ecliptic is inclined and verges towards the horizon; and conversely, when the ecliptic has a greater elevation (above the horizon), it makes a greater eastern angle, when the beginning of Libra is emerging and the beginning of Cancer is in the middle of the heavens; since these three circles, the equator, the ecliptic, and the horizon, coincide in one common section at the poles of the meridian circle, and the arcs of the meridian circle intercepted by them show how great the angle of rising should be judged to be.

But in order that the way of taking the measurements of the other parts of the ecliptic may be clear, again let *ABC* be the meridian circle, let *BED* be the semicircle of the horizon, let *AEC* be the semi-circle of the ecliptic, and let any degree of the ecliptic be rising at *E*.

Our problem is to find how great angle *AEB* is, according as four right angles are equal to 360°. Therefore since *E* is given as the rising degree, there are given by the foregoing the degree which is in the middle of the heavens and the arc *AE*.
And since

$$\text{angle } ABE = 90°;$$
$$\text{ch. } 2\ AE : \text{ch. } 2\ AB = \text{dmt. sph.} : \text{ch. } 2\ AEB.$$
[40ᵃ] Therefore too
$$\text{angle } AEB \text{ is given.}$$

But if the degree given is not rising but the degree in the middle of the heavens—and let it be *A*—nevertheless angle *AEB* will be the measure of the eastern angle or angle of rising. For with *E* as pole, let *FGH* the quadrant of a great circle be described, and let the quadrants *EAG* and *EBH* be completed. Therefore since

$$\text{meridian altitude } AB \text{ is given,}$$
and
$$AF = 90° - AB,$$
and by the foregoing
$$\text{angle } FAG \text{ is given,}$$
and

$$\text{angle } FGA = 90°;$$

therefore

$$\text{arc } FG \text{ is given,}$$

and

$$90° - FG = GH,$$

which measures the sought angle of rising. Similarly, it is also made evident here how for the degree which is in the middle of the heavens the degree which is rising is given, because

$$\text{ch. } 2 \ GH : 2 \ AB = \text{dmt. sph. : ch. } 2 \ AE,$$

as in spherical triangles.

We are subjoining three sets of tables of these things. The first will be the table of ascensions in the right sphere, beginning with Aries and increasing by sixtieth parts of the ecliptic. The second will be that of ascensions in the oblique sphere, proceeding by steps of 6° in the ecliptic, from the parallel for which there is a polar elevation of 39° to the parallel which has a polar elevation of 57°—increasing the elevation by 3°s each time. The remaining table contains the angles made with the horizon and proceeds through the ecliptic by steps of 6° beneath the same seven segments. These tables have been set up in accordance with the least obliquity of the ecliptic, namely 23°28', which is approximately right for our age.

TABLE OF THE ASCENSIONS OF THE SIGNS IN THE REVOLUTION OF THE RIGHT SPHERE

Ecliptic Signs	Deg.	Ascensions Times	Min.	One Degree Times	Min.	Ecliptic Signs	Deg.	Ascensions Times	Min.	One Degree Times	Min.
Aries ♈	6	5	30	0	55	Libra ♎	6	185	30	0	55
	12	11	0	0	55		12	191	0	0	55
	18	16	34	0	56		18	196	34	0	56
	24	22	10	0	56		24	202	10	0	56
	30	27	54	0	57		30	207	54	0	57
Taurus ♉	6	33	43	0	58	Scorpio ♏	6	213	43	0	58
	12	39	35	0	59		12	219	35	0	59
	18	45	32	1	0		18	225	32	1	0
	24	51	37	1	1		24	231	37	1	1
	30	57	48	1	2		30	237	48	1	2
Gemini ♊	6	64	6	1	3	Sagittarius ♐	6	244	6	1	3
	12	70	29	1	4		12	250	29	1	4
	18	76	57	1	5		18	256	57	1	5
	24	83	27	1	5		24	263	27	1	5
	30	90	0	1	5		30	270	0	1	5
Cancer ♋	6	96	33	1	5	Capricornus ♑	6	276	33	1	5
	12	103	3	1	5		12	283	3	1	5
	18	109	31	1	5		18	289	31	1	5
	24	115	54	1	4		24	295	54	1	4
	30	122	12	1	3		30	302	12	1	3
Leo ♌	6	128	23	1	2	Aquarius ♒	6	308	23	1	2
	12	134	28	1	1		12	314	28	1	1
	18	140	25	1	0		18	320	25	1	0
	24	146	17	0	59		24	326	17	0	59
	30	152	6	0	58		30	332	6	0	58
Virgo ♍	6	157	50	0	57	Pisces ♓	6	337	50	0	57
	12	163	26	0	56		12	343	26	0	56
	18	169	0	0	56		18	349	0	0	56
	24	174	30	0	55		24	354	30	0	55
	30	180	0	0	55		30	360	0	0	55

TABLE OF THE ASCENSIONS IN THE OBLIQUE SPHERE

Ecliptic Signs		Elevation of the Pole of the Equator												
		39°		42°		45°		48°		51°		54°		57°
		Ascension		Ascension		Ascension		Ascension		Ascension		Ascension		Ascension
		Times	Min.	Times	Min.	Times	Min.	Times	Min.	Times	Min.	Times	Min.	Times Min.
♈	6	3	34	3	20	3	6	2	50	2	32	2	12	1 49
	12	7	10	6	44	6	15	5	44	5	8	4	27	3 40
	18	10	50	10	10	9	27	8	39	7	47	6	44	5 34
	24	14	32	13	39	12	43	11	40	10	28	9	7	7 32
	30	18	26	17	21	16	11	14	51	13	26	11	40	9 40
♉	6	22	30	21	12	19	46	18	14	16	25	14	22	11 57
	12	26	39	25	10	23	32	21	42	19	38	17	13	14 23
	18	31	0	29	20	27	29	25	24	23	2	20	17	17 2
	24	35	38	33	47	31	43	29	25	26	47	23	42	20 2
	30	40	30	38	30	36	15	33	41	30	49	27	26	23 22
♊	6	45	39	43	31	41	7	38	23	35	15	31	34	27 7
	12	51	8	48	52	46	20	43	27	40	8	36	13	31 26
	18	56	56	54	35	51	56	48	56	45	28	41	22	36 20
	24	63	0	60	36	57	54	54	49	51	15	47	1	41 49
	30	69	25	66	59	64	16	61	10	57	34	53	28	48 2
♋	6	76	6	73	42	71	0	67	55	64	21	60	7	54 55
	12	83	2	80	41	78	2	75	2	71	34	67	28	62 26
	18	90	10	87	54	85	22	82	29	79	10	75	15	70 28
	24	97	27	95	19	92	55	90	11	87	3	83	22	78 55
	30	104	54	102	54	100	39	98	5	95	13	91	50	87 46
♌	6	112	24	110	33	108	30	106	11	103	33	100	28	96 48
	12	119	56	118	16	116	25	114	20	111	58	109	13	105 58
	18	127	29	126	0	124	23	122	32	120	28	118	3	115 13
	24	135	4	133	46	132	21	130	48	128	59	126	56	124 31
	30	142	38	141	33	140	23	139	3	137	38	135	52	133 52
♍	6	150	11	149	19	148	23	147	20	146	8	144	47	143 12
	12	157	41	157	1	156	19	155	29	154	38	153	36	153 24
	18	165	7	164	40	164	12	163	41	163	5	162	24	162 47
	24	172	34	172	21	172	6	171	51	171	33	171	12	170 49
	30	180	0	180	0	180	0	180	0	180	0	180	0	180 0

TABLE OF THE ASCENSIONS IN THE OBLIQUE SPHERE

Ecliptic Signs	39° Ascension Times	Min.	42° Ascension Times	Min.	45° Ascension Times	Min.	48° Ascension Times	Min.	51° Ascension Times	Min.	54° Ascension Times	Min.	57° Ascension Times	Min.
≎ 6	187	26	187	39	187	54	188	9	188	27	188	48	189	11
12	194	53	195	19	195	48	196	19	196	55	197	36	198	23
18	202	21	203	0	203	41	204	30	205	24	206	25	207	36
24	209	49	210	41	211	37	212	40	213	52	215	13	216	48
30	217	22	218	27	219	37	220	57	222	22	224	8	226	8
♏ 6	224	56	226	14	227	38	229	12	231	1	233	4	235	29
12	232	56	234	0	235	37	237	28	239	32	241	57	244	47
18	240	31	241	44	243	35	245	40	248	2	250	47	254	2
24	247	36	249	27	251	30	253	49	256	27	259	32	263	12
30	255	36	257	6	259	21	261	52	264	47	268	10	272	14
♐ 6	262	8	264	41	267	5	269	49	272	57	276	38	281	5
12	269	50	272	6	274	38	277	31	280	50	284	45	289	32
18	276	58	279	19	281	58	248	58	288	26	292	32	297	34
24	283	54	286	18	289	0	292	5	295	39	299	53	305	5
30	290	75	293	1	295	45	298	50	302	26	306	42	311	58
♑ 6	297	0	299	24	302	6	305	11	308	45	312	59	318	11
12	303	4	305	25	308	4	311	4	314	32	318	38	323	40
18	308	52	311	8	313	40	316	33	319	52	323	47	328	34
24	314	21	316	29	318	53	321	37	324	45	328	26	332	53
30	319	30	321	30	323	45	326	19	329	11	332	34	336	38
♒ 6	324	21	326	13	328	16	330	35	333	13	336	18	339	58
12	330	0	330	40	332	31	334	36	336	58	339	43	342	58
18	333	21	334	50	336	27	338	18	340	22	342	47	345	37
24	337	30	338	48	340	3	341	46	343	35	345	38	348	3
30	341	34	342	39	343	49	345	9	346	34	348	20	350	20
♓ 6	345	29	346	21	347	17	348	20	349	32	350	53	352	28
12	349	11	349	51	350	33	351	21	352	14	353	16	354	26
18	352	50	353	16	353	45	354	16	354	52	355	33	356	20
24	356	26	356	40	356	23	357	10	357	53	357	48	358	11
30	360	0	360	0	360	0	360	0	360	0	360	0	360	0

Elevation of the Pole of the Equator

TABLE OF THE ANGLES MADE BY THE ECLIPTIC WITH THE HORIZON

Ecliptic		Elevation of the Pole of the Equator													Ecliptic
		39°		42°		45°		48°		51°		54°		57°		
		Angles		Angles		Angles		Angles		Angles		Angles		Angles		
Sign		Deg.	Min.	Deg.	Min.	Deg.	Min.	Deg.	Min.	Deg.	Min.	Deg.	Min.	Deg.	Min.	Sign
♈	0	27	32	24	32	21	32	18	32	15	32	12	32	9	32	30
	6	27	37	24	36	21	36	18	36	15	35	12	35	9	35	24
	12	27	49	24	39	21	48	18	47	15	45	12	43	9	41	18
	18	28	13	25	9	22	6	19	3	15	59	12	56	9	53	12
	24	28	45	25	40	22	34	19	29	16	23	13	18	10	13	6 ♓
	30	29	27	26	15	23	11	20	5	16	56	13	45	10	31	30
♉	6	30	19	27	9	23	59	20	48	17	34	14	20	11	2	24
	12	31	21	28	9	24	56	20	41	18	23	15	3	11	40	18
	18	32	35	29	20	26	3	22	43	19	21	15	56	12	26	12
	24	34	5	30	43	27	23	24	2	20	41	16	59	13	20	6 ♒
	30	35	40	32	17	28	52	25	26	21	52	18	14	14	26	30
♊	6	37	29	34	1	30	37	27	5	23	11	19	42	15	48	24
	12	39	32	36	4	32	32	28	56	25	15	21	25	17	23	18
	18	41	44	38	14	34	41	31	3	27	18	23	25	19	16	12
	24	44	8	40	32	37	2	33	22	29	35	25	37	21	26	6 ♑
	30	46	41	43	11	39	33	35	53	32	5	28	6	23	52	30
♋	6	49	18	45	51	42	15	38	35	34	44	30	50	26	36	24
	12	52	3	48	34	45	0	41	8	37	55	33	43	29	34	18
	18	54	44	51	20	47	48	44	13	40	31	36	40	32	39	12
	24	57	30	54	5	50	38	47	6	43	33	39	43	35	50	6 ♐
	30	60	4	56	42	53	22	49	54	46	21	42	43	38	56	30
♌	6	62	40	59	27	56	0	52	34	49	9	45	37	41	57	24
	12	64	59	61	44	58	26	55	7	51	46	48	19	44	48	18
	18	67	7	63	56	60	20	57	26	54	6	50	47	47	24	12
	24	68	59	65	52	62	42	59	30	56	17	53	7	49	47	6 ♏
	30	70	38	67	27	64	18	61	17	58	9	54	50	52	38	30
♍	6	72	0	68	53	65	51	62	46	59	37	56	27	53	16	24
	12	73	4	70	2	66	59	63	56	60	53	57	50	54	46	18
	18	73	51	70	50	67	49	64	48	61	46	58	45	55	44	12
	24	74	19	71	20	68	20	65	19	62	18	59	17	56	16	6 ♎
	30	74	28	71	28	68	28	65	28	62	28	59	28	56	28	0

11. On the Use of These Tables

[42^b] Now the use of these tables is clear from the demonstrations, since if we take the right ascension corresponding to the known degree of the sun and if for every equal hour measured from noon we add 15 "times" to it—not counting the 360° of the whole circle, if there is more than that—the sum of the right ascensions will show the degree of the ecliptic in the middle of the heavens at the proposed hour.

Similarly if you do the same thing in the case of the oblique ascension of your region, you will have the rising degree of the ecliptic for the hour measured from sunrise.

Moreover, in the case of certain stars which are outside the ecliptic but of which the right ascension has been established—as we taught above—by their right ascension from the beginning of Aries the degrees of the ecliptic which are in the middle of the heavens together with them are given according to the table; and by their oblique ascension the degree of the ecliptic which arises with them, according as the ascensions and parts of the ecliptic are placed in corresponding regions of the tables. It is possible to operate with the setting similarly but by means of the position opposite.

Moreover, if to the right ascension in the middle of the heavens a quadrant of a circle is added, the sum is the oblique ascension of the rising degree. Wherefore the rising degree is given by means of the degree in the middle of the heavens, and vice versa.

There follows the table of the angles of the ecliptic and the horizon, which are measured at the rising degree of the ecliptic. Hence it is understood how great the elevation of the 90th degree of the ecliptic is above the horizon; and it is particularly necessary to know that in the case of solar eclipses.

12. On the Angles and Arcs of the Circles Which Pass Through the Poles of the Horizon and Intersect the Same Circle of the Ecliptic

In what follows we shall expound the ratio of the angles and arcs made by the intersection of the ecliptic with the circles through the vertex of the horizon, in the cases wherein the intersections have some altitude above the horizon. But we spoke above concerning the meridian altitude of the sun or of any degree of the ecliptic which is in the middle of the heavens, and concerning the angle of section with the meridian, since [43^a] the meridian circle is also one of those circles which pass through the vertex of the horizon. Moreover we have already talked about the angle of the rising sign, the complementary angle to which is the angle which is comprehended by a

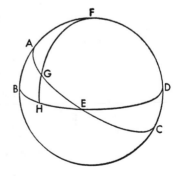

great circle passing through the vertex of the horizon and by the rising ecliptic. Therefore there remain to be considered the mean sections, that is, the mean sections of the meridian circle with the semicircles of the ecliptic and the horizon.

Let the above figure be repeated. Let G be taken as any point on the ecliptic between midday and the point of rising or setting. Through G from F the pole of the horizon let fall FGH a quadrant of a circle.

"Hour" AGE is given as the whole arc of the ecliptic between the meridian and the horizon, and by hypothesis AG is given. Similarly, because meridian altitude AB is given, and meridian angle FAG is given; therefore AF is given. And by what has been shown concerning spherical triangles, arc FG is given. And hence altitude of G is given, because 90° − FG = GH. And meridian angle FAG is given. And those are what we were looking for.

En route, we have taken from Ptolemy these truths about the angles and intersections of the ecliptic, and have referred ourselves to the geometry of spherical triangles. If anyone wishes to pursue this study at length, he can find by himself more utilities than we have given examples of.

13. On the Rising and Setting of the Stars

The rising and setting of the stars also seems to depend upon the daily revolution, not only the simple risings and settings of which we have just spoken but also those which occur in the morning or evening; because although their occurrence is affected by the course of the annual revolution, it will be better to speak of them here.

The ancient mathematicians distinguish the true risings and settings from the apparent. The morning rising of the star is true when the star rises at the same time as the sun; and the morning setting is true, when the star sets at sunrise; for morning is said to occur at the midpoint of this time. But the evening rising is true when the star rises at sunset; and the evening setting is true when the star sets at the same time as the sun; for evening is said to occur at the midpoint of this time, namely the time [43^b] between the time which is beginning and the time which ceases with night.

But the morning rising of a star is apparent when it rises first in the twilight before sunrise and begins to be apparent; and the morning setting is apparent when the star is seen to set very early before the sun rises. The evening rising is apparent when the star is seen to rise first in the evening; and the evening setting is apparent, when the star ceases to be apparent some time after sunset, and the star is occulted by the approach of the sun, until they come forth in their previous order at the morning rising.

This is true of the fixed stars and of the planets Saturn, Jupiter, and Mars. But Venus and Mercury rise and set in a different fashion. For they are not occulted by the approach of the sun, as the higher planets are; and they are not uncovered again by its departure. But, coming in front, they mingle with the radiance of the sun and free themselves. But when the higher planets have an evening rising and a morning setting, they are not obscured at any time, so as not to traverse the night with their illumination. But the lower planets remain hidden indifferently from sunset to sunrise, and cannot be seen anywhere. There is still another difference, namely that in higher planets the true morning risings and settings are prior to the apparent ones; and the evening risings and settings are posterior to the apparent, according as in the morning they precede the rising of the sun and in the evening follow its setting. But in the lower planets the apparent morning and evening risings are posterior to the true, while the apparent settings are prior to the true.

Now it can be understood from the above, where we expounded the oblique ascension of any star having a known position, how (the risings and settings) may be discerned, and together with what degree of the ecliptic the star rises or sets and at what position, or degree opposite—if the sun has become apparent by that time—the star has its true morning or evening rising or setting. The apparent risings and settings differ from the true according to the clarity and magnitude of the star, so that the stars which give a more powerful light are less dimmed by the rays of the sun than those which are less luminous. And the boundaries of occultation and apparition are determined in the lower hemisphere, between the horizon and the sun, on the arcs of circles which pass through the poles of the horizon. And the limits are 12° for the primary stars, 11° for Saturn, 10° for Jupiter, $11^1/_2$° for Mars, 5° for Venus, and 10° for Mercury. But in this whole period during which what is left of daylight yields to night—this period embraces twilight, or dusk—there are 18° of the aforesaid circle. When the sun has traversed these degrees, the smaller stars too begin to be apparent. By this distance the mathematicians determine [44ᵃ] a parallel below the horizon in the lower hemisphere, and they say that when the sun has reached this parallel, day has ended and night has begun. Therefore when we have learned with what degree of the ecliptic the star rises or sets and what the angle of section of the ecliptic with the horizon at that point is, and if then too we find as many degrees of the ecliptic between the rising degree and the sun as are sufficient to give the sun an altitude below the horizon in accord with the prescribed limits of the star in question; we shall pronounce that the first emergence or occultation of the star has taken place.

But what we have expounded, in the foregoing explanation, in the case of the altitude of the sun above the Earth agrees in all respects with its descent below the Earth. For there is no difference in the corresponding positions; and consequently those stars which are setting in the visible hemisphere are rising in the hidden hemisphere; and everything is the converse, and is easy to understand. What has been said concerning the rising and setting of the stars and the daily revolution of the terrestrial globe shall be sufficient.

14. On Investigating the Positions of the Stars and the Catalogue of the Fixed Stars

After the daily revolution of the terrestrial globe and its consequences have been expounded by us, the demonstrations relating to the annual circuit ought to follow now. But since some of the ancient mathematicians thought the phenomena of the fixed stars ought to come first as being the first beginnings of this art, accordingly we decided to act in accordance with this opinion, as among our principles and hypotheses we had assumed that the sphere of the fixed stars, to which the wanderings of all the planets are equally referred, is wholly immobile. But no one should be surprised at our following this order, although Ptolemy in his *Almagest* held that an explanation of the fixed stars could not be given, unless knowledge of the positions of the sun and moon had preceded it, and accordingly he judged that whatever had to do with the fixed stars should be put off till then. We think that this opinion must be opposed. But if you understand it of the numbers with which the apparent motion of the sun and moon is computed perhaps the opinion will stand. For Menelaus the geometer discovered the positions of many stars by means of the numbers relating to their conjunctions with the moon. [44b] But we shall do a much better job if we determine a star by the aid of instruments after examining carefully the positions of the sun and moon, as we will show how in a little while. We are even admonished by the wasted attempt of those who thought that the magnitude of the solar year could be defined simply by the equinoxes or solstices without the fixed stars. We shall never agree with them on that, so much so that there will nowhere be greater discord. Ptolemy called our attention to this: when he had evaluated the solar year in his time not without suspicion of an error which might emerge with the passage of time, he admonished posterity to examine the further certainty of the thing later on. Therefore it seemed to us to be worth the trouble to show how by means of artificial instruments the positions of the sun and moon may be determined, that is, how far distant they are from the spring equinox or some other cardinal points of the world. The knowledge of these positions will afford us some facilities for investigating the other stars, and thus we shall be able to set forth before your eyes the sphere of the fixed stars and an image of it embroidered with constellations.

Now we have set forth above with what instruments the distance of the tropics, the obliquity of the ecliptic, and the inclination of the sphere, or the altitude of the pole of the equator, may be determined. In the same way we can determine any other altitude of the sun at midday. This altitude will exhibit to us through its difference from the inclination of the sphere how great the declination of the sun from the equator is. Then by means of this declination the position of the sun at midday will become clear as measured from the solstice or the equinox. Now the sun seems to traverse approximately 1° during the space of 24 hours; $2^1/_2'$ come as the hourly allotment. Hence its position at any other definite hour will easily be determined.

But for observing the positions of the moon and stars another instrument is constructed, which Ptolemy calls the astrolabe. For let two circles, or rather four-sided rims of circles, be constructed in such a way that they may have their concave and convex surfaces at right angles to the plane sides. These rims are to be equal and similar in every respect and of a suitable size, in order not to become hard to handle through being too large, though they must be of sufficient amplitude to be divided into degrees and minutes. Their width and thickness [45a] should be at least one thirtieth of the diameter. Therefore they are to be fitted together and joined at right angles to one another, having their convex sides as it were on the surface of the same sphere, and their concave sides on the surface of another single sphere. Now one of the circles should have the relative position of the ecliptic; and the other, that of the circle which passes through both poles, *i.e.*, the poles of the equator and of the ecliptic. Therefore the circle of the ecliptic is to be divided along its sides into the conventional number of 360°, which are again to be subdivided according to the capacity of the instrument. Moreover, when quadrants on the other circle have been measured from the ecliptic, the poles of the ecliptic should be marked on it; and when a distance proportionate to the obliquity has been measured from those points, the poles of the equator are also to be marked down. When these circles are finished, two other circles should be prepared and constructed around the same poles of the ecliptic: they will move about these poles, one circle inside and one circle outside. They should be of equal thicknesses between their plane surfaces, and the width of their plane surfaces should be equal to that of the others; and they should be so constructed that at all points the concave surface of the larger will touch the convex surface of the ecliptic; and the convex surface of the smaller, the concave surface of the ecliptic. Nevertheless do not let their revolutions be impeded, but have them able to traverse freely and easily both the ecliptic together with its meridian circle and one another. Therefore we shall make holes in these circles diametrically opposite the poles of the ecliptic, and pass axles through these holes, so that by means of these axles the circles will be bound together and carried along. Moreover, the inside circle should be divided into 360° in such fashion that the single quadrant of 90° will be at the poles. Furthermore, within the concavity of the inside circle a fifth circle should be placed which can be turned in the same plane and which has an apparatus fixed to its plane surfaces which has openings diametrically opposite and reflectors or eyepieces, through which the light of the sun, as in a dioptra, can break through and go out along the diameter of the circle. And certain appliances or pointers for numbers are fitted on to this fifth circle at opposite points for the sake of observing the latitudes on the container circle. Finally, a sixth circle is to be applied which will embrace and support the whole astrolabe, which is hung on to it by means of fastenings at the poles of the equator; this last circle is to be placed upon some sort of column, or stand, and made to rest upon it perpendicular to the plane of the horizon. Moreover, the poles (of the equator) should be adjusted to the inclination of the sphere, so that the outmost circle will have a position similar to that of a natural meridian and will by no means waver from it.

Therefore after the instrument has been prepared in this way, when we wish to determine the position of some star, then in the evening or at the approach of sunset and at a time when the moon too is visible, we shall adjust the outer circle to the degree of the ecliptic, in which [45^b] we have determined—by the methods spoken of—that the sun is at that time. And we shall turn the intersection of the (ecliptic and the outer) circle towards the sun itself, until both of them—I mean the ecliptic and the outer circle which passes through its poles—cast shadows on themselves evenly.[1] Then we shall turn the inner circle towards the moon; and with the eye placed in its plane we shall mark its position on the ecliptic part of the instrument there where we shall view the moon as opposite, or as it were bisected by the same plane. That will be the position of the moon as seen in longitude. For without the moon there is no way of discovering the positions of the stars, as the moon alone among all is a partaker of both day and night. Then after nightfall, when the star whose position we are seeking is visible, we shall adjust the outer circle to the position of the moon; and thus, as we did in the case of the sun, we shall bring the position of the astrolabe into relation with the moon. Then also we shall turn the inner circle towards the star, until the star seems to be in contact with the plane surfaces of the circle and is viewed through the eyepieces which are on the little circle contained (by the inner circle). For in this way we shall have discovered the longitude and latitude of the star. When this is being

[1]*i.e.,* until the shadows intersect as two straight lines at right angles to one another.

done, the degree of the ecliptic which is in the middle of the heavens will be before the eyes; and accordingly it will be obvious at what hour the thing itself was done.[1]

For example, in the 2nd year of the Emperor Antoninus Pius, on the 9th day of Pharmuthi, the 8th month by the Egyptian calendar, Ptolemy, who was then at Alexandria and wished to observe at the time of sunset the position of the star which is in the breast of Leo and is called Basiliscus or Regulus, adjusted his astrolabe to the setting sun at 5 equatorial hours after midday. At this time the sun was at $3^1/_{24}°$ of Pisces, and by moving the inner circle he found that the moon was $92^1/_8°$ east of the sun: hence it was seen that the position of the moon was then at $5^1/_6°$ of Gemini. After half an hour—which made six hours since noon—when the star had already begun to be apparent and 4° of Gemini was in the middle of the heavens, he turned the outer circle of the instrument to the already determined position of the moon. Proceeding with the inner circle, he took the distance of the star from the moon as $57^1/_{10}°$ to the east. Accordingly the moon had been found at $92^1/_8°$ from the setting sun, as was said—which placed the moon at $5^1/_6°$ of Gemini; but it was correct for the moon to have moved $^1/_4°$ in the space of half an hour; since the hourly allotment in the movement of the moon is more or less $^1/_2°$; but on account of the then subtractive parallax of the moon it must have been slightly less than $^1/_4°$, [46ᵃ] that is to say, about $^1/_6°$: hence the moon was at $5^1/_3°$ of Gemini. But when we have discussed the parallaxes of

[1]Legend:
1. Circle through poles of ecliptic
2. Ecliptic
3. Outer circle
4. Inner circle
5. Little circle
6. Meridian circle
A and A_1. Poles of equator
BCC_1B_1. Axis of ecliptic
D. Zenith, or pole of horizon

The astrolabe is constructed as an image of the Ptolemaic heavens, or as a "smaller world." Accordingly the astrolabe in operation is an imitation of the revolving heavens on a reduced scale.

The astrolabe is set up with the meridian circle (6) fixed in the meridian line and with the northern and southern poles of the equator (A and A_1) pointing towards the celestial poles above and below the horizon, as the meridian does not change during the course of the daily revolution. The degrees of the celestial ecliptic are marked off on the ecliptic circle (2), with the solstices or equinoxes at the intersections of the ecliptic (2) and the circle through the poles of the ecliptic (1). The outer circle (3) is turned to that point on the ecliptic where the position of the sun is computed to be, and then this intersection of the outer circle and the ecliptic is turned towards the sun itself, until each circle casts its shadow in the form of a straight line intersecting the other shadow at right angles. Now as the revolution of the outer circle around the axis of the equator makes the axis of the ecliptic, the circle through the poles of the ecliptic, the inner circle, and the little circle swing round the axis of the equator, and as the pole of the ecliptic revolves around the pole of the equator during the daily revolution; the turning towards the sun of the intersection of the two circles serves to bring the yearly and daily movement of the sun into proper ratio with one another; and the cruciform shadow is a sign that the wooden ecliptic occupies the relative position in the astrolabe that the celestial ecliptic occupies at this moment of the daily revolution. The inner circle (4) can now be turned towards the moon in order to mark on the ecliptic the lunar longitude, and the little circle (5) can be wheeled around in the plane of the inner circle, in order to mark the lunar latitude on the graduated inner circle.

COPERNICUS' ASTROLABE

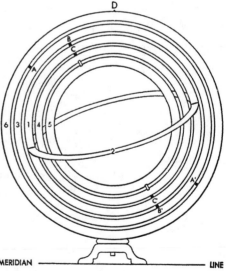

MERIDIAN ——————————— LINE

the moon, the difference will not appear to have been so great; and hence it will be evident enough that the position of the moon viewed was more than $5^1/_3$° but a little less than $5^2/_5$°. The addition of $57^1/_{10}$° to this locates the position of the star at 2°30' of Leo at a distance of about $32^1/_2$° from the summer solstice of the sun and with a northern latitude of $1/_6$°. This was the position of Basiliscus; and consequently the way was laid open to the other fixed stars. This observation of Ptolemy's was made in the year of Our Lord 139 by the Roman calendar, on the 24th day of February, in the 1st year of the 229th Olympiad.

That most outstanding of mathematicians took note of what position at that time each of the stars had in relation to the spring equinox, and catalogued the constellations of the celestial animals. Thus he helps us not a little in this our enterprise and relieves us of some difficult enough labour, so that we, who think that the positions of the stars should not be referred to the equinoxes which change with time but that the equinoxes should be referred to the sphere of the fixed stars, can easily draw up a description of the stars from any other unchanging starting-point. We decided to begin this description with the Ram as being the first sign, and with its first star, which is in its head—so that in this way a configuration which is absolute and always the same will be possessed by those stars which shine together as if fixed and clinging perpetually and at the same time to the throne which they have seized. But by the marvellous care and industry of the ancients the stars were distributed into forty-eight constellations with the exception of those which the circle of the always hidden stars removed from the fourth climate, which passes approximately through Rhodes; and in this way the unconstellated stars remained unknown to them. According to the opinion of Theo the Younger in the *Aratean Treatise* the stars were not arranged in the form of images for any other reason except that their great multitude might be divided into parts and that they might be designated separately by certain names in accordance with an ancient enough custom, since even in Hesiod and Homer we read the names of the Pleiades, Hyas, Arcturus, and Orion. Accordingly in the description of the stars according to longitude we shall not employ the "twelve divisions," or dodekatemoria, which are measured from the equinoxes or solstices, but the simple and conventional number of degrees. We shall follow Ptolemy as to the rest with the exception of a few cases, where we have either found some corruption or a different state of affairs. We shall however teach you in the following book how to find out what their distances are from those cardinal points (*i.e.*, the equinoxes).

CATALOGUE OF THE SIGNS AND OF THE STARS
AND FIRST THOSE OF THE NORTHERN REGION

Constellations	Longitude			Latitude		Magnitude
	Deg.	Min.		Deg.	Min.	
URSA MINOR, OR THE LITTLE BEAR, OR CYNOSURA						
The (star) at the tip of the tail	53	30	N	66	0	3
The (star) to the east in the tail	55	50	N	70	0	4
The (star) at the base of the tail	69	20	N	74	0	4
The more southern (star) on the western side of the quadrilateral	83	0	N	75	20	4
The northern (star) on the same side	87	0	N	77	40	4
The more southern of the stars on the eastern side	100	30	N	72	40	2
The more northern on the same side	109	30	N	74	50	2
7 stars: 2 of second magnitude, 1 of third, 4 of fourth						
The most southern unconstellated star near the Cynosure, in a straight line with the eastern side	103	20	N	71	10	4
URSA MAJOR, OR THE GREAT BEAR						
The star in the muzzle	78	40	N	39	50	4
The western star in the two eyes	79	10	N	43	0	5
The star to the east of that	79	40	N	43	0	5
The more western star of the two in the forehead	79	30	N	47	10	5
The star to the east in the forehead	81	0	N	47	0	5
The western star in the right ear	81	30	N	50	30	5
The more western of the two in the neck	85	50	N	43	50	4
The eastern	92	50	N	44	20	4

NORTHERN SIGNS

	Longitude			Latitude		Magnitude
The more northern of the two in the breast	94	20	N	44	0	4
The more southern	93	20	N	42	0	4
The star at the knee of the left foreleg	89	0	N	35	0	3
The more northern of the two in the left forefoot	89	50	N	29	0	3
The more southern	88	40	N	28	30	3
At the knee of the right foreleg	89	0	N	36	0	4
The star below the knee	101	10	N	33	30	4
The star on the shoulder	104	0	N	49	0	2
The star on the flanks	105	30	N	44	30	2
The star at the base of the tail	116	30	N	51	0	3
The star in the left hind leg	117	20	N	46	30	2
The more western of the two in the left hind foot	106	0	N	29	38	3
The star to the east of that	107	30	N	28	15	3
[47ª] The star in the hollow of the left leg	115	0	N	35	15	4
The more northern of the two which are in the right hind foot	123	10	N	25	50	3

NORTHERN SIGNS

Constellations	Longitude Deg.	Min.		Latitude Deg.	Min.	Magnitude
The more southern	123	40	N	25	0	3
The first of the three in the tail after the base	125	30	N	53	30	2
The middle star	131	20	N	55	40	2
The star which is last and at the tip of the tail	143	10	N	54	0	2
27 stars: 6 of second magnitude, 8 of third, 8 of fourth, and 5 of fifth						
UNCONSTELLATED STARS NEAR THE GREAT BEAR						
The star to the south of the tail	141	10	N	39	45	3
The more obscure star to the west	133	30	N	41	20	5
The star between the forefeet of the Bear and the head of the Lion	98	20	N	17	15	4
The star more to the north than that one	96	40	N	19	10	4
The last of the three obscure stars	99	30	N	20	0	0 obscure
The one to the west of that	95	30	N	22	45	obscure
The one more to the west	94	30	N	23	15	obscure
The star between the forefeet and the Twins	100	20	N	22	15	obscure
8 unconstellated stars: 1 of third magnitude, 2 of fourth, 1 of fifth, 4 obscure						
DRACO, OR THE DRAGON						
The star in the tongue	200	0	N	76	30	4
On the jaws	215	10	N	78	30	4 greater
Above the eye	216	30	N	75	40	3
In the cheek	229	40	N	75	20	4
Above the head	233	30	N	75	30	3
The most northern star in the first curve of the neck	258	40	N	82	20	4
The most southern	295	50	N	78	15	4
The star in between	262	10	N	80	20	4
The star to the east of them at the second curve	282	50	N	81	10	4
The more southern star on the western side of the quadrilateral	331	20	N	81	40	4
The more northern star on the same side	343	50	N	83	0	4
The more northern star on the eastern side	1	0	N	78	50	4
The more southern on the same side	346	10	N	77	50	4
The more southern star in the triangle at the third curve	4	0	N	80	30	4
The more western of the other two in the triangle	15	0	N	81	40	5
The star to the east	19	30	N	80	15	5
<The star to the east> in the triangle to the west	66	20	N	83	30	4
The more southern of the remaining two in the same triangle	43	40	N	83	30	4

NORTHERN SIGNS

Constellations	Longitude			Latitude		Magnitude
	Deg.	Min.		Deg.	Min.	
[47b] The star which is more northern than the two above	35	10	N	84	50	4
Of the small stars west of the triangle, the more eastern	200	0	N	87	30	6
The more western	195	0	N	86	50	6
The most southern of the three which are in a straight line towards the east	152	30	N	81	15	5
The one in the middle	152	50	N	83	0	5
The most northern	151	0	N	84	50	3
The more northern of the two which follow towards the west	153	20	N	78	0	3
The more southern	156	30	N	74	40	4 greater
The star to the west of them, in the coil of the tail	156	0	N	70	0	3
The more western of the two rather distant from that one	120	40	N	64	40	4
The star to the east of it	124	30	N	65	30	3
The star to the east in the tail	192	30	N	61	15	3
At the tip of the tail	186	30	N	56	15	3
Therefore 31 stars: 8 of third magnitude, 16 of fourth, 5 of fifth, 2 of sixth						
CEPHEUS						
In the right foot	28	40	N	75	40	4
In the left foot	26	20	N	64	15	4
On the right side beneath the belt	0	40	N	71	10	4
The star which touches the top of the right shoulder	340	0	N	69	0	3
The star which touches the right joint of the elbow	332	40	N	72	0	4
The star to the east which touches the same elbow	333	20	N	74	0	4
The star on the chest	352	0	N	65	30	5
On the right arm	1	0	N	62	30	4 greater
The most southern of the three on the tiara	339	40	N	60	15	5
The one in the middle	340	40	N	61	15	4
The most northern	342	20	N	61	30	5
11 stars: 1 of third magnitude, 7 of fourth, 3 of fifth						
Of the two unconstellated stars, the one to the west of the tiara	337	0	N	64	0	5
The one to the east of the tiara	344	40	N	59	30	4
BOÖTES, OR ARCTURUS						
The more western of the three in the left hand	145	40	N	58	40	5
The middle one of the three, the more southern	147	30	N	58	20	5
The more eastern of the three	149	0	N	60	10	5
The star in the left joint of the elbow	143	0	N	54	40	5

NORTHERN SIGNS

Constellations	Longitude			Latitude		Magnitude
	Deg.	Min.		Deg.	Min.	
On the left shoulder	163	0	N	49	0	3
On the head	170	0	N	53	50	4 greater
On the right shoulder	179	0	N	48	40	4
[48ª] The more southern of the two on the crook	179	0	N	53	15	4
The star more to the north, at the tip of the crook	178	20	N	57	30	4
The more northern of the two under the shoulder and on the spear	181	0	N	46	10	4 greater
The more southern	181	50	N	45	30	5
At the extremity of the right hand	181	35	N	41	20	5
The more western of the two in the palm	180	0	N	41	40	5
The one to the east	180	20	N	42	30	5
At the extremity of the handle of the crook	181	0	N	40	20	5
On the right leg	183	20	N	40	15	3
The more eastern of the two in the belt	169	0	N	41	40	4
The more western	168	20	N	42	10	4 greater
At the right heel	178	40	N	28	0	3
The more northern of the three on the left ham	164	40	N	28	0	3
The middle one of the three	163	50	N	26	30	4
The more southern of them	164	50	N	25	0	4
22 stars: 4 of third magnitude, 9 of fourth, 9 of fifth						
The unconstellated star between the thighs, which they call Arcturus	170	20	N	31	30	1
CORONA BOREALIS, OR THE NORTHERN CROWN						
The brilliant star in the crown	188	0	N	44	30	2 greater
The most western of all	185	0	N	46	10	4 greater
The eastern star towards the north	185	10	N	48	0	5
The eastern star more to the north	193	0	N	50	30	6
The star to the south-east of the brilliant one	191	30	N	44	45	4
The next star to the east	190	30	N	44	50	4
The star farther to the east	194	40	N	46	10	4
The most eastern of all in the crown	195	0	N	49	20	4
8 stars: 1 of second magnitude, 5 of fourth, 1 of fifth, 1 of sixth						
ENGONASI, OR THE KNEELING MAN						
On the head	221	0	N	37	30	3
At the right arm-pit	207	0	N	43	0	3
On the right arm	205	0	N	40	10	3
In the right flank	201	20	N	37	10	4
On the left shoulder	220	20	N	49	30	4 greater
[48ᵇ] In the left flank	231	0	N	42	0	4

NORTHERN SIGNS

Constellations	Longitude			Latitude		Magnitude
	Deg.	Min.		Deg.	Min.	
<The more eastern> of the three in the left palm	238	50	N	52	50	4 greater
The more northern of the remaining two	235	0	N	54	0	4 greater
The more southern	234	50	N	53	0	4
On the right side	207	10	N	56	10	3
On the left side	213	30	N	53	30	4
On the <lower part of the> left buttock	213	20	N	56	10	5
At the beginning of the left leg	214	30	N	58	30	5
The most western of the three in the left ham	217	20	N	59	50	3
The more eastern	218	40	N	60	20	4
The most eastern	219	40	N	61	15	4
At the left knee	237	10	N	61	0	4
On the <upper part of the> left buttock	225	30	N	69	20	4
The most western of the three in the left foot	188	40	N	70	15	6
The middle star	220	10	N	71	15	6
The most eastern of the three	223	0	N	72	0	6
At the beginning of the right leg	207	0	N	60	15	4 greater
The more northern on the right ham	198	50	N	63	0	4
At the right knee	189	0	N	65	30	4 greater
The more southern of the two under the right knee	186	40	N	63	40	4
The more northern	183	30	N	64	15	4
On the right shin	184	30	N	60	0	4
At the extremity of the right foot, the same as the tip of Boötes' crook	178	20	N	57	30	4
Besides that last one, 28 stars: 6 of third magnitude, 17 of fourth, 2 of fifth, 3 of sixth						
The unconstellated star to the south of the right arm	26	0	N	38	10	5

LYRA, OR THE LYRE

The brilliant star which is called Lyra or Fidicula	250	40	N	62	0	1
The more northern of the two adjacent stars	253	40	N	62	40	4 greater
The more southern	253	40	N	61	0	4 greater
The star which is at the centre of the beginning of the horns	262	0	N	60	0	4
The more northern of the two which are next and to the east	265	20	N	61	20	4
The more southern	265	0	N	60	20	4
The more northern of the two westerly stars on the cross-piece	254	20	N	56	10	3
The more southern	254	10	N	55	0	4 smaller
The more northern of the two easterly stars on the cross-piece	257	30	N	55	20	3
The more southern	258	20	N	54	45	4 smaller

Northern Signs

Constellations	Longitude			Latitude		Magnitude
	Deg.	Min.		Deg.	Min.	
10 stars: 1 of first magnitude, 2 of third magnitude, 7 of fourth						
[49ª] CYGNUS, OR THE SWAN						
At the mouth	267	50	N	49	20	3
On the head	272	20	N	50	30	5
In the middle of the neck	279	20	N	54	30	4 greater
In the breast	291	50	N	56	20	3
The brilliant star in the tail	302	30	N	60	0	2
In the elbow of the right wing	282	40	N	64	40	3
The most southern of the three in the flat of the wing	285	50	N	69	40	4
The middle star	284	30	N	71	30	4 greater
The last of the three, and at the tip of the wing	310	0	N	74	0	4 greater
At the elbow of the left wing	294	10	N	49	30	3
In the middle of the left wing	298	10	N	52	10	4 greater
At the tip of the same	300	0	N	74	0	3
In the left foot	303	20	N	55	10	4 greater
At the left knee	307	50	N	57	0	4
The more western of the two in the right foot	294	30	N	64	0	4
The more eastern	296	0	N	64	30	4
The nebulous star at the right knee	305	30	N	63	45	5
17 stars: 1 of second magnitude, 5 of third, 9 of fourth, 2 of fifth						
AND TWO UNCONSTELLATED STARS NEAR THE SWAN						
The more southern of the two under the left wing	306	0	N	49	40	4
The more northern	307	10	N	51	40	4
CASSIOPEIA						
On the head	1	10	N	45	20	4
On the breast	4	10	N	46	45	3 greater
On the girdle	6	20	N	47	50	4
Above the seat, at the hips	10	0	N	49	0	3 greater
At the knees	13	40	N	45	30	3
On the leg	20	20	N	47	45	4
At the extremity of the foot	355	0	N	48	20	4
On the left arm	8	0	N	44	20	4
On the left forearm	7	40	N	45	0	5
On the right forearm	357	40	N	50	0	6
At the foot of the chair	8	20	N	52	40	4
At the middle of the settle	1	10	N	51	40	3 smaller
At the extremity	27	10	N	51	40	6
13 stars: 4 of third magnitude, 6 of fourth, 1 of fifth, 2 of sixth						

NORTHERN SIGNS

Constellations	Longitude			Latitude		Magnitude
	Deg.	Min.		Deg.	Min.	
[49^b] PERSEUS						
The nebulous star at the extremity of the right hand	21	0	N	40	30	nebulous
On the right forearm	24	30	N	37	30	4
On the right shoulder	26	0	N	34	30	4 smaller
On the left shoulder	20	50	N	32	20	4
On the head, or a nebula	24	0	N	34	30	4
On the shoulder-blades	24	50	N	31	10	4
The brilliant star on the right side	28	10	N	30	0	2
The most western of the three on the same side	28	40	N	27	30	4
The middle one	30	20	N	27	40	4
The remaining one of the three	31	0	N	27	30	3
On the left forearm	24	0	N	27	0	4
The brilliant star in the left hand and in the head of Medusa	23	0	N	23	0	2
The easterly star on the head of the same	22	30	N	21	0	4
The more western on the head of the same	21	0	N	21	0	4
The most western	20	10	N	22	15	4
On the right knee	38	10	N	28	15	4
The one to the west of this one at the knee	37	10	N	28	10	4
The more western of the two on the belly	35	40	N	25	10	4
The more eastern	37	20	N	26	15	4
On the right hip	37	30	N	24	30	5
On the right calf	39	40	N	28	45	5
On the left hip	30	10	N	21	40	4 greater
On the left knee	32	0	N	19	50	3
On the left calf	31	40	N	14	45	3 greater
On the left heel	24	30	N	12	0	3 smaller
On the top part of the left foot	29	40	N	11	0	3 greater
26 stars: 2 of second magnitude, 5 of third, 16 of fourth, 2 of fifth, 1 nebulous						
UNCONSTELLATED STARS AROUND PERSEUS						
To the east of the left hand	34	10	N	31	0	5
To the north of the right hand	38	20	N	31	0	5
To the west of Medusa's head	18	0	N	20	40	obscure
3 stars: 2 of fifth magnitude, 1 obscure						
[50^a] AURIGA, OR THE CHARIOTEER						
The more southern of the two on the head	55	50	N	30	0	4
The more northern	55	40	N	30	50	4
The brilliant star on the left shoulder, which is called Capella	48	20	N	22	30	1
On the right shoulder	56	10	N	20	0	2
On the right forearm	54	30	N	15	15	4
On the palm of the right hand	56	10	N	13	30	4 greater
On the left forearm	45	20	N	20	40	4 greater
The star to the west of the Haedi	45	30	N	18	0	4 smaller

NORTHERN SIGNS

Constellations	Longitude Deg.	Min.		Latitude Deg.	Min.	Magnitude
The star on the palm of the left hand which is to the east of the Haedi	46	0	N	18	0	4 greater
On the left calf	53	10	N	10	10	3 small
On the right calf and at the tip of the northern horn of Taurus	49	0	N	5	0	3 greater
At the ankle	49	20	N	8	30	5
On the buttocks	49	40	N	12	20	5
The small star on the left foot	24	0	N	10	20	6
14 stars: 1 of first magnitude, 1 of second, 2 of third, 7 of fourth, 2 of fifth, 1 of sixth						
OPHIUCHUS, OR THE SERPENT-HOLDER						
On the head	228	10	N	36	0	3
The more western of the two on the right shoulder	231	20	N	27	15	4 greater
The more eastern	232	20	N	26	45	4
The more western of the two on the left shoulder	216	40	N	33	0	4
The more eastern	218	0	N	31	50	4
At the left elbow	211	40	N	34	30	4
The more western of the two in the left hand	208	20	N	17	0	4
The more eastern	209	20	N	12	30	3
At the right elbow	220	0	N	15	0	4
The more western in the right hand	205	40	N	18	40	4 smaller
The more eastern	207	40	N	14	20	4
At the right knee	224	30	N	4	30	3
On the right shin	227	0	N	2	15	3 greater
The most western of the four on the right foot	226	20	S	2	15	4 greater
The more easterly	227	40	S	1	30	4 greater
The next to the east	228	20	S	0	20	4 greater
The most easterly	229	10	S	1	45	5 greater
The star which touches the heel	229	30	S	1	0	5
[50b] At the left knee	215	30	N	11	50	3
The most northern of the three in a straight line on the lower part of the left leg	215	0	N	5	20	5 greater
The middle one	214	0	N	3	10	5
The most southern of the three	213	10	N	1	40	5 greater
The star on the left heel	215	40	N	0	40	5
The star touching the hollow of the left foot	214	0	S	0	45	4
24 stars: 5 of third magnitude, 13 of fourth, 6 of fifth						
UNCONSTELLATED STARS AROUND OPHIUCHUS						
The most northern of the three to the east of the right shoulder	235	20	N	28	10	4
The middle one	236	0	N	26	20	4
The most southern of the three	233	40	N	25	0	4
Another one, farther to the east of the three	237	0	N	27	0	4

NORTHERN SIGNS

Constellations	Longitude			Latitude		Magnitude
	Deg.	Min.		Deg.	Min.	
A star separate from the four, to the north	238	0	N	33	0	4
Therefore 5 unconstellated stars: all of fourth magnitude						
SERPENS OPHIUCHI, OR THE SERPENT						
On the quadrilateral, the star in the cheeks	192	10	N	38	0	4
The star touching the nostrils	201	0	N	40	0	4
On the temples	197	40	N	35	0	3
At the beginning of the neck	195	20	N	34	15	3
At the middle of the quadrilateral, and on the jaws	194	40	N	37	15	4
To the north of the head	201	30	N	42	30	4
At the first curve of the neck	195	0	N	29	15	3
The most northern of the three to the east	198	10	N	26	30	4
The middle one	197	40	N	25	20	3
The most southern of the three	199	40	N	24	0	3
The star to the west of the left hand of Ophiuchus	202	0	N	16	30	4
The star to the east of the same hand	211	30	N	16	15	5
The star to the east of the right hip	227	0	N	10	30	4
The more southern of the two to the east of that	230	20	N	8	30	4 greater
The more northern	231	10	N	10	30	4
To the east of the right hand, in the coil of the tail	237	0	N	20	0	4
Farther east in the tail	242	0	N	21	10	4
At the tip of the tail	251	40	N	27	0	4 greater
18 stars: 5 of third magnitude, 12 of fourth, 1 of fifth						
[51ª] SAGITTA, OR THE ARROW						
At the head	273	30	N	39	20	4
The most eastern of the three on the shaft	270	0	N	39	10	6
The middle one	269	10	N	39	50	5
The most western of the three	268	0	N	39	0	5
At the notch	266	40	N	38	45	5
5 stars: 1 of fourth magnitude, 3 of fifth, 1 of sixth						
AQUILA, OR THE EAGLE						
In the middle of the head	270	30	N	26	50	4
On the neck	268	10	N	27	10	3
The brilliant star on the shoulder-blades, which is called Aquila	267	10	N	29	10	2 greater
The star to the north which is very near	268	0	N	30	0	3 smaller
The more western on the left shoulder	266	30	N	31	30	3
The more eastern	269	20	N	31	30	5
The star to the west in the right shoulder	263	0	N	28	40	5
The star to the east	264	30	N	26	40	5 greater

NORTHERN SIGNS

Constellations	Longitude			Latitude		Magnitude
	Deg.	Min.		Deg.	Min.	
The star in the tail, which touches the milky circle	265	30	N	26	30	3
9 stars: 1 of second magnitude, 4 of third, 1 of fourth, 3 of fifth						
UNCONSTELLATED STARS AROUND AQUILA						
The more western star south of the head	272	0	N	21	40	3
The more eastern	272	20	N	29	10	3
Away from the right shoulder and to the south-west	259	20	N	25	0	4 greater
To the south	261	30	N	20	0	3
Farther south	263	0	N	15	30	5
West of all	254	30	N	18	10	3
6 unconstellated stars: 4 of third magnitude, 1 of fourth, and 1 of fifth						
DELPHINUS, OR THE DOLPHIN						
The most western of the three in the tail	281	0	N	29	10	3 smaller
The more northern of the two remaining	282	0	N	29	0	4 smaller
The more southern	282	0	N	26	40	4
The more southern on the western side of the rhomboid	281	50	N	32	0	3 smaller
The more northern on the same side	283	30	N	33	50	3 smaller
The more southern on the eastern side	284	40	N	32	0	3 smaller
The more northern on the same side	286	50	N	33	10	3 smaller
The most southern of the three between the tail and the rhombus	280	50	N	34	15	6
The more western of the other two to the north	280	50	N	31	50	6
The more eastern	282	20	N	31	30	6
10 stars: 5 of third magnitude, 2 of fourth, 3 of sixth						
[51b] EQUI SECTIO, OR THE SECTION OF THE HORSE						
The more western of the two on the head	289	40	N	20	30	obscure
The more eastern	292	20	N	20	40	obscure
The more western of the two at the mouth	289	40	N	25	30	obscure
The more eastern	291	21	N	25	0	obscure
4 stars: all obscure						
PEGASUS, OR THE WINGED HORSE						
Within the open mouth	298	40	N	21	30	3 greater
The more northern of the two close together on the head	302	40	N	16	50	3
The more southern	301	20	N	16	0	4
The more southern of the two on the mane	314	40	N	15	0	5
The more northern	313	50	N	16	0	5
The more western of the two on the neck	312	10	N	18	0	3
The more eastern	313	50	N	19	0	4

NORTHERN SIGNS

Constellations	Longitude			Latitude		Magnitude
	Deg.	Min.		Deg.	Min.	
On the left pastern	305	40	N	36	30	4 greater
On the left knee	311	0	N	34	15	4 greater
On the right pastern	317	0	N	41	10	4 greater
The more western of the two close together on the breast	319	30	N	29	0	4
The more eastern	320	20	N	29	30	4
The more northern of the two on the right knee	322	20	N	35	0	3
The more southern	321	50	N	24	30	5
The more northern of the two beneath the wing, on the body	327	50	N	25	40	4
The more southern	328	20	N	25	0	4
At the shoulder-blades and juncture of the wing	350	0	N	19	40	2 smaller
On the right shoulder and at the beginning of the leg	325	30	N	31	0	2 smaller
At the tip of the wing	335	30	N	12	30	2 smaller
At the navel, and on the head of Andromeda too	341	10	N	26	0	2 smaller
20 stars: 4 of second magnitude, 4 of third, 9 of fourth, 3 of fifth						
ANDROMEDA						
On the shoulder-blades	348	40	N	24	30	3
On the right shoulder	349	40	N	27	0	4
On the left shoulder	347	40	N	23	0	4
The most southern of the three on the right arm	347	0	N	32	0	4
The most northern	348	0	N	33	30	4
The middle one of the three	348	20	N	32	20	5
The most southern of the three on the top of the right hand	343	0	N	41	0	4
The middle star	344	0	N	42	0	4
[52a] The most northern of the three	345	30	N	44	0	4
On the left arm	347	30	N	17	30	4
At the left elbow	349	0	N	15	50	3
The most southern of the three on the girdle	357	10	N	25	20	3
The middle one	355	10	N	30	0	3
The most northern	355	20	N	32	30	3
On the left foot	10	10	N	23	0	3
On the right foot	10	30	N	37	10	4 greater
To the south of those two	8	30	N	35	20	4 greater
The more northern of the two under the hamstrings	5	40	N	29	0	4
The more southern	5	20	N	28	0	4
At the right knee	5	30	N	35	30	5
The more northern of the two on the flowing robe	6	0	N	34	30	5
The more southern	7	30	N	32	30	5

NORTHERN SIGNS

Constellations	Longitude			Latitude		Magnitude
	Deg.	Min.		Deg.	Min.	
The unconstellated star west of the right hand	5	0	N	44	0	3
23 stars: 7 of third magnitude, 12 of fourth,						
4 of fifth						
TRIANGULUM, OR THE TRIANGLE						
At the vertex of the triangle	4	20	N	16	30	3
The most western of the three on the base	9	20	N	20	40	3
The middle one	9	30	N	20	20	4
The most eastern of the three	10	10	N	19	0	3
4 stars: 3 of third magnitude, 1 of fourth						

Therefore in the northern region there are 360 stars, all in all: 3 of first magnitude, 18 of second, 81 of third, 177 of fourth, 58 of fifth, 13 of sixth, 1 nebulous, and 9 obscure.

THE SIGNS AND STARS WHICH ARE IN THE MIDDLE AND AROUND THE ECLIPTIC

Constellations	Longitude			Latitude		Magnitude
	Deg.	Min.		Deg.	Min.	
ARIES, OR THE RAM						
The star which is first of all and the more western of the two on the horn	0	0	N	7	20	3 smaller
The more eastern on the horn	1	0	N	8	20	3
The more northern of the two in the opening of the jaws	4	20	N	7	40	5
The more southern	4	50	N	6	0	5
On the neck	9	50	N	5	30	5
On the kidneys	10	50	N	6	0	6
At the beginning of the tail	14	40	N	4	50	5
The most western of the three on the tail	17	10	N	1	40	4
The middle one	18	40	N	2	30	4
[52b] The most eastern	20	20	N	1	50	4
On the hips	13	0	N	1	10	5
On the ham	11	20	S	1	30	5
At the tip of the hind foot	8	10	S	5	15	4 greater
13 stars: 2 of third magnitude, 4 of fourth, 6 of fifth, 1 of sixth						
UNCONSTELLATED STARS AROUND ARIES						
The brilliant star over the head	3	50	N	10	0	3 greater
The very northerly star above the back	15	0	N	10	10	4
The most northern of the remaining three small stars	14	40	N	12	40	5
The middle one	13	0	N	10	40	5
The most southern	12	30	N	10	40	5
5 stars: 1 of third magnitude, 1 of fourth, 3 of fifth						
TAURUS, OR THE BULL						
The most northern of the four in the section	19	40	S	6	0	4
The next after that	19	20	S	7	15	4
The third	18	0	S	8	30	4

IN THE MIDDLE, AND AROUND THE ECLIPTIC

Constellations	Longitude			Latitude		Magnitude
	Deg.	Min.		Deg.	Min.	
The fourth and most southern	17	50	S	9	15	4
On the right shoulder	23	0	S	9	30	5
In the breast	27	0	S	8	0	3
At the right knee	30	0	S	12	40	4
On the right pastern	26	20	S	14	50	4
At the left knee	35	30	S	10	0	4
On the left pastern	36	20	S	13	30	4
Of the five called Hyades and on the face, the one at the nostrils	32	0	S	5	45	3 smaller
Between that star and the northern eye	33	40	S	4	15	3 smaller
Between that same star and the southern eye	34	10	S	8	50	3 smaller
The brilliant star, in the very eye, called Palilicius by the Romans	36	0	S	5	10	1
On the northern eye	35	10	S	3	0	3 smaller
The star south of the horn between the base and the ear	40	30	S	4	0	4
The more southern of the two on the same horn	43	40	S	5	0	4
The more northern	43	20	S	3	30	5
At the extremity of the same	50	30	S	2	30	3
To the north of the base of the horn	49	0	S	4	0	4
At the extremity of the horn and on the right foot of Auriga	49	0	N	5	0	3
The more northern of the two in the north ear	35	20	N	4	30	5
The more southern	35	0	N	4	30	5 Apogee of Venus: 48°20'
[53ª] The more western of the two small stars on the neck	30	20	N	0	40	5
The more eastern	32	20	N	1	0	6
The more southern on the western side of the quadrilateral on the neck	31	20	N	5	0	5
The more northern on the same	32	10	N	7	10	5
The more southern on the eastern side	35	20	N	3	0	5
The more northern on the same side	35	0	N	5	0	5
The northern limit of the western side of the Pleiades	25	30	N	4	30	5
The southern limit of the same side	25	50	N	4	40	5
The very narrow limit of the eastern side of the Pleiades	27	0	N	5	20	5
A small star of the Pleiades, separated from the limits	26	0	N	3	0	5
32 stars, apart from that which is at the tip of the northern horn: 1 of first magnitude, 6 of third, 11 of fourth, 13 of fifth, 1 of sixth						
UNCONSTELLATED STARS AROUND TAURUS						
Between the foot and below the shoulder	18	20	S	17	30	4

IN THE MIDDLE, AND AROUND THE ECLIPTIC

Constellations	Longitude			Latitude		Magnitude
	Deg.	Min.		Deg.	Min.	
The most western of the three to the south						
of the horn	43	20	S	2	0	5
The middle one	47	20	S	1	45	5
The most eastern of the three	49	20	S	2	0	5
The more northern of the two under the tip						
of the same horn	52	20	S	6	20	5
The more southern	52	20	S	7	40	5
The most western of the five under the						
northern horn	50	20	N	2	40	5
The next to the east	52	20	N	1	0	5
The third and to the east	54	20	N	1	20	5
The more northern of the remaining two	55	40	N	3	20	5
The more southern	56	40	N	1	15	5
11 unconstelled stars: 1 of fourth magnitude,						
10 of fifth						
GEMINI, OR THE TWINS						
On the head of the western Twin, Castor	76	40	N	9	30	2
The reddish star on the head of the eastern						
Twin, Pollux	79	50	N	6	15	2
At the left elbow of the western Twin	70	0	N	10	0	4
On the left arm	72	0	N	7	20	4
At the shoulder-blades of the same Twin	75	20	N	5	30	4
On the right shoulder of the same	77	20	N	4	50	4
On the left shoulder of the eastern Twin	80	0	N	2	40	4
On the right side of the western Twin	75	0	N	2	40	5
On the left side of the eastern Twin	76	30	N	3	0	5
[53ᵇ] At the left knee of the western Twin	66	30	N	1	30	3
At the left knee of the eastern	71	35	S	2	30	3
On the left groin of the same	75	0	S	0	30	3
At the hollow of the right knee of the same	74	40	S	0	40	3
The more western star in the foot of the western						
Twin	60	0	S	1	30	4 greater
The more eastern star in the same foot	61	30	S	1	15	4
At the extremity of the foot of the western Twin	63	30	S	3	30	4
On the top of the foot of the eastern Twin	65	20	S	7	30	3
On the bottom of the foot of the same	68	0	S	10	30	4
18 stars: 2 of second magnitude, 5 of third,						
9 of fourth, 2 of fifth						
UNCONSTELLATED STARS AROUND GEMINI						
The star west of the top of the foot of the						
western Twin	57	30	S	0	40	4
The brilliant star to the west of the knee of						
the same	59	50	N	5	50	4 greater
To the west of the left knee of the eastern Twin	68	30	S	2	15	5
The most northern of the three east of the						
right hand of the eastern Twin	81	40	S	1	20	5

IN THE MIDDLE, AND AROUND THE ECLIPTIC

Constellations	Longitude Deg.	Min.		Latitude Deg.	Min.	Magnitude
The middle one	79	40	S	3	20	5
The most southern of the three, and in the neighbourhood of the right arm	79	20	S	4	30	5
The brilliant star to the east of the three	84	0	S	2	40	4
7 unconstellated stars; 3 of fourth magnitude, 4 of fifth						
CANCER, OR THE CRAB						
The nebulous star in the breast, which is called Praeses	93	40	N	0	40	nebulous
The more northern of the two west of the quadrilateral	91	0	N	1	15	4 smaller
The more southern	91	20	S	1	10	4 smaller
The more northern of the two to the east, which are called the Asses	93	40	N	2	40	4 greater
The southern Ass	94	40	S	0	10	4 greater
On the claws or the southern arm	99	50	S	5	30	4
On the northern arm	91	40	N	11	50	4
At the extremity of the northern foot	86	0	N	1	0	5
At the extremity of the southern foot	90	30	S	7	30	4 greater
9 stars: 7 of fourth magnitude, 1 of fifth, 1 nebulous						
UNCONSTELLATED STARS AROUND CANCER						
Above the elbow of the southern claw	103	0	S	2	40	4 smaller
East of the extremity of the same claw	105	0	S	5	40	4 smaller
[54ᵃ] The more western of the two above the little cloud	97	20	N	4	50	5
The more eastern	100	20	N	7	15	5
4 unconstellated stars: 2 of fourth magnitude, 2 of fifth						
LEO, OR THE LION						
At the nostrils	101	40	N	10	0	4
At the opening of the jaws	104	30	N	7	30	4
The more northern of the two on the head	107	40	N	12	0	3
The more southern	107	30	N	9	30	3 greater
The most northern of the three on the neck	113	30	N	11	0	3 Apogee of Mars: 109°50'
The middle one	115	30	N	8	30	2
The most southern of the three	114	0	N	4	30	3
At the heart, the star called Basiliscus or Regulus	115	50	N	0	10	1
The more southern of the two on the breast	116	50	S	1	50	4
A little to the west of the star at the heart	113	20	S	0	15	5
At the knee of the right foreleg	110	40		0	0	5
On the right pad	117	30	S	3	40	6
At the knee of the left foreleg	122	30	S	4	10	4
On the left pad	115	50	S	4	15	4

IN THE MIDDLE, AND AROUND THE ECLIPTIC

Constellations	Longitude			Latitude		Magnitude
	Deg.	Min.		Deg.	Min.	
At the left arm-pit	122	30	S	0	10	4
The most western of the three on the belly	120	20	N	4	0	6
The more northern of the two to the east	126	20	N	5	20	6
The more southern	125	40	N	2	20	6
The more western of the two on the loins	124	40	N	12	15	5
The more eastern	127	30	N	13	40	2
The more northern of the two on the rump	127	40	N	11	30	5
The more southern	129	40	N	9	40	3
At the hips	133	40	N	5	50	3
At the hollow of the knee	135	0	N	1	15	4
On the lower part of the leg	135	0	S	0	50	4
On the hind foot	134	0	S	3	0	5
At the tip of the tail	137	50	N	11	50	1 smaller
27 stars: 2 of first magnitude, 2 of second, 6 of third, 8 of fourth, 5 of fifth, 4 of sixth						
UNCONSTELLATED STARS AROUND LEO						
The more western of the two above the back	119	20	N	13	20	5
The more eastern	121	30	N	15	30	5
The most northern of the three below the belly	129	50	N	1	10	4 smaller
[54b] The middle one	130	30	S	0	30	5
The most southern of the three	132	20	S	2	40	5
The star farthest north between the extremities of Leo and the nebulous complex called Coma Berenices	138	10	N	30	0	luminous
The more western of the two to the south	133	50	N	25	0	obscure
The star to the east, in the shape of an ivy leaf	141	50	N	25	30	obscure
8 unconstellated stars: 1 of fourth magnitude, 4 of fifth, 1 luminous, 2 obscure						
VIRGO, OR THE VIRGIN						
The more southwestern of the two on the top of the head	139	40	N	4	15	5
The more northeastern	140	20	N	5	40	5
The more northern of the two on the face	144	0	N	8	0	5
The more southern	143	30	N	5	30	5
At the tip of the left and southern wing	142	20	N	6	0	3
The most western of the four on the left wing	151	35	N	1	10	3
The next to the east	156	30	N	2	50	3
The third	160	30	N	2	50	5
The last and most eastward of the four	164	20	N	1	40	4
On the right side beneath the girdle	157	40	N	8	30	3
The most western of the three on the right and northern wing	151	30	N	13	50	5
The more southern of the two remaining	153	30	N	11	40	6 Apogee of Jupiter: 154°20'

IN THE MIDDLE, AND AROUND THE ECLIPTIC

Constellations	Longitude Deg.	Min.		Latitude Deg.	Min.	Magnitude
The more northern of them, called Vindemiator	155	30	N	15	10	3 greater
On the left hand, called Spica	170	0	S	2	0	1
Beneath the girdle and on the right buttock	168	10	N	8	40	3
The more northern of the two on the western side of the quadrilateral on the left hip	169	40	N	2	20	5
The more southern	170	20	N	0	10	6
The more northern of the two on the eastern side	173	20	N	1	30	4
The more southern	171	20	N	0	20	5
At the left knee	175	0	N	1	30	5
On the posterior side of the right hip	171	20	N	8	30	5
On the flowing robe, in the middle	180	0	N	7	30	4
More to the south	180	40	N	2	40	4
More to the north	181	40	N	11	40	4 Apogee of Mercury: 183°20'
On the left and southern foot	183	20	N	0	30	4
On the right and southern foot	186	0	N	9	50	3

26 Stars: 1 of first magnitude, 7 of third, 6 of fourth, 10 of fifth, 2 of sixth

UNCONSTELLATED STARS AROUND VIRGO

Constellations	Longitude Deg.	Min.		Latitude Deg.	Min.	Magnitude
[55ª] The most western of the three in a straight line under the left arm	158	0	S	3	30	5
The middle one	162	20	S	3	30	5
The most eastern	165	35	S	3	20	5
The most western of the three in a straight line under Spica	170	30	S	7	20	6
The middle one, which is also a double star	171	30	S	8	20	5
The most eastern of the three	173	20	S	7	50	6

6 unconstellated stars: 4 of fifth magnitude, 2 of sixth

CHELAE, OR THE CLAWS

Constellations	Longitude Deg.	Min.		Latitude Deg.	Min.	Magnitude
The bright one of the two at the extremity of the southern claw	191	20	N	0	40	2 greater
The more obscure star to the north	190	20	N	2	30	5
The bright one of the two at the extremity of the northern claw	195	30	N	8	30	2
The more obscure star to the west of that	191	0	N	8	30	5
In the middle of the southern claw	197	20	N	1	40	4
In the same claw, but to the west	194	40	N	1	15	4
At the middle of the northern claw	200	50	N	3	45	4
In the same claw, but to the east	206	20	N	4	30	4

8 stars: 2 of second magnitude, 4 of fourth, 2 of fifth

In the Middle, and Around the Ecliptic

Constellations	Longitude Deg.	Min.		Latitude Deg.	Min.	Magnitude
UNCONSTELLATED STARS AROUND THE CHELAE						
The most western of the three north of the northern claw	199	30	N	9	0	5
The more southern of the two to the east	207	0	N	6	40	4
The more northern	207	40	N	9	15	4
The most eastern of the three between the claws	205	50	N	5	30	6
The more northern of the remaining two to the west	203	40	N	2	0	4
The more southern	204	30	N	1	30	5
The most western of the three beneath the southern claw	196	20	S	7	30	3
The more northern of the remaining two to the east	204	30	S	8	10	4
The more southern	205	20	S	9	40	4
9 unconstellated stars: 1 of third magnitude, 5 of fourth, 2 of fifth, 1 of sixth						
SCORPIO, OR THE SCORPION						
The most northern of the three bright stars on the forehead	209	40	N	1	20	3 greater
The middle one	209	0	S	1	40	3
The most southern of the three	209	0	S	5	0	3
More to the south and in the foot	209	20	S	7	50	3
The more northern of the two adjacent bright stars	210	20	N	1	40	4
The more southern	210	40	N	0	30	4
The most western of the three bright stars on the body	214	0	S	3	45	3
The reddish star in the middle, called Antares	216	0	S	4	0	2 greater
The most eastern of the three	217	50	S	5	30	3
[55b] The more western of the two at the extremity of the foot	212	40	S	6	10	5
The more eastern	213	50	S	6	40	5
At the first vertebra of the body	221	50	S	11	0	3
At the second vertebra	222	10	S	15	0	4
The more northern of the double at the third	223	20	S	18	40	4
The more southern of the double	223	30	S	18	0	3
At the fourth vertebra	226	30	S	19	30	3 Apogee of Saturn: 226°30'
At the fifth	231	30	S	18	50	3
At the sixth vertebra	233	50	S	16	40	3
At the seventh, and next to the sting	232	20	S	15	10	3
The more eastern of the two on the sting	230	50	S	13	20	3
The more western	230	20	S	13	30	4
21 stars: 1 of second magnitude, 13 of third, 5 of fourth, 2 of fifth						

IN THE MIDDLE, AND AROUND THE ECLIPTIC

Constellations	Longitude			Latitude		Magnitude
	Deg.	Min.		Deg.	Min.	
UNCONSTELLATED STARS AROUND SCORPIO						
The nebulous star to the east of the sting	234	30	S	12	15	nebulous
The more western of the two north of the sting	228	50	S	6	10	5
The more eastern	232	50	S	4	10	5
3 unconstellated stars: 2 of fifth magnitude, 1 nebulous						
SAGITTARIUS, OR THE ARCHER						
At the head of the arrow	237	50	S	6	30	3
In the palm of the left hand	241	0	S	6	30	3
On the southern part of the bow	241	20	S	10	50	3
The more southern of the two to the north	242	20	S	1	30	3
More northward, at the extremity of the bow	240	0	N	2	50	4
On the left shoulder	248	40	S	3	10	3
To the west and on the dart	246	20	S	3	50	4
The nebulous double star in the eye	248	30	N	0	45	nebulous
The most western of the three on the head	249	0	N	2	10	4
The middle one	251	0	N	1	30	4 greater
The most eastward	252	30	N	2	0	4
The most southern of the three on the northern garment	254	40	N	2	50	4
The middle one	255	40	N	4	30	4
The most northern	256	10	N	6	30	4
The obscure star east of the three	259	0	N	5	30	6
The most northern of the two on the southern garment	262	50	N	5	0	5
The more southern	261	0	N	2	0	6
On the right shoulder	255	40	S	1	50	5
[56a] At the right elbow	258	10	S	2	50	5
At the shoulder-blades	253	20	S	2	30	5
At the foreshoulder	251	0	S	4	30	4 greater
Beneath the arm-pit	249	40	S	6	45	3
On the pastern of the left foreleg	251	0	S	23	0	2
At the knee of the same leg	250	20	S	18	0	2
On the pastern of the right foreleg	240	0	S	13	0	3
At the left shoulder blade	260	40	S	13	30	3
At the knee of the right foreleg	260	0	S	20	10	3
The more western on the northern side of the quadrilateral at the beginning of the tail	261	0	S	4	50	5
The more eastern on the same side	261	10	S	4	50	5
The more western on the southern side	261	50	S	5	50	5
The more eastern on the same side	263	0	S	6	50	5
31 stars: 2 of second magnitude, 9 of third, 9 of fourth, 8 of fifth, 2 of sixth, 1 nebulous.						

IN THE MIDDLE, AND AROUND THE ECLIPTIC

Constellations	Longitude			Latitude		Magnitude
	Deg.	Min.		Deg.	Min.	
CAPRICORNUS, OR THE GOAT						
The most northern of the three on the western						
horn	270	40	N	7	30	3
The middle one	271	0	N	6	40	6
The most southern of the three	270	40	N	5	0	3
At the extremity of the eastern horn	272	20	N	8	0	6
The most southern of the three at the opening						
of the jaws	272	20	N	0	45	6
The more western of the two remaining	272	0	N	1	45	6
The more eastern	272	10	N	1	30	6
Under the right eye	270	30	N	0	40	5
The more northern of the two on the neck	275	0	N	4	50	6
The more southern	275	10	S	0	50	5
At the right knee	274	10	S	6	30	4
At the left knee, which is bent	275	0	S	8	40	4
On the left shoulder	280	0	S	7	40	4
The more western of the two contiguous stars						
below the belly	283	30	S	6	50	4
The more eastern	283	40	S	6	0	5
The most eastern of the three in the middle of						
the body	282	0	S	4	15	5
The more southern of the two remaining to						
the west	280	0	S	7	0	5
The more northern	280	0	S	2	50	5
The more western of the two on the back	280	0		0	0	4
The more eastern	284	20	S	0	50	4
The more western of the two on the southern						
part of the spine	286	40	S	4	45	4
[56b] The more eastern	288	20	S	4	30	4
The more western of the two at the base of						
the tail	288	40	S	2	10	3
The more eastern	289	40	S	2	0	3
The more western of the four in the northern						
part of the tail	290	10	S	2	20	4
The most northern of the remaining three	292	0	S	5	0	5
The middle one	291	0	S	2	50	5
The most northern, at the extremity of the tail	292	0	N	4	20	5
28 stars: 4 of third magnitude, 9 of fourth,						
9 of fifth, 6 of sixth						
AQUARIUS, OR THE WATER-BOY						
On the head	293	40	N	15	45	5
The brighter of the two on the right shoulder	299	40	N	11	0	3
The more obscure	298	30	N	9	40	5
On the left shoulder	290	0	N	8	50	3
Under the arm-pit	290	40	N	6	15	5
The most eastern of the three under the left						
hand and on the coat	280	0	N	5	30	3
The middle one	279	30	N	8	0	4

IN THE MIDDLE, AND AROUND THE ECLIPTIC

Constellations	Longitude			Latitude		Magnitude
	Deg.	Min.		Deg.	Min.	
The most western of the three	278	0	N	8	30	3
At the right elbow	302	50	N	8	45	3
The farthest north on the right hand	303	0	N	10	45	3
The more western of the two remaining to						
the south	305	20	N	9	0	3
The more eastern	306	40	N	8	30	3
The more western of the two adjacent stars						
on the right hip	299	30	N	3	0	4
The more eastern	300	20	N	2	10	5
On the right buttock	302	0	S	0	50	4
The more southern of the two on the left						
buttock	295	0	S	1	40	4
The more northern	295	30	N	4	0	6
The more southern of the two on the right						
shin	305	0	S	6	30	3
The more northern	304	40	S	5	0	4
On the left hip	301	0	S	5	40	5
The more southern of the two on the left shin	300	40	S	10	0	5
The northern star beneath the knee	302	10	S	9	0	5
The first star in the fall of water from the hand	303	20	N	2	0	4
More to the south-east	308	10	N	0	10	4
To the east at the first bend in the water	311	0	S	1	10	4
To the east of that	313	20	S	0	30	4
In the second and southern bend	313	50	S	1	40	4
The more northern of the two to the east	312	30	S	3	30	4
The more southern	312	50	S	4	10	4
Farther off to the south	314	10	S	8	15	5
[57a] Eastward, the more western of the						
two adjacent	316	0	S	11	0	5
The more eastern	316	30	S	10	50	5
The most northern of the three at the third						
bend in the water	315	0	S	14	0	5
The middle one	316	0	S	14	45	5
The most eastern of the three	316	30	S	15	40	5
The most northern of three in a similar figure						
to the east	310	20	S	14	10	4
The middle one	310	50	S	15	0	4
The most southern of the three	311	40	S	15	45	4
The most western of the three at the last bend						
in the water	305	10	S	14	50	4
The more southern of the two to the east	306	0	S	15	20	4
The more northern	306	30	S	14	0	4
The last in the water, and in the mouth of the						
southern Fish	300	20	S	23	0	1
42 stars: 1 of first magnitude, 9 of third,						
18 of fourth, 13 of fifth, 1 of sixth						

IN THE MIDDLE, AND AROUND THE ECLIPTIC

Constellations	Longitude			Latitude		Magnitude
	Deg.	Min.		Deg.	Min.	
UNCONSTELLATED STARS AROUND AQUARIUS						
The most western of the three east of the						
bend in the water	320	0	S	15	30	4
The more northern of the two remaining	323	0	S	14	20	4
The more southern	322	20	S	18	15	4
3 stars: greater than fourth magnitude						
PISCES, OR THE FISH						
In the mouth of the western Fish	315	0	N	9	15	4
The more southern of the two on the occiput	317	30	N	7	30	4 greater
The more northern	321	30	N	9	30	4
The more western of the two on the back	319	20	N	9	20	4
The more eastern	324	0	N	7	30	4
The more western one on the belly	319	20	N	4	30	4
The more eastern	323	0	N	2	30	4
On the tail of the same Fish	329	20	N	6	20	4
On the fishing-line, the first star from the tail	334	20	N	5	45	6
To the east of that	336	20	N	2	45	6
The most western of the three bright stars to						
the east	340	30	N	2	15	4
The middle one	343	50	N	1	10	4
The most eastern	346	20	S	1	20	4
The more northern of the two small stars on						
the curvature	345	40	S	2	0	6
The more southern	346	20	S	5	0	6
The most western of the three after the curvature	350	20	S	2	20	4
The middle one	352	0	S	4	40	4
The most eastern one	354	0	S	7	45	4
[57b] At the knot of the two fishing-lines	356	0	S	8	30	3
In the northern line, west of the knot	354	0	S	4	20	4
The most southern of the three to the east	353	30	N	1	30	5
The middle one	353	40	N	5	20	3
The most northern of the three and the last in						
the line	353	50	N	9	0	4
THE EASTERN FISH						
The more northern of the two in the mouth	355	20	N	21	45	5
The more southern	355	0	N	21	30	5
The most eastern of the three small stars on						
the head	352	0	N	20	0	6
The middle one	351	0	N	19	50	6
The most western of the three	350	20	N	23	0	6
The most western of the three on the southern						
fin, near the left elbow of Andromeda	349	0	N	14	20	4
The middle one	349	40	N	13	0	4
The most eastern of the three	351	0	N	12	0	4
The more northern of the two on the belly	355	30	N	17	0	4
The more southern	352	40	N	15	20	4

IN THE MIDDLE, AND AROUND THE ECLIPTIC

Constellations	Longitude			Latitude		Magnitude
	Deg.	Min.		Deg.	Min.	
On the eastern fin, near the tail	353	20	N	11	45	4
34 stars: 2 of third magnitude, 22 of fourth, 3 of fifth, 7 of sixth						
UNCONSTELLATED STARS AROUND PISCES						
The more western on the northern side of the quadrilateral under the western Fish	324	30	S	2	40	4
The more eastern	325	35	S	2	30	4
The more western on the southern side	324	0	S	5	50	4
The more eastern	325	40	S	5	30	4
4 unconstellated stars: of fourth magnitude						

Therefore, all in all, there are 348 stars in the zodiac: 5 of first magnitude, 9 of second, 65 of third, 132 of fourth, 105 of fifth, 27 of sixth, 3 nebulous, 2 obscure; and, over and above the count, the Coma, which we said above was called Coma Berenices by Conon the mathematician.

THE STARS OF THE SOUTHERN REGION

Constellations	Longitude			Latitude		Magnitude
	Deg.	Min.		Deg.	Min.	
CETUS, OR THE WHALE						
At the extremity of the nose	11	0	S	7	45	4
The most eastern of the three in the jaws	11	0	S	11	20	3
The middle one, in the middle of the mouth	6	0	S	11	30	3
The most western of the three, on the cheek	3	50	S	14	0	3
In the eye	4	0	S	8	10	4
Northward, in the hair	5	30	S	6	20	4
[58ª] Westward, in the mane	1	0	S	4	10	4
The more northern on the western side of the quadrilateral in the breast	355	20	S	24	30	4
The more southern	356	40	S	28	0	4
The more northern of the two to the east	0	0	S	25	10	4
The more southern	0	0	S	27	30	3
The middle one of the three on the body	345	20	S	25	20	3
The most southern	346	20	S	30	30	4
The most northern of the three	348	20	S	20	0	3
The more eastern of the two at the tail	343	0	S	15	20	3
The more western	338	20	S	15	40	3
The more northern on the eastern side of the quadrilateral in the tail	335	0	S	11	40	5
The more southern	334	0	S	13	40	5
The more northern of the two remaining to the west	332	40	S	13	0	5
The more southern	332	20	S	14	0	5
At the northern extremity of the tail	327	40	8	9	30	3
At the southern extremity of the tail	329	0	S	20	20	3
22 stars: 10 of third magnitude, 8 of fourth, 4 of fifth						

SOUTHERN SIGNS

Constellations	Longitude			Latitude		Magnitude
	Deg.	Min.		Deg.	Min.	
ORION						
The nebulous star on the head	50	20	S	16	30	nebulous
The bright, reddish star on the right shoulder	55	20	S	17	0	1
On the left shoulder	43	40	S	17	30	2 greater
East of that star	48	20	S	18	0	4 smaller
At the right elbow	57	40	S	14	30	4
On the right forearm	59	40	S	11	50	6
The more eastern on the southern side of the quadrilateral in the right hand	59	50	S	10	40	4
The more western	59	20	S	9	45	4
The more eastern on the northern side	60	40	S	8	15	6
The more western on the same side	59	0	S	8	15	6
The more western of the two on the club	55	0	S	3	45	5
The more eastern	57	40	S	3	15	5
The most eastern of the four in a straight line on the back	50	50	S	19	40	4
More western	49	40	S	20	0	6
Still more western	48	40	S	20	20	6
Most western	47	30	S	20	30	5
The most northern of the nine on the shield	43	50	S	8	0	4
The second	42	40	S	8	10	4
The third	41	20	S	10	15	4
The fourth	39	40	S	12	50	4
The fifth	38	30	S	14	15	4
The sixth	37	50	S	15	50	3
[58b] The seventh	38	10	S	17	10	3
The eighth	38	40	S	20	20	3
The last and most southern	39	40	S	21	30	3
The most western of the three bright stars on the sword-belt	48	40	S	24	10	2
The middle one	50	40	S	24	50	2
The most eastern of the three in a straight line	52	40	S	25	30	2
On the hilt of the sword	47	10	S	25	50	3
The most northern of the three on the sword	50	10	S	28	40	4
The middle one	50	0	S	29	30	3
The most southern one	50	20	S	29	50	3 smaller
The more eastern of the two at the tip of the sword	51	0	S	30	30	4
The more western	49	30	S	30	50	4
On the left foot, the bright star which belongs to Fluvius too	42	30	S	31	30	1
On the left shin	44	20	S	30	15	4 greater
At the right heel	46	40	S	31	10	4
At the right knee	53	30	S	33	30	3

38 stars: 2 of first magnitude, 4 of second,
8 of third, 15 of fourth, 3 of fifth, 5 of sixth,
and 1 nebulous

SOUTHERN SIGNS

Constellations	Longitude			Latitude		Magnitude
	Deg.	Min.		Deg.	Min.	
FLUVIUS, OR THE RIVER						
After the left foot of Orion, and at the beginning of Fluvius	41	40	S	31	50	4
The most northern star within the bend of Orion's leg	42	10	S	28	15	4
The more eastern of the two after that	41	20	S	29	50	4
The more western	38	0	S	28	15	4
The more eastern of the next two	36	30	S	25	15	4
The more western	33	30	S	25	20	4
The most eastern of the three after them	29	40	S	26	0	4
The middle one	29	0	S	27	0	4
The most western of the three	26	18	S	27	50	4
The most eastern of the four after the interval	20	20	S	32	50	3
More western	18	0	S	31	0	4
Still more western	17	30	S	28	50	3
The most western of all four	15	30	S	28	0	3
Again similarly, the most eastward of the four	10	30	S	25	30	3
More westward	8	10	S	23	50	4
Still more westward	5	30	S	23	10	3
The most westward of the four	3	50	S	23	15	4
The star in the bend of Fluvius which touches the breast of Cetus	358	30	S	32	10	4
East of that	359	10	S	34	50	4
The most westward of the three to the seat	2	10	S	38	30	4
[59ª] The middle one	7	10	S	38	10	4
The most eastward of the three	10	50	S	39	0	5
The more northern of the two on the western side of the quadrilateral	14	40	S	41	30	4
The more southern	14	50	S	42	30	4
The more western on the eastern side	15	30	S	43	20	4
The most eastward of those four	18	0	S	43	20	4
The more northern of the two contiguous stars towards the east	27	30	S	50	20	4
The more southern	28	20	S	51	45	4
The more eastern of the two at the bend	21	30	S	53	50	4
The more western	19	10	S	53	10	4
The most eastern of the three in the remaining space	11	10	S	53	0	4
The middle one	8	10	S	53	30	4
The most western of the three	5	10	S	52	0	4
The bright star at the extremity of the river	353	30	S	53	30	1
34 stars; 1 of first magnitude, 5 of third, 27 of fourth, 1 of fifth						
LEPUS, OR THE RABBIT						
The more northern one on the western side of the quadrilateral at the ears	43	0	S	35	0	5
The more southern	43	10	S	36	30	5
The more northern one on the eastern side	44	40	S	35	30	5

SOUTHERN SIGNS

Constellations	Longitude			Latitude		Magnitude
	Deg.	Min.		Deg.	Min.	
The more southern	44	40	S	36	40	5
At the chin	42	30	S	39	40	4 greater
At the extremity of the left forefoot	39	30	S	45	15	4 greater
In the middle of the body	48	50	S	41	30	3
Beneath the belly	48	10	S	44	20	3
The more northern of the two on the hind feet	54	20	S	44	0	4
The more southern	52	20	S	45	50	4
On the loins	53	20	S	38	20	4
At the tip of the tail	56	0	S	38	10	4
12 stars: 2 of third magnitude, 6 of fourth, 4 of fifth						
CANIS, OR THE DOG						
The very bright star called Canis, in the mouth	71	0	S	39	10	1 very great
On the ears	73	0	S	35	0	4
On the head	74	40	S	36	30	5
The more northern of the two on the neck	76	40	S	37	45	4
The more southern	78	40	S	40	0	4
On the breast	73	50	S	42	30	5
The more northern of the two at the right knee	69	30	S	41	15	5
The more southern	69	20	S	42	30	5
At the extremity of the forefoot	64	20	S	41	20	3
[59b] The more western of the two on the left knee	68	0	S	46	30	5
The more eastern	69	30	S	45	50	5
The more eastern of the two on the left shoulder	78	0	S	46	0	4
The more western	75	0	S	47	0	5
On the left hip	80	0	S	48	45	3 smaller
Beneath the belly between the thighs	77	0	S	51	30	3
In the hollow of the right foot	76	20	S	55	10	4
At the extremity of the same foot	77	0	S	55	40	3
At the tip of the tail	85	30	S	50	30	3 smaller
18 stars: 1 of first magnitude, 5 of third, 5 of fourth, 7 of fifth						
UNCONSTELLATED STARS AROUND CANIS						
North of the head of the Dog	72	50	S	25	15	4
The most southern in a straight line under the hind feet	63	20	S	60	30	4
The more northern	64	40	S	58	45	4
Still more northern	66	20	S	57	0	4
The last and farthest north of the four	67	30	S	56	0	4
The most western of the three westward as it were in a straight line	50	20	S	55	30	4
The middle one	53	40	S	57	40	4
The most eastern of the three	55	40	S	59	30	4

SOUTHERN SIGNS

Constellations	Longitude			Latitude		Magnitude
	Deg.	Min.		Deg.	Min.	
The more western of the two bright stars						
beneath them	52	20	S	59	40	2
The more western	49	20	S	57	40	2
The remaining star, more southern	45	30	S	59	30	4
11 stars: 2 of second magnitude, 9 of fourth						
CANICULA, OR PROCYON, OR THE LITTLE BITCH						
On the neck	78	20	S	14	0	4
The bright star on the thigh, that is,						
Προκύων or Canicula, the Dog-star	82	30	S	16	10	1
2 stars: 1 of first magnitude, 1 of fourth						
ARGO, OR THE SHIP						
The more western of the two at the extremity						
of the Ship	93	40	S	42	40	5
The more eastern	97	40	S	43	20	3
The more northern of the two on the stern	92	10	S	45	0	4
The more southern	92	10	S	46	0	4
West of the two	88	40	S	45	30	4
The bright star in the middle of the shield	89	40	S	47	15	4
The most western of the three beneath the shield	88	40	S	49	45	4
The most eastern	92	40	S	49	50	4
The middle one of the three	91	40	S	49	15	4
At the extremity of the rudder	97	20	S	49	50	4
The more northern of the two on the stern keel	87	20	S	53	0	4
The more southern	87	20	S	58	30	3
[60ᵃ] The most northern on the cross-bank						
of the stem	93	30	S	55	30	5
The most western of the three on the same						
cross-bank	95	30	S	58	30	5
The middle one	96	40	S	57	15	4
The most eastern	99	50	S	57	45	4
The bright star to the east on the cross-bank	104	30	S	58	20	2
The more western of the two obscure stars						
beneath that	101	30	S	60	0	5
The more eastern	104	20	S	59	20	5
The more western of the two east of the						
aforesaid bright star	106	30	S	56	40	5
The more eastern	107	40	S	57	0	5
The most northern of the three on the small						
shields and at the foot of the mast	119	0	S	51	30	4 greater
The middle one	119	30	S	55	30	4 greater
The most southern of the three	117	20	S	57	10	4
The more northern of the two contiguous						
stars beneath them	122	30	S	60	0	4
The more southern	122	20	S	61	15	4

SOUTHERN SIGNS

Constellations	Longitude			Latitude		Magnitude
	Deg.	Min.		Deg.	Min.	
The more southern of the two in the middle						
of the mast	113	30	S	51	30	4
The more northern	112	40	S	49	0	4
The more western of the two at the top part						
of the sail	111	20	S	43	20	4
The more eastern	112	20	S	43	30	4
Below the third star east of the shield	98	30	S	54	30	2 smaller
In the section of the bridge	100	50	S	51	15	2
Between the oars in the keel	95	0	S	63	0	4
The obscure star east of that	102	20	S	64	30	6
The bright star, east of that and below the						
cross-bank	113	20	S	63	50	2
The bright star to the south, more within						
the keel	121	50	S	69	40	2
The most western of the three to the east of that	128	30	S	65	40	3
The middle one	134	40	S	65	50	3
The most eastern	139	20	S	65	50	2
The more western of the two in the section	144	20	S	62	50	3
The more eastern	151	20	S	62	15	3
The more western in the northwestern oar	57	20	S	65	50	4 greater
The more eastern	73	30	S	65	40	3 greater
The more western one in the remaining oar,						
Canopus	70	30	S	75	0	1
The remaining star east of that	82	20	S	71	50	3 greater
45 stars: 1 of first magnitude, 6 of second,						
8 of third, 22 of fourth, 7 of fifth, 1 of sixth						
HYDRA						
Of the two more western of the five on the						
head, the more southern, at the nostrils	97	20	S	15	0	4
The more northern of the two, and in the eye	98	40	S	13	40	4
On the occiput, the more northern of the two						
to the east	99	0	S	11	30	4
[60ᵇ] The more southern, and at the jaws	98	50	S	14	45	4
East of all those and on the cheeks	100	50	S	12	15	4
The more western of the two at the beginning						
of the neck	103	40	S	11	50	5
The more eastern	106	40	S	13	30	4
The middle one of the three at the curve of						
the neck	111	40	S	15	20	4
East of that	114	0	S	14	50	4
The most southern	111	40	S	17	10	4
The obscure and northern star of the two						
contiguous to the south	112	30	S	19	45	6
The bright one and to the south-east	113	20	S	20	30	2
The most western of the three after the curve						
in the neck	119	20	S	26	30	4

SOUTHERN SIGNS

Constellations	Longitude			Latitude		Magnitude
	Deg.	Min.		Deg.	Min.	
The most eastern	124	30	S	23	15	4
The middle one	122	0	S	24	0	4
The most western of the three in a straight line	131	20	S	24	30	3
The middle one	133	20	S	23	0	4
The most eastern one	136	20	S	23	10	3
The more northern of the two beneath the base of the Cup	144	50	S	25	45	4
The more southern	145	40	S	30	10	4
East of them, the most western of the three on the triangle	155	30	S	31	20	4
The most southern	157	50	S	34	10	4
The most eastern of the same three	159	30	S	31	40	3
East of the Crow, near the tail	173	20	S	13	30	4
At the extremity of the tail	186	50	S	17	30	4
25 stars: 1 of second magnitude, 3 of third, 19 of fourth, 1 of fifth, 1 of sixth						
UNCONSTELLATED STARS AROUND HYDRA						
South of the head	96	0	S	23	15	3
East of those on the neck	124	20	S	26	0	3
2 unconstellated stars: of third magnitude						
CRATER, OR THE CUP						
On the base of the Cup and in Hydra too	139	40	S	23	0	4
The more southern of the two in the middle of the Cup	146	0	S	19	30	4
The more northern of them	143	30	S	18	0	4
On the southern rim of the Cup	150	20	S	18	30	4 greater
On the northern part of the rim	142	40	S	13	40	4
On the southern part of the stem	152	30	S	16	30	4 smaller
On the northern part	145	0	S	11	50	4
7 stars: of fourth magnitude						
[61ª] CORVUS, OR THE CROW						
On the beak, and in Hydra too	158	40	S	21	30	5
On the neck	157	40	S	19	40	5
In the breast	160	0	S	18	10	5
On the right wing, the western wing	160	50	S	14	50	3
The more western of the two on the eastern wing	160	0	S	12	30	3
The more eastern	161	20	S	11	45	4
At the extremity of the foot, and in Hydra too	163	50	S	18	10	3
7 stars: 5 of third magnitude, 1 of fourth, 1 of fifth						
CENTAURUS, OR THE CENTAUR						
The most southern of the four on the head	183	50	S	21	20	5
The more northern	183	20	S	13	50	5

SOUTHERN SIGNS

Constellations	Longitude			Latitude		Magnitude
	Deg.	Min.		Deg.	Min.	
The more western of the two in the middle	182	30	S	20	30	5
The more eastern and last of the four	182	20	S	20	0	5
On the left and western shoulder	179	30	S	25	30	3
On the right shoulder	189	0	S	22	30	3
On the left forearm	182	30	S	17	30	4
The more northern of the two on the western side of the quadrilateral on the shield	191	30	S	22	30	4
The more southern	192	30	S	23	45	4
Of the remaining two, the one at the top of the shield	195	20	S	18	15	4
The more southern	196	50	S	20	50	4
The most western of the three on the right side	186	40	S	28	20	4
The middle one	187	20	S	29	20	4
The most eastern	188	30	S	28	0	4
On the right arm	189	40	S	26	30	4
On the right elbow	196	10	S	25	15	3
At the extremity of the right hand	200	50	S	24	0	4
The bright star at the junction of the human body	191	20	S	33	30	3
The more eastern of the two obscure stars	191	0	S	31	0	5
The more western	189	50	S	30	20	5
At the beginning of the back	185	30	S	33	50	5
West of that, on the horse's back	182	20	S	37	30	5
The most eastern of the three on the loins	179	10	S	40	0	3
The middle one	178	20	S	40	20	4
The most western of the three	176	0	S	41	0	5
The more western of the two contiguous stars on the right hip.	176	0	S	46	10	2
The more eastern	176	40	S	46	45	4
On the breast, beneath the horse's wing	191	40	S	40	45	4
[61ᵇ] The more western of the two under the belly	179	50	S	43	0	2
The more eastern	181	0	S	43	45	3
In the hollow of the right hind foot	183	20	S	51	10	2
On the pastern of the same	188	40	S	51	40	2
In the hollow of the left <hind> foot	188	40	S	55	10	4
Under the muscle of the same foot	184	30	S	55	40	4
On top of the right forefoot	181	40	S	41	10	1
At the left knee	197	30	S	45	20	2
The unconstellated star below the right thigh	188	0	S	49	10	3
37 stars: 1 of first magnitude, 5 of second, 7 of third, 15 of fourth, 9 of fifth						
BESTIA QUAM TENET CENTAURUS, OR THE BEAST HELD BY THE CENTAUR—THE WOLF						
At the top of the hind foot and in the hand of the Centaur	201	20	S	24	50	3

SOUTHERN SIGNS

Constellations	Longitude Deg.	Min.		Latitude Deg.	Min.	Magnitude
On the hollow of the same foot	199	10	S	20	10	3
The more western of the two on the foreshoulder	204	20	S	21	15	4
The more eastern	207	30	S	21	0	4
In the middle of the body	206	20	S	25	10	4
On the belly	203	30	S	27	0	5
On the hip	204	10	S	29	0	5
The more northern of the two at the beginning of the hip	208	0	S	28	30	5
The more southern	207	0	S	30	0	5
The upmost part of the loins	208	40	S	33	10	5
The most southern of the three at the extremity of the tail	195	20	S	31	20	5
The middle one	195	10	S	30	0	4
The most northern of the three	196	20	S	29	20	4
The more southern of the two at the throat	212	10	S	17	0	4
The more northern	212	40	S	15	20	4
The more western of the two at the opening of the jaws	209	0	S	13	30	4
The more eastern	210	0	S	12	50	4
The more southern of the two on the forefoot	240	40	S	11	30	4
The more northern	239	50	S	10	0	4
19 stars: 2 of third magnitude, 11 of fourth, 6 of fifth						
ARA OR THURIBULUM, THE ALTAR OR THE CENSER						
The more northern of the two at the base	231	0	S	22	40	5
The more southern	233	40	S	25	45	4
At the center of the altar	229	30	S	26	30	4
[62ª] The most northern of the three on the hearth	224	0	S	30	20	5
The more southern of the remaining two contiguous stars	228	30	S	34	10	4
The more northern	228	20	S	33	20	4
In the midst of the flames	224	10	S	34	10	4
7 Stars: 5 of fourth magnitude, 2 of fifth						
CORONA AUSTRINA, OR SOUTHERN CROWN						
The more western star on the outer periphery	242	30	S	21	30	4
East of that on the crown	245	0	S	21	0	5
East of that too	246	30	S	20	20	5
Farther east of that also	248	10	S	20	0	4
East of that and west of the knee of Sagittarius	249	30	S	18	30	5
The bright star to the north on the knee	250	40	S	17	10	4
The more northern	250	10	S	16	0	4
Still more northern	249	50	S	15	20	4

Southern Signs

Constellations	Longitude			Latitude		Magnitude
	Deg.	Min.		Deg.	Min.	
The more eastern of the two on the northern part of the periphery	248	30	S	15	50	6
The more western	248	0	S	14	50	6
Some distance west of those	245	10	S	14	40	5
Still west of that	243	0	S	15	50	5
The last star, more towards the south	242	30	S	18	30	5
13 stars: 5 of fourth magnitude, 6 of fifth, 2 of sixth						
PISCIS AUSTRINUS, OR THE SOUTHERN FISH						
In the mouth, and the same as at the extremity of Aqua	300	20	S	23	0	1
The most western of the three on the head	294	0	S	21	20	4
The middle one	297	30	S	22	15	4
The most eastern	299	0	S	22	30	4
At the gills	297	40	S	16	15	4
On the southern and dorsal fin	288	30	S	19	30	5
The more eastern of the two in the belly	294	30	S	15	10	5
The more western	292	10	S	14	30	4
The most eastern of the three on the northern fin	288	30	S	15	15	4
The middle one	285	10	S	16	30	4
The most western of the three	284	20	S	18	10	4
At the extremity of the tail	289	20	S	22	15	4
11 stars beside the first: 9 of fourth magnitude, 2 of fifth						
[62b] UNCONSTELLATED STARS AROUND PISCIS AUSTRINUS						
The most western of the bright stars west of Piscis	271	20	S	22	20	3
The middle one	274	30	S	22	10	3
The most eastern of the three	277	20	S	21	0	3
The obscure star west of that	275	20	S	20	50	5
The more southern of the two remaining to the north	277	10	S	16	0	4
The more northern	277	10	S	14	50	4
6 stars: 3 of third magnitude, 2 of fourth, 1 of fifth						

In the southern region 316 stars: 7 of first magnitude, 18 of second, 60 of third, 167 of fourth, 54 of fifth, 9 of sixth, and 1 nebulous. And so there are altogether 1024 stars: 15 of first magnitude, 45 of second, 206 of third, 476 of fourth, 217 of fifth, 49 of sixth, 11 obscure, and 5 nebulous.

BOOK THREE

1. ON THE PRECESSIONS OF THE SOLSTICES AND EQUINOXES

[63ᵃ] Having depicted the appearance of the fixed stars in relation to the annual revolution, we must pass on; and we shall treat first of the change of the equinoxes, by reason of which even the fixed stars are believed to move. Now we find that the ancient mathematicians made no distinction between the "turning" or natural year, which begins at an equinox or solstice, and the year which is determined by means of some one of the fixed stars. That is why they thought the Olympic years, which they measured from the rising of Canicula, were the same as the years measured from the summer solstice, since they did not yet know the distinction between the two.

But Hipparchus of Rhodes, a man of wonderful acumen, was the first to call attention to the fact that there was a difference in the length of these two kinds of year. While making careful observations of the magnitude of the year, he found that it was longer as measured from the fixed stars than as measured from the equinoxes or solstices. Hence he believed that the fixed stars too possessed a movement eastward, but one so slow as not to be immediately perceptible. But now through the passage of time, the movement has become very evident. By it we discern a rising and setting of the signs and stars which are already far different from those risings and settings described by the ancients; and we see that the twelve parts of the ecliptic have receded from the signs of the fixed stars by a rather great interval, although in the beginning they agreed in name and in position.

Moreover, an irregular movement has been found; and wishing to assign the cause for its irregularity, astronomers have brought forward different theories. Some maintained that there was a sort of swinging movement of the suspended world—like the movement in latitude which we find in the case of the planets—and that back and forth within fixed limits as far out as the world has gone forward in one direction it will come back again in the other at some time,[1] and that the extent of its digression from the middle on either side was not more than 8°. But this already outdated theory can no longer hold, especially because [63ᵇ] it is already clear enough that the head of the constellation of Aries has become more than three times 8° distant from the spring equinox—and similarly for other stars—and no trace of a regression has been perceived during so many ages. Others indeed have opined that the sphere of the fixed stars moves forward but does so by irregular steps; and nevertheless they have failed to define any fixed mode of movement.

[1] *i.e.*, the sphere of the world has rotated westward and will at some time rotate eastward the same distance.

Moreover, there is an additional surprise of nature, in that the obliquity of the ecliptic does not appear so great to us as before Ptolemy—as we said above.

For the sake of a cause for these facts some have thought up a ninth sphere and others a tenth: they thought these facts could be explained through those spheres; but they were unable to produce what they had promised. Already an eleventh sphere has begun to see the light of day; and in talking of the movement of the Earth we shall easily prove that this number of circles is superfluous.

For, as we have already set out separately in Book I, the two revolutions, that is, of the annual declination and of the centre of the Earth, are not altogether equal, namely because the restoration of the declination slightly anticipates the period of the centre, whence it necessarily follows that the equinoxes seem to arrive before their time—not that the sphere of the fixed stars is moved eastward, but rather that the equator is moved westward, as it is inclined obliquely to the plane of the ecliptic in proportion to the amount of deflexion of the axis of the terrestrial globe. For it seems more accurate to say that the equator is inclined obliquely to the ecliptic than that the ecliptic, a greater circle, is inclined to the equator, a smaller. For the ecliptic, which is described by the distance between the sun and the Earth during the annual circuit, is much greater than the equator, which is described by the daily movement of the Earth around its axis. And in this way the common sections of the equator and the oblique ecliptic are perceived, with the passage of time, to get ahead, while the stars are perceived to lag behind. But the measure of this movement and the ratio of its irregularity were hidden from our predecessors, because the period of revolution was not yet known on account of its surprising slowness—I mean that during the many ages after it was first noticed by men, it has advanced through hardly a fifteenth part of a circle, or 24°. Nevertheless, we shall state things with as much certitude as possible, with the aid of what we have learned concerning these facts from the history of observations down to our own time.

2. HISTORY OF THE OBSERVATIONS CONFIRMING THE IRREGULAR PRECESSION OF THE EQUINOXES AND SOLSTICES

[64ª] Accordingly in the 36th year of the first of the seventy-six-year periods of Callippus, which was the 30th year after the death of Alexander the Great, Timochares the Alexandrian, who was the first to investigate the positions of the fixed stars, recorded that Spica, which is in the constellation of Virgo, had an angular elongation of $82^1/_3°$ from the point of summer solstice with a southern latitude of 2°; and that the star in the forehead of Scorpio which is the most northward of the three and is first in the order of formation of the sign had a latitude of $1^1/_3°$ and a longitude of 32° from the autumn equinox.

And again in the 48th year of the same period he found that Spica in Virgo had a longitude of $82^1/_2°$ from the summer solstice but had kept the same latitude.

Now Hipparchus in the 50th year of the third period of Callippus, in the 196th year since the death of Alexander, found that the star called Regulus, which is in the breast of Leo, was $29^5/_6°$ to the east of the summer solstice.

Next Menelaus, the Roman geometer, in the first year of Trajan's reign, *i.e.*, in the 99th year since the birth of Christ, and in the 422nd year since the death of Alexander, recorded that Spica in Virgo had a longitude of $86^1/_4°$ from the (summer) solstice and that the star in the forehead of Scorpio had a longitude of $35^{11}/_{12}°$ from the autumn equinox.

Following them, Ptolemy, in the second year of the reign of Antoninus Pius, in the 462nd year since the death of Alexander, discovered that Regulus in Leo had a longitude of $32^1/_2°$ from the (summer) solstice; Spica, $86^1/_2°$; and that the star in the forehead of Scorpio had a longitude of $36^1/_3°$ from the autumn equinox, with no change in latitude—as was set forth above in drawing up the tables. And we have passed these things in review, just as they were recorded by our predecessors.

After a great lapse of time, however, in the 1202nd year after the death of Alexander, came the observations of al-Battani the Harranite; and we may place the utmost confidence in them. In that year Regulus, or Basiliscus, was seen to have attained a longitude of 44°5' from the (summer) solstice; and the star in the forehead of Scorpio, one of 47°50' [64ᵇ] from the autumn equinox. The latitude of these stars stayed completely the same, so that there is no longer any doubt on that score.

Wherefore in the year of Our Lord 1525, in the year after leap-year by the Roman calendar and 1849 Egyptian years after the death of Alexander, we were taking observations of the often mentioned Spica, at Frauenburg, in Prussia. And the greatest altitude of the star on the meridian circle was seen to be approximately 27°. We found that the latitude of Frauenburg was $54°19^1/_2°$. Wherefore its declination from the equator stood to be 8°40'. Hence its position became known as follows:

For we have described the meridian circle *ABCD* through the poles of the ecliptic and the equator. Let *AEC* be the diameter and common section with the equator; and *BED* is the diameter and common section with the ecliptic. Let *F* be the north pole of the ecliptic and *FEG* its axis; and let *B* be the beginning of Capricorn and *D* of Cancer. Now let

arc *BH* = 2°,

which is the southern latitude of the star. And from point *H* let *HL* be drawn parallel to *BD*; and let *HL* cut the axis of the ecliptic at *I* and the equator at *K.* Moreover, let

$$\text{arc } MA = 8°40',$$

in proportion to the southern declination of the star; and from point M let MN be drawn parallel to AC.

MN will cut HIL the parallel to the ecliptic; therefore let MN cut HIL at point O; and if the straight line OP is drawn at right angles to MN and AC, then

$$OP = {}^1/_2 \text{ ch. } 2 \; AM.$$

But the circles having the diameters FG, HL, and MN are perpendicular to plane $ABCD$; and by Euclid's *Elements*, XI, 19, their common sections are at right angles to the same plane in points O and I. Hence by XI, 6, they (the common sections) are parallel to one another. And since I is the centre of the circle whose diameter is HL, therefore line OI will be equal to half the chord subtending twice an arc in a circle of diameter HL—an arc similar to the arc which measures the longitude of the star from the beginning of Libra, and this arc is what we are looking for. It is found in this way:

Since the exterior angle is equal to its interior and opposite,

$$\text{angle } AEB = \text{angle } OKP$$

and

$$\text{angle } OPK = 90°.$$

Accordingly

$$[65^a] \quad OP : OK = {}^1/_2 \text{ ch. } 2 \; AB : BE = {}^1/_2 \text{ ch. } 2 \; AH : HIK.$$

For the lines comprehend triangles similar to OPK.

But

$$\text{arc } AB = 23°28{}^1/_2',$$

and

$$\begin{aligned}{}^1/_2 \text{ ch. } 2 \; AB &= 39{,}832, \\ \text{where } BE &= 100{,}000.\end{aligned}$$

And

$$\begin{aligned}\text{arc } ABH &= 25°28{}^1/_2', \\ {}^1/_2 \text{ ch. } 2 \; ABH &= 43{,}010, \\ \text{arc } MA &= 8°40',\end{aligned}$$

which is the declination, and

$$ {}^1/_2 \text{ ch. } 2 \; MA = 15{,}069.$$

It follows from this that

$$\begin{aligned}HIK &= 107{,}978, \\ OK &= 37{,}831,\end{aligned}$$

and by subtraction

$$HO = 70{,}147.$$

But

$$HOI = {}^{1}/_{2} \text{ ch. } HGL$$

and

$$\text{arc } HGL = 176°.$$

Then

$$HOI = 99,939,$$
$$\text{where } BE = 100,000.$$

And therefore by subtraction,

$$OI = HOI - HO = 29,792.$$

But in so far as $HOI = $ radius $= 100,000$,

$$OI = 29,810$$
$$\fallingdotseq {}^{1}/_{2} \text{ ch. 2 arc } 17°21'.$$

This was the distance of Spica in (the constellation) Virgo from the beginning of Libra; and the position of the star was here. Moreover, ten years before, in 1515, we found that it had a declination of 8°36'; and its position was 17°14' distant from the beginning of the Balances.

Now Ptolemy recorded that it had a declination of only $^{1}/_{2}°$. Therefore its position was at 26°40' of the (zodiacal sign) Virgo, which seems to be more or less true in comparison with the previous observations.

Hence it appears clearly enough that during nearly the whole period of 432 years from Timochares to Ptolemy the equinoxes and solstices were moved according to a precession of 1° per 100 years—if a constant ratio is set up between the time and the amount of precession, which added up to $4^{1}/_{3}°$. For in the 266 years between Hipparchus and Ptolemy the longitude of Basiliscus in Leo from the summer solstice moved $2^{2}/_{3}°$, so that here too, by taking the time into comparison, there is found a precession of 1° per 100 years.

Moreover, because during the 782 mean years between the observation of Menelaus and that of al-Battani the first star in the forehead of Scorpio had a change in longitude of 11°55', it will certainly seem that 1° should be assigned not to 100 years but rather to 66 years; but for the 741 years after Ptolemy, 1° to only 65 years.

If finally the remaining space of 645 years is compared with the difference of 9°11' given by our observation, there will be 71 years allotted to 1°.

From this it is clear that the precession of the equinoxes was slower [65b] during the 400 years before Ptolemy than during the time between Ptolemy and al-Battani, and that the precession in this middle period was speedier than in the time from al-Battani to us.

Moreover, there is found a difference in the movement of obliquity, since Aristarchus of Samos found that the obliquity of the ecliptic and the equator was 23°51'20", just as Ptolemy did; al-Battani, 23°35'; 190 years later Arzachel the

Spaniard, 23°34'. And similarly after 230 years Prophatius the Jew found that the obliquity was approximately 2' smaller. And in our time it has not been found greater than 23°28$^1/_2$'. Hence it is also clear that the movement was least from the time of Aristarchus to that of Ptolemy and greatest from that of Ptolemy to that of al-Battani.

3. THE HYPOTHESES BY MEANS OF WHICH THE MUTATION OF THE EQUINOXES AND OF THE OBLIQUITY OF THE ECLIPTIC AND THE EQUATOR ARE SHOWN

Accordingly it seems clear from this that the solstices and equinoxes change around in an irregular movement. No one perhaps will bring forward a better reason for this than that there is a certain deflexion of the axis of the Earth and the poles of the equator. For that seems to follow upon the hypothesis of the movement of the Earth, since it is clear that the ecliptic remains perpetually unchangeable—the constant latitudes of the fixed stars bear witness to that—while the equator moves. For if the movement of the axis of the Earth were simply and exactly in proportion to the movement of the centre, there would not appear at all any precession of the equinoxes and solstices, as we said; but as these movements differ from one another by a variable difference, it was necessary for the solstices and equinoxes to precede the positions of the stars in an irregular movement.

The same thing happens in the case of the movement of declination, which changes the obliquity of the ecliptic irregularly—although this obliquity should be assigned more rightly to the equator.

For this reason you should understand two reciprocal movements belonging wholly to the poles, like hanging balances, since the poles and circles in a sphere imply one another mutually and are in agreement. Therefore there will be one movement which changes the inclination of those circles [66ª] by moving the poles up and down in proportion to the angle of section. There is another which alternately increases and decreases the solstitial and equinoctial precessions by a movement taking place crosswise. Now we call these movements "librations," or "swinging movements," because like hanging bodies swinging over the same course between two limits, they become faster in the middle and very slow at the extremes. And such movements occur very often in connection with the latitudes of the planets, as we shall see in the proper place.

They differ moreover in their periods, because the irregular movement of the equinoxes is restored twice during one restoration of obliquity. But as in every apparent irregular movement, it is necessary to understand a certain mean, through which the ratio of irregularity can be determined; so in this case too it was quite necessary to consider the mean poles and the mean equator and also the mean equinoxes and points of solstice. The poles and the terrestrial equator, by being deflected in opposite directions away from these mean poles, though within fixed limits, make those regular movements appear to be irregular. And so these two librations competing with one another make the poles of the earth in the passage of time describe certain lines similar to a twisted garland.

But since it is not easy to explain these things adequately with words, and still

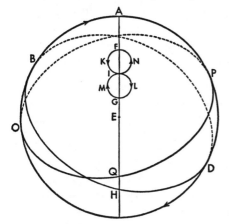

less—I fear—to have them grasped by the hearing, unless they are also viewed by the eyes, therefore let us describe on a sphere circle *ABCD* which is the ecliptic. Let the north pole (of the ecliptic) be *E*, the beginning of Capricornus *A*, that of Cancer *C*, that of Aries *B*, and that of Libra *D*. And through points *A* and *C* and pole *E* let circle *AEC* be drawn. And let the greatest distance between the north poles of the ecliptic and of the equator be *EF*, and the least *EG*. Similarly let *I* be the pole in the middle

position, and around *I* let the equator *BHD* be described, and let that be called the mean equator; and let *B* and *D* be called the mean equinoxes.

Let the poles of the equator, the equinoxes, and the equator be all carried around *E* by an always regular movement westward, *i.e.*, counter to the order of the signs in the sphere of the fixed stars, and with a slow movement, as I said. Now let there be understood two reciprocal movements of the terrestrial poles like hanging bodies—one of them between the limits *F* and *G*, which will be called the movement of anomaly,[1]— *i.e.*, irregularity—of declination; the other from westward to eastward, and from eastward to westward. This second movement, which has twice the velocity of the first, we shall call the anomaly of the equinoxes. As both of these movements belong to the poles of the earth, they deflect the poles in a surprising way.

For first with *F* as the north pole of the earth, [66ᵇ] the equator described around the pole will pass through the same sections *B* and *D*, *i.e.*, through the poles of circle *AFEC*. But it will make greater angles of obliquity in proportion to arc *FI*. Now the second movement supervening does not allow the terrestrial pole, which was about to cross from the assumed starting point *F* to the mean obliquity at *I*, to proceed in a straight line along *FI*, but draws it aside in a circular movement towards its farthest eastward latitude, which is at *K*. The intersection of the apparent equator *OQP* described around this position will not be in *B* but to the east of it in *O*, and the precession of the equinoxes will be decreased in proportion to arc *BO*. Changing its direction and

---

[1]The term *anomaly* will be used to designate a regular movement the compounding of which with the principal regular movement being considered makes that principal movement appear irregular.

moving westwards, the pole is carried by the two simultaneously competing movements to the mean position *I*. And the apparent equator is in all respects identical with the regular or mean equator. Crossing there, the pole of the earth moves westward and separates the apparent equator from the mean equator and increases the precession of the equinoxes up to the other limit *L*. There changing its direction again, it subtracts what it had just added to the precession of the equinoxes, until, when situated at point *G*, it causes the least obliquity at the same common section *B*, where once more the movement of the equinoxes and solstices will appear very slow, in approximately the same way as at *F*. At this time the irregularity of the equinoxes stands to have completed its revolution, since it has passed from the mean through both extremes and back to the mean; while the movement of obliquity in going from greatest declination to least has completed only half its circuit. Moving on from there, the pole advances eastward to the farthest limit *M*; and, after reversing its direction there becomes one with the mean pole *I*; and once more it proceeds westward and after reaching the limit *N* finally [67ᵃ] completes what we called the twisted line *FKILGMINF*. And so it is clear that in one cycle of obliquity the pole of the Earth reaches the westward limit twice and the eastward limit twice.

4. HOW THE RECIPROCAL MOVEMENT OR MOVEMENT OF LIBRATION IS COMPOSED OF CIRCULAR MOVEMENTS

Accordingly we shall make clear exactly how this movement agrees with the appearances. In the meantime someone will ask how the regularity of these librations is to be understood, since it was said in the beginning that the celestial movement was regular, or composed of regular and circular movements. But here in either case of libration two movements are apparent as one movement between two limits, and the two limits necessarily make a cessation of movement intervene. For we acknowledge that there are twin movements, which are demonstrated from regular movements in this way.

Let there be the straight line *AB*, and let it be cut into four equal parts at points *C*, *D*, and *E*. Let the homocentric circles *ADB* and *CDE* be described around *D* in the same plane. And in the selfsame plane *ADB* and *CDE*, let any point *F* be taken on the circumference of the inner circle; and with *F* as centre and radius equal to *FD* let circle *GHD* be described. And let it cut the straight line *AB* at point *H*; and let the diameter *DFG* be drawn. We have to show that when the twin movements of circles *GHD* and *CFE* compete

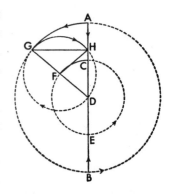

with one another, the movable point *H* proceeds back and forth along the same straight line *AB* by a reciprocal motion.

This will take place if we understand that *H* is moved in a different direction from *F* and through twice the distance, since the angle same *CDF* which is situated at the centre of circle *CFE* and on the circumference of circle *GHD* comprehends both arcs of the equal circles: arc *FC* and arc *GH* which is twice arc *FC*.

It is laid down that at some time upon the coincidence of the straight lines *ACD* and *DFG* the moving point *H* will be at *G*, which will then coincide with *A*; and *F* will be at *C*. Now, however, the centre *F* has moved towards the right along *CF*, and *H* has moved along the circumference to the left twice the distance *CF* [67ᵇ] or vice versa; accordingly *H* will be deflected along line *AB*; otherwise the part would be greater than the whole, as it is easy to see. But *H* has moved away from its first position along the length *AH* made by the bent line *DFH*, which is equal to *AD*; and *H* has moved for a distance by which the diameter *DFG* exceeds chord *DH*. And in this way *H* will be made to arrive at centre *D*, which will be the point of tangency of circle *DHG* with straight line *AB*, namely when *GD* is at right angles to *AB*; and then *H* will reach the other limit at *B*, and from that position it will move back again according to the same ratio.

Therefore it is clear that movement along a straight line is compounded of two circular movements which compete with one another in this way; and that a reciprocal and irregular movement is composed of regular movements; as was to be shown. Moreover it follows from this that the straight line *GH* will always be at right angles to *AB*; for lines *DH* and *HG*, being in a semicircle, will always comprehend a right angle. And accordingly

$$GH = {}^1/_2 \text{ ch. } 2 \ AG;$$

and

$$DH = {}^1/_2 \text{ ch. } 2 \ (90° - AG),$$

because circle *AGB* has twice the diameter of circle *HGD*.

5. A Demonstration of the Irregularity of the Equinoctial Precession and the Obliquity

For this reason some call this movement of the circle a movement in width, *i.e.*, along the diameter. But they determine its periodicity and its regularity by means of the circumference, and its magnitude by means of the chords subtending. On that account it is easily shown that the movement appears irregular and faster at the centre and slower [68ᵃ] at the circumference.

For let there be the semicircle *ABC* with centre *D* and diameter *ADC*, and let it be bisected at point *B*. Now let equal arcs *AE* and *BF* be taken, and from points *F* and *E* let *EG* and *FK* be drawn perpendicular to *ADC*. Therefore, since

$$2 \, DK = 2 \text{ ch. } BF,$$

and

$$2 \, EG = 2 \text{ ch. } AE,$$

then

$$DK = EG.$$

But by Euclid's *Elements*, III, 7,

$$AG < GE;$$

hence

$$AG < DK.$$

But *GA* and *KD* will take up equal time because

$$\text{arc } AE = \text{arc } BF;$$

therefore the movement in the neighbourhood of an *A* will (appear to) be slower than in the neighbourhood of the centre *D*.

Having shown this, let us take *L* as the centre of the Earth, so that the straight line *DL* is perpendicular to plane *ABC* of the semicircle; and with *L* as centre and through points *A* and *C* let arc *AMC* of a circle be described; and let *LDM* be drawn in a straight line. Accordingly the pole of the semicircle *ABC* will be at *M*, and *ADC* will be the common section of the circles. Let *LA* and *LC* be joined; and similarly *LK* and *LG* too. And let *LK* and *LG* extended in straight lines cut arc *AMC* at *N* and *O*. Therefore, since angle *LDK* is right, angle *LKD* is acute. Wherefore too the line *LK* is longer than *LD*, and all the more is side *LG* greater than side *LK*, and *LA* than *LG* in the obtuse triangles. Therefore the circle described with *L* as centre and *LK* as radius will fall beyond *LD*, but will cut *LG* and *LA*; let it be described, and let it be *PKRS*. And since

$$\text{trgl. } LDK < \text{sect. } LPK,$$

while

$$\text{trgl. } LGA > \text{sect. } LRS,$$

on that account

$$\text{trgl. } LDK : \text{sect. } LPK < \text{trgl. } LGA : \text{sect. } LRS.$$

Hence, alternately also,

$$\text{trgl. } LDK : \text{trgl. } LGA < \text{sect. } LPK : \text{sect. } LRS.$$

And by Euclid's *Elements*, VI, 1,

$$\text{trgl. } LDK : \text{trgl. } LGA = \text{base } DK : \text{base } AG.$$

But

$$\text{sect. } LPK : \text{sect. } LRS = \text{angle } DLK : \text{angle } RLS = \text{arc } MN : \text{arc } OA.$$

Therefore

$$\text{base } DK : \text{base } GA < \text{arc } MN : \text{arc } OA.$$

But we have already shown that

$$DK > GA.$$

All the more then

[68ᵇ] $MN > OA.$

And arcs *MN* and *OA* are understood as having been described during equal intervals of time by the poles of the earth in accordance with the equal arcs of anomaly *AE* and *BF*—as was to be shown. But since the difference between greatest and least obliquity is so slight as not to exceed $2/_5°$, there will be no sensible difference between the curved line *AMC* and the straight line *ADC*; and so no error will arise if we work simply with line *ADC* and semicircle *ABC.*

Practically the same thing happens in the case of the other movement of the poles, which has to do with the equinoxes, since this movement does not ascend to the mean degree, as will be apparent below. Once more let there be the circle *ABCD* through the poles of the ecliptic and the mean equator. We may call it the mean colure of Cancer. Let the semicircle of the ecliptic be *DEB* and the mean equator *AEC*; and let them cut one another at point *E*, where the mean equinox will be. Now let the pole of the equator be *F*, and let the great circle *FEI* be described through it. On that account it will be the colure of the mean or regular equinoxes.

Therefore for the sake of an easier demonstration let us separate the libration of the equinoxes from the obliquity of the ecliptic. On the colure *EF* let arc *FG* be taken,

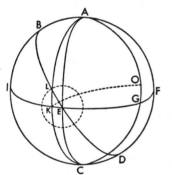

and through that distance let *G* the apparent pole of the equator be understood as removed from *F* the mean pole. And with *G* as a pole let *ALKC* the semicircle of the apparent equator be described. It will cut the ecliptic at *L*. Therefore point *L* will be the apparent equinox; and its distance from the mean equinox will be measured by arc *LE*, which is produced by arc *EK* the equal of *FG.*

But with *K* as a pole we shall describe circle *AGC*; and let it be understood that the equatorial pole during the time in which the libration *FG* takes place does not remain the "true" pole at point *G*, but, driven by another libration or swinging movement, moves away in the direction of the oblique ecliptic through arc *GO*. Therefore while ecliptic *BED* abides, the "true" equator will be changed to the "apparent" in accordance with the transposition of the pole to *O*. And similarly the movement of intersection *L* the apparent equinox will be faster in the neighbourhood of the mean (equinox) *E* and very slow in the neighbourhood of the extreme equinoxes, more or less in proportion to the swinging movement of the poles which we have already demonstrated—as was worth the trouble of our attention.

6. On the Regular Movements of the Precession of the Equinoxes and of the Inclination of the Ecliptic

[69ᵃ] Now every apparent irregular circular movement passes through four termini: there is the terminus where it appears slow and the terminus where it appears fast, as if at the extremes, and the terminus where it appears to have a mean velocity, as if at the means, since from the point which is the end of decrease in velocity and the beginning of increase it passes on to a mean velocity; and from the mean velocity it increases till it becomes fast; again after being fast it approaches a mean velocity, whence for the remainder of the cycle it changes to its former slowness.

By means of that it is possible to know in what part of the circle the position of the non-uniform movement, or irregularity, is at a given time; and too by means of these indications the restitution of the irregularity is perceptible.[1] Accordingly in a quadrisected circle let *A* be the position of greatest slowness, *B* the mean of increasing velocity, *C* the end of the increase and the beginning of the decrease, and *D* the mean of decreasing velocity. Therefore, since, as was reported above, the apparent movement of the precession of the equinoxes was found to be rather slow in the time between Timochares and Ptolemy in comparison with the other times, and because for a while it appeared regular and uniform, as is shown by the observations of Aristyllus, Hipparchus, Agrippa, and Menelaus which were made at the middle of that time; it argues that the apparent movement of the equinoxes had been simply at its slowest and at the middle of this time was at the beginning of increase in velocity, when the cessation of the decrease conjoined to the beginning of the increase by reason of mutual compensation made the movement seem uniform for the time being. Accordingly Timochares' observation must be placed in the fourth quadrant of the circle along *DA*; but Ptolemy's falls in the first quadrant along *AB*. Again, because in the second interval, the one between Ptolemy and al-Battani the Harranite the movement is found to have been faster than in the third, it is clear that the point of highest velocity was passed during the second interval of time, and the irregularity had already reached the third quadrant of the circle along *CD*, and that from the third interval down to us the restoration of the irregularity was nearly completed and has nearly returned to its starting-point with Timochares. For if we divide the cycle of 1819 years between Timochares and us into

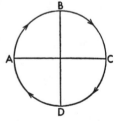

the customary 360 parts, we shall have proportionately an arc of 85$^1/_2$° for 432 years; 146°51' for 742 years; and the remaining arc of 127°39' for the remaining 645 years.

---

[1]This circle is not the circle of libration, of course, but a circle which typifies the cycle in velocity which results from compounding the libration with the regular movement of precession of the equinoxes.

We have made these determinations by an obvious and simple inference; [69ᵇ] but upon working them over with stricter calculations, to see how exactly they agree with observations, we find that the movement of irregularity during the 1819 Egyptian years has already exceeded its complete revolution by 21°24', and that the time of the period comprehends only 1717 Egyptian years. In accordance with this ratio it is discovered that the first segment of the circle has 90°35'; the second, 155°34'; while the third period of 543 years will comprehend the remaining 113°51' of the circle. Now that these things have been set up in this way, the mean movement of the precession of the equinoxes is also disclosed; and it is 23°57' for these same 1717 years, at the end of which the whole movement of irregularity was restored to its pristine status, since for the 1819 years we have an apparent movement of approximately 25°1'.

But in 102 years after Timochares—the difference between the 1717 years and the 1819 years—the apparent movement must have been about 1°4', because it is probable that the apparent movement was then a little greater than to need only 1° per 100 years, since it was decreasing without yet having reached the end of the decrease. Hence, if we subtract 1°4' from 25°1' there will remain, as we said, for the 1717 Egyptian years a mean and regular movement of 23°57', now corrected according to the apparent and irregular movement: hence the complete and regular revolution of the precession of the equinoxes arises in 25,816 years, during which time 15 cycles of irregularity are traversed and approximately $1/28$ cycle over and above.

Moreover, the movement of obliquity, whose restoration we said was twice as slow as the irregular precession of the equinoxes, accords with this ratio. For the fact that Ptolemy reports that during the 400 years between the time of Aristarchus of Samos and his own the obliquity of 23°51'20" had hardly changed at all indicates that the obliquity was then close to the limit of greatest obliquity, namely when the precession of the equinoxes was in its slowest motion. But now also when the same restoration of slowness is approaching, the inclination of the axis is not at its greatest but is near its least. In the middle of the time in between, al-Battani the Harranite, as was said, found that the inclination was 23°35'; 190 years after him Arzachel the Spaniard, 23°34'; and similarly 230 years later Prophatius the Jew found it approximately 2' less. Finally as regards our own times, we in 30 years of frequent observation found it approximately 23°28$2/5$'—from which George Peurbach and John of Monteregium, [70ᵃ] our nearest predecessors, differ slightly. Here again it is perfectly obvious that for the 900 years after Ptolemy the change in obliquity was greater than for any other interval of time.

Therefore since we already have the cycle of irregularity of precession in 1717 years, we shall also have half the period of obliquity in that time, and in 3434 years its complete restoration. Wherefore if we divide the 360° by the number of the 3434 years or 180° by 1717, the quotient will be the annual movement of simple anomaly of 6'17"24"'9"". These in turn distributed through the 365 days give a daily movement of 1"2"'2"".

Similarly, when the mean precession of equinoxes has been distributed through the 1717 years—and there was 23°57'—an annual movement of 50"12"'5"" will be the result; and this distributed through the 365 days will give a daily movement of 8"'15"".

But in order that the movements may be more in the open and may be found right at hand when there is need of them, we shall draw up their tables or canons by the continuous and equal addition of annual movement—60 parts always being carried over into the minutes or degrees, if the sum exceeds that—and for the sake of convenience we shall keep on adding until we reach the 60th year, since the same configuration of numbers returns every sixty years, only with the denominations of degrees and minutes moved up, so that what were formerly seconds become minutes and so on.[1] By this abridgement in the form of brief tables, it will be possible merely by a double entry to determine and infer the regular movements for the years in question among the 3600 years. This is also the case with the number of days.

Moreover, in our computations of the celestial movements we shall employ the Egyptian years, which alone among the legal years are found equal. For it is necessary for the measure to agree with the measured; but that is not the case with the years of the Romans, Greeks, and Persians, for intercalations are made not in any single way, but according to the will of the people. But the Egyptian year contains no ambiguity as regards the fixed number of 365 days, in which throughout twelve equal months— which they name in order by these names: Thoth, Phaophi, Athyr, Chiach, Tybi, Mechyr, Phamenoth, Pharmuthi, Pachon, Pauni, Epiphi, and Mesori—in which, I say, six periods of 60 days are comprehended evenly together with the five remaining days, which they call the intercalary days. For that reason Egyptian years are most convenient for calculating regular movements. Any other years are easily reducible to them by resolving the days.

---

[1]That is to say, the same configurations of numbers return in multiples of sixty years, because the cycle of movement is divided according to the sexagesimal system—jut as it would return in multiples of ten years if the circle were divided according to the decimal system.

REGULAR MOVEMENT OF THE PRECESSION OF THE EQUINOXES IN YEARS AND PERIODS OF SIXTY YEARS

Egyptian Years	60°	°	′	″	‴	Egyptian Years	60°	°	′	″	‴
1	0	0	0	50	12	31	0	0	25	56	14
2	0	0	1	40	24	32	0	0	26	46	26
3	0	0	2	30	36	33	0	0	27	36	38
4	0	0	3	20	48	34	0	0	28	26	50
5	0	0	4	11	0	35	0	0	29	17	2
6	0	0	5	1	12	36	0	0	30	7	15
7	0	0	5	51	24	37	0	0	30	57	27
8	0	0	6	41	36	38	0	0	31	47	38
9	0	0	7	31	48	39	0	0	32	37	51
10	0	0	8	22	0	40	0	0	33	28	3
11	0	0	9	12	12	41	0	0	34	18	15
12	0	0	10	2	25	42	0	0	35	8	27
13	0	0	10	52	37	43	0	0	35	58	39
14	0	0	11	42	49	44	0	0	36	48	51
15	0	0	12	33	1	45	0	0	37	39	3
16	0	0	13	23	13	46	0	0	38	29	15
17	0	0	14	13	25	47	0	0	39	19	27
18	0	0	15	3	37	48	0	0	40	9	40
19	0	0	15	53	49	49	0	0	40	59	52
20	0	0	16	44	1	50	0	0	41	50	4
21	0	0	17	34	13	51	0	0	42	40	16
22	0	0	18	24	25	52	0	0	43	30	28
23	0	0	19	14	37	53	0	0	44	20	40
24	0	0	20	4	50	54	0	0	45	10	52
25	0	0	20	55	2	55	0	0	46	1	4
26	0	0	21	45	14	56	0	0	46	51	16
27	0	0	22	35	26	57	0	0	47	41	28
28	0	0	23	25	38	58	0	0	48	31	40
29	0	0	24	15	50	59	0	0	49	21	52
30	0	0	25	6	2	60	0	0	50	12	5

Position of the Birth of Christ—5 32'

REGULAR MOVEMENT OF THE PRECESSION OF THE EQUINOXES IN DAYS AND PERIODS OF SIXTY DAYS

Days	60°	°	′	″	‴	Days	60°	°	′	″	‴
1	0	0	0	0	8	31	0	0	0	4	15
2	0	0	0	0	16	32	0	0	0	4	24
3	0	0	0	0	24	33	0	0	0	4	32
4	0	0	0	0	33	34	0	0	0	4	40
5	0	0	0	0	41	35	0	0	0	4	48
6	0	0	0	0	49	36	0	0	0	4	57
7	0	0	0	0	57	37	0	0	0	5	5
8	0	0	0	1	6	38	0	0	0	5	13
9	0	0	0	1	14	39	0	0	0	5	21
10	0	0	0	1	22	40	0	0	0	5	30
11	0	0	0	1	30	41	0	0	0	5	38
12	0	0	0	1	39	42	0	0	0	5	46
13	0	0	0	1	47	43	0	0	0	5	54
14	0	0	0	1	55	44	0	0	0	6	3
15	0	0	0	2	3	45	0	0	0	6	11
16	0	0	0	2	12	46	0	0	0	6	19
17	0	0	0	2	20	47	0	0	0	6	27
18	0	0	0	2	28	48	0	0	0	6	36
19	0	0	0	2	36	49	0	0	0	6	44
20	0	0	0	2	45	50	0	0	0	6	52
21	0	0	0	2	53	51	0	0	0	7	0
22	0	0	0	3	1	52	0	0	0	7	9
23	0	0	0	3	9	53	0	0	0	7	17
24	0	0	0	3	18	54	0	0	0	7	25
25	0	0	0	3	26	55	0	0	0	7	33
26	0	0	0	3	34	56	0	0	0	7	42
27	0	0	0	3	42	57	0	0	0	7	50
28	0	0	0	3	51	58	0	0	0	7	58
29	0	0	0	3	59	59	0	0	0	8	6
30	0	0	0	4	7	60	0	0	0	8	15

Position of the Birth of Christ—5 32'

MOVEMENT OF THE SIMPLE ANOMALY OF EQUINOXES IN YEARS AND PERIODS OF SIXTY YEARS

Position of the Birth of Christ–6 °45'

Egyptian Years	Longitude 60°	°	'	"	'''	Egyptian Years	Longitude 60°	°	'	"	'''
1	0	0	6	17	24	31	0	3	14	59	28
2	0	0	12	34	48	32	0	3	21	16	52
3	0	0	18	52	12	33	0	3	27	34	16
4	0	0	25	9	36	34	0	3	33	51	41
5	0	0	31	27	0	35	0	3	40	9	5
6	0	0	37	44	24	36	0	3	46	26	29
7	0	0	44	1	49	37	0	3	52	43	53
8	0	0	50	19	13	38	0	3	59	1	17
9	0	0	56	36	36	39	0	4	5	18	42
10	0	1	2	54	1	40	0	4	11	36	6
11	0	1	9	11	25	41	0	4	17	53	30
12	0	1	15	28	49	42	0	4	24	10	54
13	0	1	21	46	13	43	0	4	30	28	18
14	0	1	28	3	38	44	0	4	36	45	42
15	0	1	34	21	2	45	0	4	43	3	0
16	0	1	40	38	26	46	0	4	49	20	31
17	0	1	46	55	50	47	0	4	55	37	55
18	0	1	53	13	14	48	0	5	1	55	19
19	0	1	59	30	38	49	0	5	8	12	43
20	0	2	5	48	3	50	0	5	14	30	7
21	0	2	12	5	27	51	0	5	20	47	31
22	0	2	18	22	51	52	0	5	27	4	55
23	0	2	24	40	15	53	0	5	33	22	20
24	0	2	30	57	39	54	0	5	39	39	44
25	0	2	37	15	3	55	0	5	45	57	8
26	0	2	43	32	27	56	0	5	52	14	32
27	0	2	49	49	52	57	0	5	58	31	56
28	0	2	56	7	16	58	0	6	4	49	20
29	0	3	2	24	40	59	0	6	11	6	45
30	0	3	8	42	4	60	0	6	17	24	9

MOVEMENT OF THE SIMPLE ANOMALY OF EQUINOXES IN DAYS AND PERIODS OF SIXTY DAYS

Position of the Birth of Christ–6 °45'

Days	Longitude 60°	°	'	"	'''	Days	Longitude 60°	°	'	"	'''
1	0	0	0	1	2	31	0	0	0	32	3
2	0	0	0	2	4	32	0	0	0	33	5
3	0	0	0	3	6	33	0	0	0	34	7
4	0	0	0	4	8	34	0	0	0	35	9
5	0	0	0	5	10	35	0	0	0	36	11
6	0	0	0	6	12	36	0	0	0	37	13
7	0	0	0	7	14	37	0	0	0	38	15
8	0	0	0	8	16	38	0	0	0	39	17
9	0	0	0	9	18	39	0	0	0	40	19
10	0	0	0	10	20	40	0	0	0	41	21
11	0	0	0	11	22	41	0	0	0	42	23
12	0	0	0	12	24	42	0	0	0	43	25
13	0	0	0	13	26	43	0	0	0	44	27
14	0	0	0	14	28	44	0	0	0	45	29
15	0	0	0	15	30	45	0	0	0	46	31
16	0	0	0	16	32	46	0	0	0	47	33
17	0	0	0	17	34	47	0	0	0	48	35
18	0	0	0	18	36	48	0	0	0	49	37
19	0	0	0	19	38	49	0	0	0	50	39
20	0	0	0	20	40	50	0	0	0	51	41
21	0	0	0	21	42	51	0	0	0	52	43
22	0	0	0	22	44	52	0	0	0	53	45
23	0	0	0	23	46	53	0	0	0	54	47
24	0	0	0	24	48	54	0	0	0	55	49
25	0	0	0	25	50	55	0	0	0	56	51
26	0	0	0	26	52	56	0	0	0	57	53
27	0	0	0	27	54	57	0	0	0	58	55
28	0	0	0	28	56	58	0	0	0	59	57
29	0	0	0	29	58	59	0	0	1	0	59
30	0	0	0	31	1	60	0	0	1	2	2

7. WHAT THE GREATEST DIFFERENCE IS BETWEEN THE REGULAR AND THE APPARENT PRECESSION OF THE EQUINOXES

[72^b] Now that the mean movements have been set out in this way, we must inquire what the greatest difference is between the regular and the apparent movement of the equinoxes, or what the diameter of the small circle is, through which the movement of anomaly turns.[1] For when this is known, it will be easy to discern various other differences in the movements. As was written above, between the observation of Timochares, which came first, and that of Ptolemy in the second year of the reign of Antoninus Pius, there were 432 years; and during that time the mean movement was 6° and the apparent 4°20'. So the difference between them is 1°40'. And the movement of double[2] anomaly was 90°35'. Moreover, it seems that at the middle of this period of time or around there the apparent movement reached its peak of greatest slowness. At that time the (position of the) apparent movement necessarily agreed with the mean movement, and the true equinox and the mean equinox occurred at the same section of the circles.[3] Wherefore if we make a distribution of the movement and the time into two equal parts, there will be in each part as differences between the irregular and the regular movement $^{10}/_{12}$°, which the circle of anomaly comprehends on either side beneath an arc of 45°17$^1/_2$'. But since all these differences are very small and do not amount to 1$^1/_2$° on the ecliptic, and the straight lines are almost equal to the arcs subtended by them, and there is scarcely any diversity found in the third-minutes: we who are staying within the minutes will make no error if we employ straight lines instead of arcs.

[73^a] Let *ABC* be a part of the ecliptic and on it let the mean equinox be *B*. And with *B* as pole let there be described the semicircle *ADC*, and let it cut the ecliptic at points *A* and *C*. Moreover let *DB* be drawn from the pole of the ecliptic, it will bisect the semicircle at *D*; and let *D* be understood to be limit of greatest slowness and beginning of the increase.[4] In the quadrant *AD* let

arc *DE* = 45°17$^1/_2$';

and through point *E* from the pole of the ecliptic, let fall *EF*; and let

$$BF = 50'.$$

Our problem is to find out from this what the whole *BFA* is.

---

[1] *i.e.*, what the diameter of the small circle is, along which the libration takes place back and forth.
[2] The anomaly of precession is called the "double" anomaly because it completes two cycles for one cycle of the anomaly of obliquity.
[3] As Copernicus showed in Chapter 4, the movement of the libration, considered above, appears fastest around the centre of the circle. Hence the apparent movement itself will appear slowest when the fastest movement of libration is in opposition to the mean movement with which it is compounded. And the fastest libration is in opposition to the mean movement when the apparent equinox is swinging eastward and is in the neighbourhood of the centre of the circle or the mean equinox.
[4] Thus, circle *ADC* is the circle of libration transferred from the pole of the ecliptic to around the equinox—as in the last diagram in Chapter 5.

Accordingly it is clear that

$$2\ BF = \text{ch. } 2\ DE.$$

But

$$FB : AFB = 7107 : 10,000 = 50' : 70'.$$

Hence

$$AB = 1°10',$$

and that is the greatest difference between the mean and the apparent movement of the equinoxes, which we were seeking; and the greatest polar deflexion of 28' follows upon it.

[72^b] For with this set-up let ABC be the arc of the ecliptic, BDE the mean equatorial arc, and B the mean section of the apparent equinoxes, either Aries or Libra, and through the poles of DBE let fall BF. Now along arc ABC on both sides let

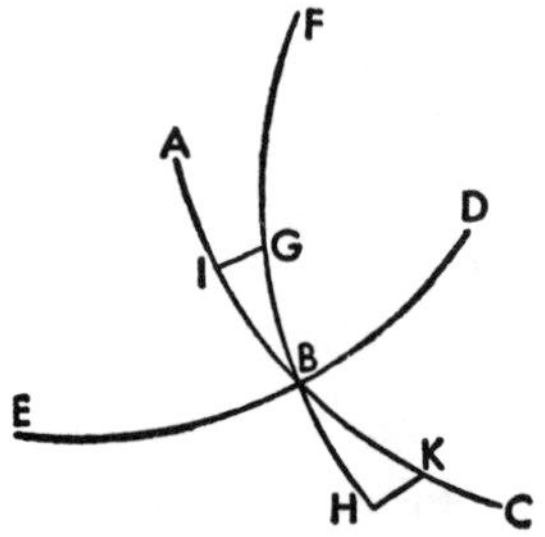

$$\text{arc } BI = \text{arc } BK = 1°10';$$

hence, by addition,

$$\text{arc } IBK = 2°20'.$$

Moreover, let there be drawn at right angles to FB extended to FBH the two arcs IG and HK of the apparent equators. Now I say "at right angles," [73^a] though the poles of IG and IK are usually outside of circle BF, since the movement of obliquity gets mixed in, as was seen in the hypothesis, but on account of the distance being very slight—for at its greatest it does not exceed 90°/350—we employ these angles as angles which are right to sense-perception. For no great error will appear on that account. Therefore in triangle IBG

$$\text{angle } IBG = 66°20',$$

since its complement, as being the angle of mean obliquity of the ecliptic,

$$\text{angle } DBA = 23°40'.$$

And

$$\text{angle } BGI = 90°.$$

Moreover,

$$\text{angle BIG} \doteqdot \text{angle IBD},$$

because they are alternate angles. And

$$\text{side } IB = 70'.$$

Therefore too

$$\text{arc } BG = 28',$$

and that is the distance between the poles of the mean and the apparent equator.

Similarly in triangle BHK,

$$\text{angle } BHK = \text{angle } IGB$$

and

$$\text{angle } HBK = \text{angle } IBG,$$

and

$$\text{side } BK = \text{side } BI.$$

Moreover,

$$BH = BG = 28'.$$

For

$$GB : IB = BH : BK;$$

and the movements will be of the same ratio in the poles as in the intersections.

8. On the Particular Differences in the Movements and the Table of Them

[73^b] Therefore since

$$\text{arc } AB = 70',$$

and since arc AB does not appear to differ from the chord subtending it lengthwise, it will not be difficult to exhibit certain other differences between the mean and the apparent movements. The Greeks call the differences προσθαφαιρέις, or "additosubtractions," and later writers "aequationes," by the subtraction or addition of which the apparent movements are made to harmonize (with the mean movements). We shall employ the Greek word as being more fitting. Therefore if

$$\text{arc } ED = 3°,$$

then in accordance with the ratio of AB to the chord BF,

$$\text{arc } BF = 4',$$

which is the additosubtraction. And if

$$ED = 6°,$$

then

$$\text{arc } BF = 7';$$

and if

$$ED = 9°,$$

then

$$BF = 11',$$

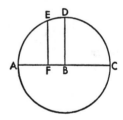

and so on.

We think we should use a similar ratio in the case of the change of obliquity also, where, as we said, a difference of 24' has been found between the greatest and the least obliquity. These 24' subtend a semicircle of simple anomaly every 1717 years, and the mean differences subtending a quadrant of a circle will be 12', where the pole of the small circle of this anomaly will be at an obliquity of 23°40'.

And in this way, as we said, we shall extract the remaining parts of difference approximately in proportion to the aforesaid as in the subjoined table. And if through these demonstrations the apparent movements can be compounded by various modes, nevertheless that mode is better whereby all the particular additosubtractions may be taken separately, that the calculus of their movements may be easier to understand and may agree better with the explanations of what has been demonstrated.

Accordingly we have drawn up a table of sixty rows, increasing by 3°'s. For in this way it will not be spread over too much space, and it will not seem to be compressed into too little—as we shall do in the case of the similar remaining tables. The table will have only four main columns, the first two of which will contain the degrees of both semicircles; and we call them the common numbers, because the obliquity of the circle of signs is taken from the simple number, and twice the number applies to the additosubtractions of the movement of the equinoxes; and the numbers have their commencement at the beginning of the increase [74a] (in velocity).[1] In the third column will be placed the additosubtractions of the equinoxes corresponding to the single 3°'s; and they are to be added to, or subtracted from, the mean movement—which we measure from the head of Aries at the spring equinox. The subtractive additosubtractions correspond to the numbers in the first semicircle of the anomaly or the first column; and the additive, to those in the second column and the second semicircle. Finally, in the last column are the minutes, which are called the differences in the proportions of obliquity and which go up to 60', since in place of the difference of 24' between greatest and least obliquity we are putting 60', and we adjust the proportional minutes to them in the same ratio in proportion to the other differences of obliquity. On that account we place 60' as corresponding to the beginning and end of the anomaly; but where the difference of obliquity is 22', as in the anomaly of 33°, we put 55' instead. In this way we put 50' in place of 20', as in the anomaly of 48°; and so on for the rest, as in the subjoined table.[2]

---

[1] *i.e.*, in the foregoing diagram, the first quadrant comprises the arc *DA*; and the fourth quadrant the arc *CD*.

[2] Thus, let line *FIG*, as in Chapter 3, represent the colors of the solstices. Point *F* is the limit of greatest obliquity of the ecliptic, point *G* that of the least; and thus the distance *FG* is 28'. The distance *KN* of the libration of the equinoxes is 2°20'. In the foregoing table the anomalies of precession and of obliquity we taken as starting at point *I* and proceeding along the route *INFKILGM*.

ADDITIONS-AND-SUBTRACTIONS OF EQUINOXES, OBLIQUITY OF THE ECLIPTIC

Common Numbers		Additions-and-Subtractions of Movement of Equinoxes		Proportional Minutes of Obliquity	Common Numbers		Additions-and-Subtractions of Movement of Equinoxes		Proportional Minutes of Obliquity
Deg.	Deg.	Deg.	Min.		Deg.	Deg.	Deg.	Min.	
3	357	0	4	60	93	267	1	10	28
6	354	0	7	60	96	264	1	10	27
9	351	0	11	60	99	261	1	9	25
12	348	0	14	59	102	258	1	9	24
15	345	0	18	59	105	255	1	8	22
18	342	0	21	59	108	252	1	7	21
21	339	0	25	58	111	249	1	5	19
24	336	0	28	57	114	246	1	4	18
27	333	0	32	56	117	243	1	2	16
30	330	0	35	56	120	240	1	1	15
33	327	0	38	55	123	237	0	59	14
36	324	0	41	54	126	234	0	56	12
39	321	0	44	53	129	231	0	54	11
42	318	0	47	52	132	228	0	52	10
45	315	0	49	51	135	225	0	49	9
48	312	0	52	50	138	222	0	47	8
51	309	0	54	49	141	219	0	44	7
54	306	0	56	48	144	216	0	41	6
57	303	0	9	46	147	213	0	38	5
60	300	1	1	45	150	210	0	35	4
63	297	1	2	44	153	207	0	32	3
66	294	1	4	42	156	204	0	28	3
69	291	1	5	41	159	201	0	25	2
72	288	1	7	39	162	198	0	21	1
75	285	1	8	38	165	195	0	18	1
78	282	1	9	36	168	192	0	14	1
81	279	1	9	35	171	189	0	11	0
84	276	1	10	33	174	186	0	7	0
87	273	1	10	32	177	183	0	4	0
90	270	1	10	30	180	180	0	0	0

9. On the Examination and Correction of That Which Was Set Forth Concerning the Precession of the Equinoxes

[75ª] But since by an inference we took the beginning of increase in the movement of anomaly as occurring in the middle of the time from the 36th year of the first period of Callippus to the 2nd year of Antoninus, and we take the order of the movement of anomaly from that beginning; it is still necessary for us to test whether we did that correctly and whether it agrees with the observations.

Let us consider again the three observations of the stars made by Timochares, Ptolemy, and al-Battani the Harranite: And it is clear that there were 432 Egyptian years in the first interval and 742 years in the second. The regular movement in the first span of time was 6°; the irregular movement 4°20'; and the movement of double anomaly 90°35', subtracting 1°40' from the regular movement. During the second interval the regular movement was 10°21', the irregular 11$^1/_2$°; and the movement of double anomaly was 155°34', adding 1°9' to the regular movement.

Now as before let the arc of the ecliptic be *ABC*, and let *B*—which is to be the mean spring equinox—be taken as a pole; let

$$\text{arc } AB = 1°10',$$

and let the small circle *ADCE* be described. But let the regular movement of *B* be understood as in the direction of *A*, *i.e.*, westward; and let *A* be the westward limit, where the irregular equinox is westernmost; and *C* the eastern limit, where the irregular equinox is easternmost. Furthermore, from the pole of the ecliptic drop *DBE* through point *B*. *DBE* together with the ecliptic will cut the small circle *ADCE* into four equal parts, since circles described through the poles of one another cut one another at right angles. However since the movement in the semicircle *ADC* is eastward, and the movement remaining in *CEA* is westward, the extreme slowness of the apparent equinox will be at *D* on account of its resistance to the forward movement of *B*; but there will be at *E* the greatest velocity for the movements moving forwards in the same direction.

Moreover on either side of *D* let

$$\text{arc } FD = \text{arc } DG = 45°17^1/_2'$$

Let *F* be the first terminus of the anomaly—the one observed by Timochares; *G* the second—the one observed by Ptolemy; and *P* the third—the one observed by al-Battani. And through these points let fall great circles *FN*, *GM*, and *OP* through the poles of the ecliptic; and they all [75ᵇ] appear in this very small circle rather much like straight lines. Therefore

$$\text{arc } FDG = 99°35',$$

where circle $ADCE = 360°$,

wherefrom

−add. $MN = 1°40'$,

where $ABC = 2°20'$.

And

arc $GCEP = 155°34'$,

wherefrom

+ add. $MBO = 109'$.

Accordingly, by subtraction,

arc $PAF = 113°51'$

wherefrom

+ add. $ON = 31'$

where $AB = 70'$.

But since by addition

arc $DGCEP = 200°51'$

and

$EP = DGCEP − 180° = 200°51'$;

therefore by the table of chords in a circle, as if a straight line,

$BO = 356$,

where $AB = 1,000$.

But

$BO \doteq 24'$,

where $AB = 70'$;

and

$MB = 50'$.

Hence, by addition,

$MBO = 74'$,

and,

$NO = MN − MBO = 26'$.

But, in the foregoing

$MBO = 69'$,

and

$NO = 31'$.

Hence NO has a deficiency of 5'; and MO has an excess of 5'. Accordingly the circle $ADCE$ must be revolved, until there is compensation on both sides.

But this will take place if

arc $DG = 42^1/_2°$,

so that by subtraction,

arc $DF = 48°5'$.

For by this both errors will seem to be corrected and everything else will be all right, since—with the beginning at D the limit of greatest slowness—

arc $DGCEPAF = 311°55'$,

which is the movement of anomaly at the first terminus; at the second terminus

arc $DG = 42^1/_2°$;

and at the third terminus

arc $DGCEP = 198°4'$.

Now since

$AB = 70'$;

at the first terminus

+ add. $BN = 52'$,

by what has been shown; at the second terminus

−add. $MB = 47^1/_2'$;

and at the third terminus again

+ add. $BO \doteqdot 21'$.

Therefore during the first interval

arc $MN = 1°40'$,

and during the second interval

arc $MBO = 1°9'$;

and they agree exactly with the observations. Moreover, by those means a simple anomaly of $155°57^1/_2°$ is made evident at the first terminus; at the second terminus, one of $21°15'$; and at the third terminus, a simple anomaly of $99°2'$—as was to be shown.

10. WHAT THE GREATEST DIFFERENCE IS BETWEEN THE INTERSECTIONS OF THE EQUATOR AND THE ECLIPTIC

[76ª] In the same way we shall confirm what we expounded concerning the change in obliquity of the ecliptic and the equator and shall find it to be correct. For we have in Ptolemy for the second year of Antoninus Pius a corrected simple anomaly of $21^1/_4°$; and a greatest obliquity of $23°51'20"$ was found to go with it. From this position down to the observation made by us there have been 1387 years, during which the movement of simple anomaly is reckoned to be $144°04'$; and at this time an obliquity of approximately $23°28^2/_5'$ is found.

In connection with this let there be drawn again arc ABC of the ecliptic, or instead of it a straight line on account of the shortness of the arc; and above it the semicircle of simple anomaly around pole B, as before. And let A be the limit of greatest declination and C the limit of least declination; and it is the difference between them which we are examining. Therefore in the small circle let arc $AE = 21°15'$,

and,

arc $ED = AD - AE = 68°45'$;

while, by calculation,

arc $EDF = 144°4'$

and,

arc $DF = EDF - ED = 75°19'$.

Drop perpendiculars EG and EK upon the diameter ABC.

Now on the great circle

arc $GK = 22'56''$,

on account of the difference in obliquities from Ptolemy to us. But on account of being like a straight line,

$$GB = \tfrac{1}{2} \text{ ch. } 2 \; ED = 932,$$

where AC the diameter's image = 2,000.

And also

$$KB = \tfrac{1}{2} \text{ ch. } 2 \; DF = 967.$$

And

$$GK = 1899,$$

where $AC = 2,000$.

But according as

$$GK = 22'56'',$$

$$AC \doteqdot 24',$$

the difference between greatest and least obliquity which we have been examining. So it is established that the greatest obliquity which occurred during the time between Timochares and Ptolemy was $23°52'$ and a least obliquity of $23°28'$ is now approaching. [76b] Hence also whatever mean inclinations of these circles there happen to be are discovered by the same mathematical reasoning we expounded in connection with the precession.

11. On Determining the Positions of the Regular Movements of the Equinoxes and of the Anomaly

With all that unfolded, it remains for us to determine the positions of the movements of the spring equinox. Some people call these positions "roots," because computations may be drawn from them for any given time. Ptolemy considered that the farthest point in history to which our knowledge of this question extends was the beginning of the reign of Nabonassar of the Chaldees, whom many people—taken in by the similarity of the names—have thought to be Nabuchodonoso, and whom the ratio of time and the computation of Ptolemy—which according to the historians falls in the reign of Shalmaneser of the Chaldees—declare to have been much later. But we, seeking better known times, have judged it sufficient if we start with the first Olympiad, which—measured from the summer solstice—is found to have preceded Nabonassar by 28 years. At this time Canicula was beginning to rise for the Greeks, and the Olympic games were being held, as Censorinus and other trustworthy authors report. Whence, according to the more exact reckoning of the times which is necessary in calculating the heavenly movements, there are 27 years and 247 days from the first Olympiad at noon on the first day of the month Hekatombaion by the Greek calendar to Nabonassar and noon of the first day of the month of Thoth by the Egyptian calendar.

From this to the death of Alexander there are 424 Egyptian years.

But from the decease of Alexander to the beginning of the years of Julius Caesar, there are 278 Egyptian years $118^1/_2$ days up to the midnight before the Kalends of January, which Julius Caesar took as the beginning of the year instituted by him; it was in his third year as Pontifex Maximus and during the consulship of Marcus Aemilius Lepidus that he instituted this year. And so the later years have been called Julian from the year as established by Julius Caesar.

And from the fourth consulship of Caesar to Octavius Augustus there are by the Roman calendar 18 years up to the Kalends of January, although it was on the 16th day before the Kalends of February that Augustus was proclaimed Emperor and son of the deified Julius Caesar by the Senate and the other citizens according to the decree of Numatius Plancus, in the seventh year of the consulship of Marcus Vipsanus and himself. But inasmuch as two years before this the Egyptians came into the power of the Romans after the fall of Antony [77a] and Cleopatra, the Egyptians reckon 15 years $246^1/_2$ days up to noon of the first day of the month Thoth, which by the Roman calendar was the 3rd day before the Kalends of September.

Accordingly from Augustus to the years of Christ, which begin similarly in January, there are 27 years by the Roman calendar but 29 years $130^1/_2$ days by the Egyptian.

From this to the 2nd year of Antoninus, when, as Claud Ptolemy says, the positions of the stars were observed by him, there are 138 Roman years 55 days. And these years add 34 days to the Egyptian reckoning.

Between the first Olympiad and that moment of time there have been altogether 913 years 101 days, for which time the regular precession of the equinoxes was 12°44', and the simple anomaly was 95°44'.

But in the second year of Antoninus, as has been narrated, the spring equinox was 6°40' to the west of the first of the stars which are in the head of Aries; and since there was a double anomaly of $42^1/_2$°, there was a subtractive difference of 48' between the regular and the apparent movement. And when this difference was restored to the 6°40' of the apparent movement, it made the mean position of the spring equinox to be at 7°28'. If to this we add the 360° of a circle, and from the sum subtract 12°44', we shall have the mean position of the spring equinox at 354°44'—that is to say, the one which was then 5°16' east of the first star of Aries—for the first Olympiad which began on noon of the first day of the month Hekatombaion among the Athenians.

In the same way if from the 21°15' of simple anomaly 95°45' are subtracted, there will remain a position of simple anomaly of 285°30' for the same beginning of the Olympiads.

And again by a series of additions of movement made in accordance with the lengths of time—the 360° are not counted where there is an excess above that—we shall have the position or root of the regular movement at the death of Alexander as 1°2', and the position of the movement of simple anomaly as 332°52'; at the beginning of the years of Caesar a mean movement of 4°55' and an anomaly of 2°2'; and at the beginning of the years of Christ a position of the mean movement at 5°32' and an anomaly of 6°45'; and in this way we shall determine the roots of movements for whatever beginnings of time are chosen.

12. ON THE COMPUTATION OF THE PRECESSION OF THE SPRING EQUINOX AND THE OBLIQUITY

[77^b] Therefore, whenever we wish to determine the position of the spring equinox, if the years from the assumed beginning to the given time are unequal, such as those of the Roman calendar, which we use commonly, we shall reduce them to equal or Egyptian years. For we do not use any other years than the Egyptian in calculating the regular movements, on account of the reason which we mentioned. In so far as the number of years is greater than a period of 60 years, we shall divide it into periods of 60 years; and when we enter the tables of movement through these 60-year periods, we shall pass over as supernumerary the first column appearing in the movements; and beginning with the second column, we shall determine the 60°'s, if there are any, together with the other degrees and minutes, which follow.[1] Next as the second entry and from the first column, as they are found, we shall take the 60°'s, degrees and minutes corresponding to the remaining years. We shall do the same thing in the case of days and the periods of 60 days, since we wish to connect the days with their regular movements according to the table of days and minutes, although in this case the minutes of days or even the days themselves are not wrongly neglected on account of the slowness of their movements, as within the daily movement there is a question only of seconds or third minutes. Therefore when we have made a sum of all these together with their root, by adding single numbers to single numbers within the same species—not counting six 60°'s, if they occur—we shall have the mean position of the spring equinox, its distance to the west of the first star of the Ram, or the distance of that star east of the equinox.

[1]That is to say, reading the column of degrees as 60°'s, the minutes as degrees, and so on.

In the same way we shall determine the anomaly too.

But we shall find placed in the last column of the table of additosubtractions and corresponding to the simple anomaly the proportional minutes: we shall set them aside and save them. Then, in the third column of the same table and corresponding to the double anomaly we shall find the additosubtraction, *i.e.*, the degrees and minutes by which the true movement differs from the mean. And if the double anomaly is less than a semicircle, we shall subtract the additosubtraction from the mean movement. But if, by having more than 180°, the double anomaly exceeds a semicircle, we shall add [78ª] the additosubtraction to the mean movement. And that which is thus the sum or remainder will comprehend the true and apparent precession of the spring equinox, or in turn the then angular elongation of the first star of Aries from the spring equinox. But if you seek the position of any other star, add its number as assigned in the catalogue of the stars.

But since things which have to do with the laboratory usually become clearer by means of some examples, let our problem be to find the true position of the spring equinox together with the obliquity of the ecliptic for the 16th day before the Kalends of May in the year of Our Lord 1525, and how great the angular distance of Spica in Virgo from the same equinox is. Therefore, it is clear that in the 1524 Roman years 106 days from the beginning of the years of Our Lord up to this time, there has been an intercalation of 381 days, *i.e.*, 1 year 16 days, which in terms of equal years make 1525 years 122 days: there are twenty-five periods of 60 years and 25 years over, and two periods of 60 days and 2 days over. But to the twenty-five periods of 60 years there correspond in the table of mean movement 20°55'2"; to the 25 years, 20'55"; to the two periods of 60 days, 16"; the remaining 2 days are in third minutes. All these together with their root—which was 5°32'—add up to 26°48', the mean precession of the spring equinox. Similarly, the movement of simple anomaly in the twenty-five periods of 60 years has been two 60°s and 37°15'3"; in the 25 years, 2°37'15"; in the two periods of 60 days, 2'4"; and in the 2 days, 2". There also, together with the root—which is 6°45'—add up to 166°40', the simple anomaly. I shall save as corresponding to this anomaly the proportional minutes found in the last column of the table of the additosubtractions; for they will come into use in investigating the obliquity; and only 1' is found in this case. Next, as corresponding to the double anomaly of 333°20', I find 32' as the additosubtraction, which is additive because the double anomaly is greater than a semicircle. And when it is added to the mean movement, there comes about a true and apparent precession of the spring equinox of 27°21'. And lastly if to that we add the 170° which is the angular distance of Spica in Virgo from the first star in Aries, I shall have its position [78ᵇ] to the east of the spring equinox at 17°21' of Libra, it was found at approximately the time of our observation.

Now the obliquity of the ecliptic and its declination have the ratio that when there are 60 proportional minutes, the differences located in the table of declinations—I mean the differences at greatest and least obliquity—are added in their entirety to the degrees of the declinations. But in this case, 1' adds only 24" to the obliquity. Wherefore the declinations of the degrees of the ecliptic placed in the table remain as they are throughout this time on account of the least obliquity already approaching as, though at some other time they would be more obviously changeable. In this way, for example, if the simple anomaly were 99°, as it was in the 1380th Egyptian year of Our Lord, there are given by it 25 proportional minutes. But 24' is the difference between greatest and least obliquity and

$$60' : 24' = 25' : 10'.$$

And the addition of 10' to 28' gives an obliquity of 23°38' for that time. If then I should wish to know the declination of any degree on the ecliptic, for example, 3° of Taurus, which is 33° distant from the equinox, I find in the table 12°32', with a difference of 12'. But

$$60' : 25' = 12' : 5';$$

and the addition of 5' to 32' gives 12°37' for 33° of the ecliptic. We can do the same thing in the case of the angles of section of the ecliptic and the equator and the right ascensions—if it is not better to make use of the ratios of spherical triangles—except that it is always necessary to add in the case of the angles of section and to subtract in the case of the right ascensions, so that all things may be corrected to accord with their time.

13. ON THE MAGNITUDE AND DIFFERENCE OF THE SOLAR YEAR

But that this is the way it is with the precession of the equinoxes and solstices—the precession being due to the inclination of the Earth's axis, as we said—will also be confirmed by the annual movement of the centre of the Earth, as it affects the appearance of the sun, which we must now discuss. It follows of absolute necessity that the magnitude of the year, when referred to one of the equinoxes or solstices, is found unequal on account of the irregular change of the termini. For these things imply one another mutually.

Wherefore we must separate and distinguish [79ª] the seasonal year from the sidereal year. For we call that the natural year which times the four seasonal changes of the year for us; and that the sidereal, the revolutions of which are referred to some one of the fixed stars. Now the observations of the ancients make clear in many ways that the natural year, which is also called the revolving year, is unequal. For Callippus, Aristarchus of Samos, and Archimedes of Syracuse determined the year as containing a quarter of a day in addition to the 365 whole days—taking the beginning of the year at the summer solstice, after the Athenian manner.

But Claud Ptolemy, realizing that the apprehension of the solstices was detailed and difficult, did not rely upon their observations very much and went over rather to Hipparchus, who left after him records not so much of the solar solstices as of the equinoxes in Rhodes and reported that there was some small deficiency in the quarter-day; and afterwards Ptolemy decided that the deficiency was $1/300$th part of a day—as follows. For he took the autumn equinox observed as accurately as possible by Hipparchus at Alexandria in the 177th year after the death of Alexander the Great, at midnight of the third intercalary day by the Egyptian calendar—which the fourth intercalary day follows. Then Ptolemy compared it with the equinox as observed by himself at Alexandria in the third year of Antoninus, which was the 463rd year since the death of Alexander, on the 9th day of Athyr, the third month of the Egyptians, at approximately one hour after the rising of the sun. Accordingly between this observation and that of Hipparchus there were 285 Egyptian years 70 days $7^1/_5$ hours, though there should have been 71 days 6 hours, if the revolving year had a full quarter-day in addition to the whole days. Accordingly the 285 years were deficient by $19/_{20}$th of a day, whence it follows that a whole day fell out in 300 years. Moreover, he made a similar inference from the spring equinox. For what he recorded as reported by Hipparchus in the 178th year of Alexander on the 27th day of Mechir, the 6th month by the Egyptian calendar, at sunrise, he himself found in the 463rd year of Alexander on the 7th day of Pachon the 9th month by the Egyptian calendar at a little more than one hour after midday; and in the same way the 285 years were deficient by $19/_{20}$th of a day. By the aid of these indications Ptolemy determined the revolving year as having 365 days 14 min. (of a day) 48 sec. (or 5 hours 55 min. 12 sec.).[1]

Afterwards al-Battani in Arata, Syria, [79b] in the 1206th year after the death of Alexander observed the autumn equinox with no less diligence and found that it occurred after the 7th day of the month Pachon, approximately $7^2/_5$ hours later in the night, i.e., $4^3/_5$ hours before the light of the 8th day. Accordingly, comparing his own observation with that of Ptolemy made in the third year of Antoninus one hour after sunrise at Alexandria—which is 10° to the west of Arata—he corrected Ptolemy's observation for the meridian at Arata and found the equinox must have occurred at $1^2/_3$ hours after sunrise. Accordingly in the period of 743 equal years the sum of the quarter-days amounted to 178 extra days and $17^3/_5$ hours instead of $185^1/_4$ days. Accordingly since there was deficiency of 7 days $^2/_5$ hours, it was seen that the quarter-day was deficient by $1/_{106}$th of a day. Therefore in accordance with the number of years he subtracted one 743rd part of the 7 days $^2/_5$ hours (which is 13 min. of an hour 36 sec.) from the quarter-day and recorded the natural year as containing 365 days 5 hours 46 min. 24 sec.

[1] i.e., Ptolemy found $1/_{300}$th part of a day lacking to a full quarter-day.

We too made observations of the autumn equinox at Frauenburg in the year of Our Lord 1515 on the 18th day before the Kalends of October: but according to the Egyptian calendar it was the 1840th year after the death of Alexander on the 6th day of the month Phaophi, half an hour after sunrise. But since Arata is about 25° to the east of this spot—which makes $1^2/_3$ hours—therefore during the time between our equinox and that of al-Battani there were 633 Egyptian years and 153 days $6^3/_4$ hours in place of 158 days 6 hours. But between the observation made by Ptolemy at Alexandria and the place and date of our observation, there were 1376 Egyptian years 332 days $^1/_2$ hour. For there is about an hour's difference between us and Alexandria. Therefore during the 633 years between al-Battani and us there have fallen out 4 days $23^3/_4$ hours, or 1 day per 128 years; but during the 1376 years after Ptolemy approximately 12 days, i.e., 1 day per 115 years, and again the year has become unequal on both sides.

[80ª] Moreover, we determined the spring equinox, which occurred in the year of Our Lord 1516, $4^1/_3$ hours after midnight on the 5th day before the Ides of March; and since the spring equinox of Ptolemy—the meridian of Alexandria being corrected for ours—there have been 1376 Egyptian years 332 days $16^1/_3$ hours, in which it is apparent that the distances between the spring and autumn equinoxes are unequal. And so it is of much importance that the solar year as determined in this way should be equal. For the fact that at the autumnal equinoxes between Ptolemy and us, as was shown, in accordance with the equal distribution of years, the quarter-day should be deficient in the 115th part of a day makes the equinox come half a day later than al-Battani's. And the period from al-Battani to us, where the quarter-day must have been deficient in the 128th part of a day, is not consonant with Ptolemy, but the date precedes by a full day the equinox observed by him, and the equinox of Hipparchus by two days. Similarly the time of al-Battani's equinox as measured from Ptolemy's precedes the equinox of Hipparchus by 2 days.

Therefore the equality of the solar year is more correctly measured from the sphere of the fixed stars, as Thebites ben Chora was the first to find; and its magnitude is 365 days 15 minutes (of a day) 23 seconds (which are approximately 6 hours 9 min. 12 sec.) according to a probable argument taken from the fact that the year appears longer in the slower passage of the equinoxes and solstices than in the faster and in accordance with a fixed proportion; and that could not be the case, if there were no equality with reference to the sphere of the fixed stars. Wherefore Ptolemy is not to be listened to in that part where he thinks that it is absurd and irrelevant to measure the annual regularity of the sun through its restitutions with reference to some one of the fixed stars and that this is no more fitting than if someone were to take Jupiter or Saturn as the measure of that regularity. And so there is a ready reason why the seasonal year was longer before Ptolemy and after him became shorter, by a variable difference.

But also in the case of the astral or sidereal year an error can come about, but nevertheless a very slight one and far less than the one which we have already described; and it occurs because this same movement of the centre of the Earth around the sun appears irregular by reason of a twofold irregularity. [80^b] The first and simple irregularity relates to the annual restoration; the other, which varies the first by changing it around, is perceptible not immediately but after a long stretch of time; and accordingly it is not simple or easy to know the ratio of the equality of the year. For if anyone wishes to determine it simply in relation to the fixed distance of some star having a known position—which can be done by using an astrolabe and with the help of the moon, in the way we described in the case of Basiliscus in Leo—he will not avoid error completely, unless at that time the sun on account of the movement of the Earth either has no additosubtraction or else obtains similar and equal additosubtractions at both termini. But unless this happens and unless there is some difference made manifest in accordance with the irregularity, an equal circuit will certainly not seem to have taken place in equal times. But if in both termini the total difference is subtracted or added proportionally, the job will be perfect. Furthermore, the apprehension of the difference requires a prior knowledge of the mean movement, which we are seeking for that reason; and we are versed in this business as in the Archimedean quadrature of the circle.

Nevertheless in order to arrive at the resolution of this knotty problem some time—we find four causes altogether for the appearance of irregularity. The *first* is the irregular precession of the equinoxes, which we have expounded; the *second* is that whereby the sun seems to traverse unequal arcs on the ecliptic, which occurs approximately annually; the *third* is the one which varies this irregularity which we call the second. There remains the *fourth*, which changes the highest and lowest apsides[1] of the centre of the Earth, as will appear below. Of all these only the second was marked by Ptolemy; and it by itself could not produce the inequality of the year but contributes to it through being involved in the others.

But for demonstrating the difference between the regular and the apparent movement of the sun the most accurate ratio of the year does not seem necessary; and it seems to be enough if in the demonstration we take as the magnitude of the year the 365$^1/_4$ days, in which the movement of the first irregularity is completed, since that which stands out so little, when taken on the total circle, vanishes utterly when taken on a lesser magnitude. But on account of the excellence of the order and the facility in teaching we are here expounding first the regular movements of the annual revolution of the centre of the Earth by means of necessary demonstrations. And then we shall build up the regular movements together with the difference between the regular and the apparent movement.

---

[1]The apsides are the positions of greatest and least altitudinal distance of a planet from the sun.

14. On the Regular and Mean Movements of the Revolutions of the Centre of the Earth

[81ª] We find that the magnitude of the year and its equality is only 1 second 10 thirds greater than Thebith ben Chora recorded it to be, so that it contains 365 days 15 minutes 24 seconds 10 third-minutes—which amounts to 6 hours 9 minutes 40 seconds, and its fixed equality with reference to the sphere of the fixed stars is disclosed.

Therefore, when we have multiplied the 360° of a circle by 365 days and have divided the sum by 365 days 15 minutes 24 seconds 10 third-minutes, we shall have the movement of an Egyptian year as 359°44'49"7"'4"" and the movement during 60 similar years—not counting the total circles—will be 344°49'7"4"'. Again, if we divide the annual movement by 365 days, we shall have a daily movement of 59'8"11"'22"".

But if we add to these the mean and regular precession of the equinoxes, we shall compose another regular annual movement in seasonal years of 359°45'39"19"'9"" and a daily movement 59'8"19"'37"". And for this reason we can call the former movement of the sun—to use the common expression—the regular and simple movement; and the latter, the regular and composite movement. And we shall set them out in tables, as we did with the precession of the equinoxes. The regular movement of the anomaly of the sun is added to them; but we shall speak of that later.

TABLE OF THE MEAN AND SIMPLE MOVEMENT OF THE SUN IN YEARS AND PERIODS OF SIXTY YEARS

Egyptian Years	Movement 60°	°	'	"	'''		Egyptian Years	Movement 60°	°	'	"	'''
1	5	59	44	49	7		31	5	52	9	22	39
2	5	59	29	38	14		32	5	51	54	11	46
3	5	59	14	27	21		33	5	51	39	0	53
4	5	58	59	16	28		34	5	51	23	50	0
5	5	58	44	5	35		35	5	51	8	39	7
6	5	58	28	54	42		36	5	50	53	28	14
7	5	58	13	43	49		37	5	50	38	17	21
8	5	57	58	32	56		38	5	50	23	6	28
9	5	57	43	22	3		39	5	50	7	55	35
10	5	57	28	11	10		40	5	49	52	44	42
11	5	57	13	0	17		41	5	49	37	33	49
12	5	56	57	49	24		42	5	49	22	22	56
13	5	56	42	38	31		43	5	49	7	12	3
14	5	56	27	27	38		44	5	48	52	1	10
15	5	56	12	16	46		45	5	48	36	50	18
16	5	55	57	5	53		46	5	48	21	39	25
17	5	55	41	55	0		47	5	48	6	28	32
18	5	55	26	44	7		48	5	47	51	17	39
19	5	55	11	33	14		49	5	47	36	6	46
20	5	54	56	22	21		50	5	47	20	55	53
21	5	54	41	11	28		51	5	47	5	45	0
22	5	54	26	0	35		52	5	46	50	34	7
23	5	54	10	49	42		53	5	46	35	23	14
24	5	53	55	38	49		54	5	46	20	12	21
25	5	53	40	27	56		55	5	46	5	1	28
26	5	53	25	17	3		56	5	45	49	50	35
27	5	53	10	6	10		57	5	45	34	39	42
28	5	52	54	55	17		58	5	45	19	28	49
29	5	52	39	44	24		59	5	45	4	17	56
30	5	52	24	33	32		60	5	44	49	7	4

Position of the Birth of Christ—272 °31'

TABLE OF THE REGULAR AND SIMPLE MOVEMENT OF THE SUN IN DAYS AND PERIODS OF SIXTY DAYS

Days	Movement 60°	°	'	"	'''		Days	Movement 60°	°	'	"	'''
1	0	0	59	8	11		31	0	30	33	13	52
2	0	1	58	16	22		32	0	31	32	22	3
3	0	2	57	24	34		33	0	32	31	30	15
4	0	3	56	32	45		34	0	33	30	38	26
5	0	4	55	40	56		35	0	34	29	46	37
6	0	5	54	49	8		36	0	35	28	54	49
7	0	6	53	57	19		37	0	36	28	3	0
8	0	7	53	5	30		38	0	37	27	11	11
9	0	8	52	13	42		39	0	38	26	19	23
10	0	9	51	21	53		40	0	39	25	27	34
11	0	10	50	30	5		41	0	40	24	35	45
12	0	11	49	38	16		42	0	41	23	43	57
13	0	12	48	46	27		43	0	42	22	52	8
14	0	13	47	54	39		44	0	43	22	0	20
15	0	14	47	2	50		45	0	44	21	8	31
16	0	15	46	11	1		46	0	45	20	16	42
17	0	16	45	19	13		47	0	46	19	24	54
18	0	17	44	27	24		48	0	47	18	33	5
19	0	18	43	35	35		49	0	48	17	41	16
20	0	19	42	43	47		50	0	49	16	49	28
21	0	20	41	51	58		51	0	50	15	57	39
22	0	21	41	0	9		52	0	51	15	5	50
23	0	22	40	8	21		53	0	52	14	14	2
24	0	23	39	16	32		54	0	53	13	22	13
25	0	24	38	24	44		55	0	54	12	30	25
26	0	25	37	32	55		56	0	55	11	38	36
27	0	26	36	41	6		57	0	56	10	46	47
28	0	27	35	49	18		58	0	57	9	54	59
29	0	28	34	57	29		59	0	58	9	3	10
30	0	29	34	5	41		60	0	59	8	11	22

Position of the Birth of Christ—272 °31'

TABLE OF THE REGULAR COMPOSITE MOVEMENT OF THE SUN IN YEARS AND PERIODS OF SIXTY YEARS

Center label (vertical): *Position of the Birth of Christ—278 2'*

Egyptian Years	60°	°	'	"	'''	Egyptian Years	60°	°	'	"	'''
1	5	59	45	39	19	31	5	52	35	18	53
2	5	59	31	18	38	32	5	52	21	58	12
3	5	59	16	57	57	33	5	52	6	37	31
4	5	59	2	37	16	34	5	51	52	16	51
5	5	58	48	16	35	35	5	51	38	56	10
6	5	58	33	55	54	36	5	51	23	35	29
7	5	58	19	35	14	37	5	51	9	14	48
8	5	58	5	14	33	38	5	50	55	54	7
9	5	57	50	53	52	39	5	50	40	33	26
10	5	57	36	33	11	40	5	50	26	12	46
11	5	57	22	12	30	41	5	50	11	52	5
12	5	57	7	51	49	42	5	49	57	31	24
13	5	56	53	31	8	43	5	49	43	10	43
14	5	56	39	10	28	44	5	49	28	50	2
15	5	56	24	49	47	45	5	49	14	29	21
16	5	56	10	29	6	46	5	49	0	8	40
17	5	55	56	8	25	47	5	48	45	48	0
18	5	55	41	47	44	48	5	48	31	27	19
19	5	55	27	27	3	49	5	48	17	6	38
20	5	55	13	6	23	50	5	48	2	45	57
21	5	54	58	45	42	51	5	47	48	25	16
22	5	54	44	25	1	52	5	47	34	4	35
23	5	54	30	4	20	53	5	47	19	43	54
24	5	54	15	43	39	54	5	47	5	23	14
25	5	54	1	22	58	55	5	46	51	2	33
26	5	53	47	2	17	56	5	46	36	41	52
27	5	53	32	41	37	57	5	46	22	21	11
28	5	53	18	20	56	58	5	46	8	0	30
29	5	53	4	0	15	59	5	45	53	39	49
30	5	52	48	39	34	60	5	45	39	19	9

TABLE OF THE REGULAR COMPOSITE MOVEMENT OF THE SUN IN DAYS AND PERIODS OF SIXTY DAYS

Center label (vertical): *Position of the Birth of Christ—278 2'*

Days	60°	°	'	"	'''	Days	60°	°	'	"	'''
1	0	0	59	8	19	31	0	30	33	18	8
2	0	1	58	16	39	32	0	31	32	26	27
3	0	2	57	24	58	33	0	32	31	34	47
4	0	3	56	33	18	34	0	33	30	43	6
5	0	4	55	41	38	35	0	34	29	51	26
6	0	5	54	49	57	36	0	35	29	59	46
7	0	6	53	58	17	37	0	36	28	8	5
8	0	7	53	6	36	38	0	37	27	16	25
9	0	8	52	14	56	39	0	38	26	24	45
10	0	9	51	23	16	40	0	39	25	33	4
11	0	10	50	31	35	41	0	40	24	41	24
12	0	11	49	39	55	42	0	41	23	49	43
13	0	12	48	48	15	43	0	42	23	58	3
14	0	13	47	56	34	44	0	43	22	6	23
15	0	14	47	4	54	45	0	44	21	14	42
16	0	15	46	13	13	46	0	45	20	23	2
17	0	16	45	21	33	47	0	46	19	31	21
18	0	17	44	29	53	48	0	47	18	39	41
19	0	18	43	38	12	49	0	48	17	48	1
20	0	19	42	46	32	50	0	49	17	56	20
21	0	20	41	54	51	51	0	50	16	4	40
22	0	21	41	3	11	52	0	51	15	13	0
23	0	22	40	11	31	53	0	52	14	21	19
24	0	23	39	19	50	54	0	53	13	29	39
25	0	24	38	28	10	55	0	54	12	37	58
26	0	25	37	36	30	56	0	55	11	46	18
27	0	26	36	44	49	57	0	56	10	54	38
28	0	27	35	53	9	58	0	57	10	2	57
29	0	28	35	1	28	59	0	58	9	11	17
30	0	29	34	9	48	60	0	59	8	19	37

TABLE OF THE REGULAR MOVEMENT OF ANOMALY[1] OF THE SUN IN YEARS AND PERIODS OF SIXTY YEARS

Egyptian Years	Movement					Position	Egyptian Years	Movement				
	60°	°	′	″	‴			60°	°	′	″	‴
1	5	59	44	24	46		31	5	51	56	48	11
2	5	59	28	49	33		32	5	51	41	12	58
3	5	59	13	14	20		33	5	51	25	37	45
4	5	58	57	39	7		34	5	51	10	2	32
5	5	58	42	3	54		35	5	50	54	27	19
6	5	58	26	28	41		36	5	50	38	52	6
7	5	58	10	53	27		37	5	50	23	16	52
8	5	57	55	18	14		38	5	50	7	41	39
9	5	57	39	43	1		39	5	49	52	6	26
10	5	57	24	7	48		40	5	49	36	31	13
11	5	57	8	32	35		41	5	49	20	56	0
12	5	56	52	57	22		42	5	49	5	20	47
13	5	56	37	22	8		43	5	48	49	45	33
14	5	56	21	46	55		44	5	48	34	10	20
15	5	56	6	11	42		45	5	48	18	35	7
16	5	55	50	36	29		46	5	48	2	59	54
17	5	55	35	1	16		47	5	47	47	24	41
18	5	55	19	26	3		48	5	47	31	49	28
19	5	55	3	50	49		49	5	47	16	14	14
20	5	54	48	15	36		50	5	47	0	39	1
21	5	54	32	40	23		51	5	46	45	3	48
22	5	54	17	5	10		52	5	46	29	28	35
23	5	54	1	29	57		53	5	46	13	53	22
24	5	53	45	54	44		54	5	45	58	18	9
25	5	53	30	19	30		55	5	45	42	42	55
26	5	53	14	44	17		56	5	45	26	7	42
27	5	52	59	9	4		57	5	45	11	32	29
28	5	52	43	33	51		58	5	44	55	57	16
29	5	52	27	58	38		59	5	44	40	22	3
30	5	52	12	23	25		60	5	44	24	46	50

Position of the Birth of Christ—211 °19′

MOVEMENT OF ANOMALY OF THE SUN IN DAYS AND PERIODS OF SIXTY DAYS

Days	Movement					Position	Days	Movement				
	60°	°	′	″	‴			60°	°	′	″	‴
1	0	0	59	8	7		31	0	30	33	11	48
2	0	1	58	16	14		32	0	31	32	19	55
3	0	2	57	24	22		33	0	32	31	28	3
4	0	3	56	32	29		34	0	33	30	36	10
5	0	4	55	40	36		35	0	34	29	44	17
6	0	5	54	48	44		36	0	35	28	52	25
7	0	6	53	56	51		37	0	36	28	0	32
8	0	7	53	4	58		38	0	37	27	8	39
9	0	8	52	13	6		39	0	38	26	16	47
10	0	9	51	21	13		40	0	39	25	24	54
11	0	10	50	29	21		41	0	40	24	33	2
12	0	11	49	37	28		42	0	41	23	41	8
13	0	12	48	45	35		43	0	42	22	49	16
14	0	13	47	53	43		44	0	43	21	57	24
15	0	14	47	1	50		45	0	44	21	5	31
16	0	15	46	9	57		46	0	45	20	13	38
17	0	16	45	18	5		47	0	46	19	21	46
18	0	17	44	26	12		48	0	47	18	29	53
19	0	18	43	34	19		49	0	48	17	38	0
20	0	19	42	42	27		50	0	49	16	46	8
21	0	20	41	50	34		51	0	50	15	54	15
22	0	21	40	58	42		52	0	51	15	2	23
23	0	22	40	6	49		53	0	52	14	10	30
24	0	23	39	14	56		54	0	53	13	18	37
25	0	24	38	23	4		55	0	54	12	26	45
26	0	25	37	31	11		56	0	55	11	34	52
27	0	26	36	39	18		57	0	56	10	42	59
28	0	27	35	47	26		58	0	57	9	51	7
29	0	28	34	55	33		59	0	58	8	59	14
30	0	29	34	3	41		60	0	59	8	7	22

Position of the Birth of Christ—211 °19′

[1] Any regular movement which, when compounded with a mean movement, causes an appearance of irregularity is called a movement of anomaly. In this case, the regular movement of anomaly is the movement of the eccentric circle, or the first epicycle.

15. THEOREMS PREREQUISITE FOR DEMONSTRATING THE APPARENT IRREGULARITY OF THE MOVEMENT OF THE SUN

[84ᵇ] But for the sake of making a better determination of the apparent irregular movement of the sun we shall now demonstrate more clearly that—with the sun occupying the central position in the world and with the Earth revolving around it as around a centre—if, as we said, there is a distance between the Earth and the sun which cannot be perceived in relation to the immensity of the sphere of the fixed stars; then the sun will be seen to have a regular motion with reference to any point or star in the same sphere (of the fixed stars).

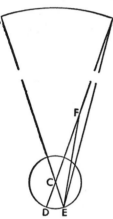

For let *AB* be the greatest circle in the world in the plane of the ecliptic. Let *C* be its centre, and let the sun be situated there. And in accordance with the distance *CD* between the sun and the Earth—in comparison with which the depth of the world is immense—let the circle *CDE*, in which the annual revolution of the centre of the Earth is located, be described in the same plane of the ecliptic: I say that the sun will seem to have a regular motion with reference to any point or star taken on circle *AB*.

Let some point be taken; and let it be *A*. And to *A* let the view of the sun from the Earth—which is at *D*—be extended as *DCA*. Now let the Earth be moved anywhere through arc *DE*; and let *AE* and *DE* be drawn from *E* the position of the Earth. Therefore the sun will now be seen from *E* at point *B*. And since *AC* is immense in comparison with *CD* or *CE* its equal, *AE* too will be immense in comparison with *CE*. For let any point *F* be taken on *AC*, and let *EF* be joined. Therefore since two straight lines from the termini *C* and *E* of the base fall outside triangle *EFC* on point *A*; by the converse of Euclid's *Elements*, I, 21,

angle *FAE* < angle *EFC*.

Wherefore the straight lines extended to immensity comprehend at last an angle *CAE* so acute that it is no longer perceptible; and

angle *CAE* = angle *BCA* − angle *AEC*.

Moreover, on account of the slightness of the difference between them angles *BCA* and *AEC* seem to be equal; and lines *AC* and *AE* seem to be parallel; and the sun seems to have [85ᵃ] a regular motion with reference to any point on the sphere of the fixed stars, just as if it were revolving around the centre *E*, as was to be shown.

But its irregular movement is demonstrated, because the movement of the centre of the Earth in its annual revolution is not absolutely around the centre of the sun.

That can be understood in two ways, either through an eccentric circle, *i.e.*, one whose centre is not the centre of the sun, or through an epicycle on a homocentric circle.

Now it is made clear through an eccentric in this way. For let *ABCD* be an eccentric circle in the plane of the ecliptic; and let its centre *E* be no very slight distance away from the centre of the sun or world. Let the centre of the world be *F*; and let *AEFD* be the diameter (of circle *ABCD*) passing through both centres. And let its apogee be at *A*—which is called the highest apsis by the Romans—the place farthest removed from the centre of the world, and *D* the perigee, which is nearest (to the centre of the world) and is the lowest apsis.

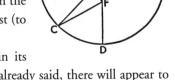

Therefore while the Earth is moved regularly in its orbital circle *ABCD* around its centre *E*, as has been already said, there will appear to be an irregular movement around *F*.

For let

arc *AB* = arc *CD*;

and let the straight lines *BE, CB, BF,* and *CF* be drawn.

Angle *AEB* = angle *CED*,

because angles *AEB* and *CED* are intercepting equal arcs around the centre *E*. But angle *CFD* is the angle of sight, and

ext. angle *CFD* > int. angle *CED*.

But

angle *AEB* = angle *CED*.

Hence

angle *CFD* > angle *AEB*.

But also

ext. angle *AEB* > int. angle *AFB*;

and hence by so much more

angle *CFD* > angle *AFB*.

But an equal time produces both angle *CFD* and angle *AFB* because

arc *AB* = arc *CD*.

Therefore the movement will appear regular from around *E* and irregular from around *F*.

Moreover, it is possible to see the same thing more simply, because arc *AB* is farther away from *F* than arc *CD* is. For by Euclid, III, 7, lines *AF* and *BF* by which arc *AB* is intercepted are longer than *CF* and *DF* by which arc *CD* is intercepted, and, as is shown in optics, equal magnitudes which are nearer appear greater than the ones farther away. And so what was proposed in the case of the eccentric circle is manifest.

The same thing will also be made clear by means of an epicycle on a homocentric circle. For let the centre of the homocentric circle *ABCD* and the centre of the world where the sun is be at *E*; and in the same plane let *A* be the centre of epicycle *FG*. And through both centres let the straight line *CEAF* be drawn. Let *F* be the apogee of the epicycle; and *I*, the perigee. Therefore it is clear that there is regularity [85ᵇ] in *A*, but apparent irregularity in epicycle *FG*. For if the movement of *A* takes place in the direction of *B*, *i.e.*, eastward, while the movement of the centre of the Earth is from its apogee *F* westward; then in the perigee—which is *I*—*E* will appear to be moving faster, because the two movements of *A* and *I* are in the same direction. But in the apogee, which is *F*, point *E* will seem to be moved more slowly, namely because it is moved only by the excelling movement out of two contraries; and the Earth situated at *G* is to the west of the regular movement but at *K* is to the east of it, and the distance of the

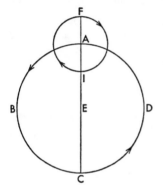

Earth from the regular movement is measured by arcs *AK* and *AG*, in accordance with which the sun will seem to move irregularly.

But whatever things take place by means of the epicycle can happen in the same way by means of the eccentric circle, which the transit of the planet in the epicycle describes equal to the homocentric circle and in the same plane; and the centre of the eccentric circle is at a distance from the centre of the homocentric circle equal to the radius of the epicycle. And all this occurs in three ways, since, if the epicycle on the homo-centric circle and the planet on the epicycle made similar revolutions but with movements opposite to one another, the movement of the planet will trace a fixed eccentric circle, *i.e.*, one whose apogee and perigee possess unchanging locations.

In this way let *ABC* be the homocentric circle, and the centre of the world *D*, the diameter *ADC*. And let us put down that, when the epicycle is at *A*, the planet is in the apogee of the epicycle, which is at *G*, and its radius is in the straight line *DAG*. Now let arc *AB* of the homocentric circle be taken; and with centre *B* and radius equal to *AG*, let the epicycle *EF* be described, and let *BD* and *BE* be extended in a straight line; and let the arc *EF* be similar to arc *AB*, but let arc *EF* be taken in the opposite direction. And let the planet or Earth be in *F*. Let *BF* be joined. And on line *AD* let

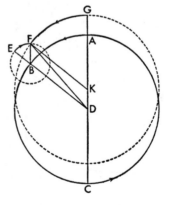

$$DK = BF.$$

Therefore, since

$$\text{angle } EBF = \text{angle } BDA,$$

and for those reasons

$$BF = DK$$

and BF is parallel to DK; and since, if straight lines are joined to equal and parallel straight lines, they are also equal and parallel by Euclid, I, 33; and since

$$DK = AG$$

[86ª] and AK is their common annex;

$$GAK = AKD$$

and therefore

$$GAK = KF.$$

Therefore the circle described with centre K and radius KAG will pass through F. By means of a movement compounded of AB and EF point F describes this circle as eccentric and equal to the homocentric and accordingly fixed too. For when the epicycle makes proportionally equal revolutions with the homocentric circle, the apsides of the eccentric circle so described necessarily remain in the same place.

But if the centre and the circumference of the epicycle make proportionately unequal revolutions, then the movement of the planet will not designate a fixed eccentric circle but one whose centre and apsides are carried westward or eastward, according as the movement of the planet is faster or slower than the centre of its epicycle. In this way if

$$\text{angle } EBF > \text{angle } BDA,$$

let

$$\text{angle } BDM = \text{angle } EBF.$$

It will similarly be shown that if on line DM there be taken DL equal to BF, the circle described with L as centre and with radius LMN equal to AD will pass through planet F. Hence it is clear that by the composite movement of the planet there is described arc NF of the eccentric circle, whose apogee meanwhile travels from point G westward along arc GN.

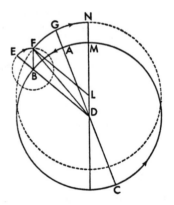

On the contrary, if the movement of the planet in the epicycle were slower, then the centre of the eccentric circle should follow it eastward, whither the centre of the epicycle is carried; that is if

$$\text{angle } EBF = \text{angle } BDM > \text{angle } BDA,$$

it is clear that what we have spoken of will take place.

From all that it is clear that the same irregularity of appearance is always produced

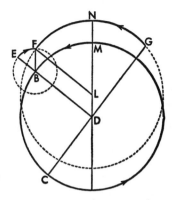

whether by means of an epicycle on a homocentric circle or by means of an eccentric circle equal to the homocentric; and they by no means differ from one another, provided the distance between the centres (of the homocentric and the eccentric) is equal to the radius of the epicycle. Accordingly it is not easy to determine which of them exists in the heavens. Indeed Ptolemy, where he understood simple irregularity and certain and immutable locations for the apsides—as he thought was the case in the sun—judged that the scheme of eccentricity was sufficient. But to the moon and to the five planets which wander in two or more different ways [86^{b}] he applied eccentric circles carrying epicycles.

From this moreover it is easily demonstrated that the greatest difference between regularity and appearance is seen at the time when the planet appears in the mean position between the highest and the lowest apsis in the case of the eccentric circle, but in the case of the epicycle at its point of contact (with the circle carrying the epicycle), as in Ptolemy.

In the case of the eccentric circle thus: For let there be the circle *ABCD* around the centre *E*; and *AEC* the diameter through *F* the sun, which is off centre. Now let line *BFD* be drawn through *F* at right angles to the diameter; and let *BE* and *ED* be joined. Let *A* be the apogee and *C* the perigee; and let *B* and *D* be the means appearing between them.

I say that no angle greater than angle *B* or *D* can be constructed with its vertex on the circumference and line *EF* as its base.

For let points *G* and *H* be taken on either side of *B*; and let *GD*, *GE*, and *GF* be joined, and also *HE*, *HF*, and *HD*. Since line *FG* is nearer the centre than line *DF*,

$$\text{line } FG > \text{line } DF.$$

And therefore

$$\text{angle } GDF > \text{angle } DGF.$$

But

$$\text{angle } EDG = \text{angle } EGD,$$

because sides *EG* and *ED* falling upon the base are equal. And therefore

$$\text{angle } EDF > \text{angle } EGF.$$

But
$$\text{angle } EDF = \text{angle } EBF.$$
Similarly too
$$\text{line } DF > \text{line } FH;$$
and
$$\text{angle } FHD > \text{angle } FDH.$$
But
$$\text{angle } EHD = \text{angle } EDH,$$
because
$$\text{line } EH = \text{line } ED.$$
Therefore, by subtraction,
$$\text{angle } EDF > \text{angle } EHF.$$
But
$$\text{angle } EDF = \text{angle } EBF.$$
Therefore no angle greater than the angles at points B and D will ever be constructed with line EF as base. And so the greatest difference between regularity and appearance is found in the mean position between apogee and perigee.

16. On the Apparent Irregularity of the Sun

These things have been demonstrated generally; and they are applicable not only to the apparent movements of the sun but also to the irregularity of the other planets. Now we shall investigate what relates to the sun and the Earth, first in respect to what has been handed down to us by Ptolemy and the other ancients, and then in respect to what modern times and experience have taught us. Ptolemy found [87^a] that $94^1/_2$ days were comprehended between the spring equinox and the summer solstice, and $92^1/_2$ between the solstice and the autumn equinox. Therefore in accordance with the ratio of time during the first interval there was a mean and regular movement of 93°9'; during the second interval, one of 91°11'.

Let $ABCD$ be the circle of the year as divided in this way; and let E be the centre. Let
$$\text{arc } AB = 93°9'$$
for the first period of time; and let
$$\text{arc } BC = 91°11'$$
for the second. Let the spring equinox be viewed from A; the summer solstice from B; the autumn equinox from C; and the remaining winter solstice from D. Let AC and BD be joined.

AC and BD cut one another at right angles at F, where we set up the sun.

Therefore since

$$\text{arc } ABC > 180°$$

and too

$$\text{arc } AB > \text{arc } BC;$$

Ptolemy understood from this that the centre of the circle was located between lines
BF and *FA*, and the apogee between the spring equinox and the summer tropic of the
sun. Now let *IEG*, which will cut *BFD* in *L*, be drawn through centre *E* parallel to
AFC, and let *HEK*, which will cut *AF* in *M*, be drawn parallel to *BFD*. In this way there
will be constructed the right parallelogram whose diameter *FE* extended in the straight
line *FEN* will indicate the Earth's greatest distance in length from the sun and the position of the apogee in *N*.

Therefore since

$$\text{arc } ABC = 184°19',$$

and

$$\text{arc } AH = {}^1/_2 \text{ arc } ABC = 92°9{}^1/_2';$$
$$\text{arc } HB = \text{arc } AGB - \text{arc } AH = 59°.$$

Again

$$\text{arc } AG = \text{arc } AH - 90° = 2°10'.$$

Now

$$LF = {}^1/_2 \text{ ch. } 2 \text{ } AG = 377,$$
$$\text{where radius} = 10,000.$$

But

$$EL = {}^1/_2 \text{ ch. } 2 \text{ } BH = 172.$$

And, as two sides of triangle *ELF* are given,

$$\text{side } EF = 414 \fallingdotseq {}^1/_{24} \text{ radius } NE$$
$$\text{where radius } NE = 10,000.$$

But

$$EF : EL = NE : {}^1/_2 \text{ ch. } 2 \text{ } NH.$$

Therefore

$$\text{arc } NH = 24{}^1/_2°.$$

And thus

$$\text{angle } NEH \text{ is given,}$$

and

$$\text{angle } NEH = \text{angle } LFE,$$

which is the angle of apparent movement. By such an interval therefore did the highest apsis before Ptolemy precede the summer solstice of the sun. But

$$\text{arc } IK - 90°,$$

and

$$[87^b] \quad \text{arc } IC = \text{arc } AG$$

and

$$\text{arc } DK = \text{arc } HB.$$

Hence

$$\text{arc } CD = \text{arc } IK - (\text{arcs } IC + DK) = 86°51'$$

and

$$\text{arc } DA = \text{arc } CDA - \text{arc } CD = 88°49'.$$

But to the 86°51' there correspond $88^1/_8$ days; and to the 88°49', 90 days and 3 hours—the eighth part of a day. During these periods the sun on account of the regular movement of the Earth seemed to cross from the autumn equinox to the winter solstice and for the remainder of the year to return from the winter solstice to the spring equinox. Indeed Ptolemy testifies that he found these things no different from what were reported by Hipparchus before him. Wherefore he judged that for the remainder of time the highest apsis would be $24^1/_2°$ before the summer tropic and that the eccentricity of—as I said—a 24th part of the radius would remain perpetually.

But now it is found that both of them have changed by a manifest difference. Al-Battani noted it as being 93 days 35 minutes (of a day) from the spring equinox to the summer solstice, and 186 days 37 minutes to the autumn, from which by Ptolemy's rule he elicited an eccentricity of not more than 346 parts whereof the radius has 10,000. Arzachel the Spaniard agrees with him in the ratio of eccentricity but reported an apogee 12°10' west of the solstice, and al-Battani viewed it as 7°43' west of the same solstice. By these tokens it has been grasped that there still remains another irregularity in the movement of the centre of the Earth, as has been attested by the observations of our time also. For during the ten and more years in which we applied our intelligence to investigating these things and especially in the year of Our Lord 1515, we found that there were 186 days $5^1/_2$ minutes from the spring equinox to the autumnal. And so as not to deceive ourselves in determining the solstices—which some suspected had happened in the case of our predecessors—we took certain other positions of the sun into consideration in this business which were not difficult to observe even in comparison with the equinoxes, such as the mean positions in the signs of Taurus, Leo, Scorpio, and Aquarius. Therefore we found that there were 45 days 16 minutes from the autumn equinox to the middle point of Scorpio, and 178 days $53^1/_2$ minutes to the spring equinox. Now the regular movement during the first interval was 44°37'; and during the second interval 176°19'.

[88ª] Now that these preparations have been made, let circle *ABCD* be repeated; and let *A* be the point from which the sun was seen at the spring equinox; *B* the point at which the autumn equinox was viewed; and *C* the midpoint of Scorpio. Let *AB* and *CD*, which cut one another at *F* the centre of the sun, be joined; and let arc *AC* be subtended.

Therefore, since

$$\text{arc } CB = 44°37',$$
$$\text{angle } BAC = 44°37',$$
where 2 rt. angles = 360°.

And

$$\text{angle } BFC = 45°,$$
where 4 rt. angles = 360°;

and is the angle of apparent movement; but

$$\text{angle } BFC = 90°,$$
where 2 rt. angles = 360°.

Hence,

$$\text{angle } ACD = 45°23',$$

because

$$\text{arc } AD = 45°23'.$$

But

$$\text{arc } ACB = 176°19',$$

and

$$\text{arc } AC = \text{arc } ACB - \text{arc } BC = 131°42',$$

and

$$\text{arc } CAD = \text{arc } AC + \text{arc } AD = 177°5'.$$

Therefore, since

$$\text{arc } ACB < 180°,$$

and

$$\text{arc } CAD < 180°,$$

it is clear that the centre of the circle is located in the remainder *BD*. And let the centre be *E*. And through *E* let the diameter *LEFG* be drawn. Let *L* be the apogee and *G* the perigee. Let *EK* be erected perpendicular to *CFD*. But the chords subtending the given arcs are also given by the table:

$$AC = 182,494$$

and

$$CFD = 199,934,$$
where diameter = 200,000.

Accordingly, as triangle *ACF* has its angles given, the ratio of the sides will be given by the first rule for plane triangles.

$$CF = 97,697,$$

according as

$$AC = 182,494;$$

and for that reason

$$FK = \frac{1}{2} CD - CF = 2,000.$$

And since

$$180° - \text{arc } CAD = 2°55';$$

and since

$$EK = \frac{1}{2} \text{ ch. } 2°55' = 2,534;$$

then, in triangle EFK, as the two sides FK and KE comprehending the right angle have been given, the triangle will have its sides and angles given:

$$EF = 323,$$
$$\text{where } EL = 10,000;$$

and

$$\text{angle } EFK = 51\frac{2}{3}°$$
$$\text{where 4 rt. angles} = 360°.$$

Therefore, by addition,

$$\text{angle } AFL = 96\frac{2}{3}°$$

and, by subtraction,

$$\text{angle } BFL = 83\frac{1}{3}°.$$

But

$$EF \doteq 1^{\text{P}}56',$$
$$\text{where } EL = 60^{\text{P}}.$$

This is the distance of the sun from the centre of the orbital circle: and it has now become approximately $\frac{1}{31}$st (of the radius of the orbital circle), [88^{b}] though to Ptolemy it seemed to he $\frac{1}{24}$th. And the apogee, which was at that time $24\frac{1}{2}°$ to the west of the summer solstice, is now $6\frac{2}{3}°$ to the east of it.

17. DEMONSTRATION OF THE FIRST AND ANNUAL IRREGULARITY OF THE SUN TOGETHER WITH ITS PARTICULAR DIFFERENCES

Therefore since many differences of the irregular movement of the sun are found, we judge that the difference which occurs annually and is more known than the rest should be deduced first.

Accordingly let circle ABC be constructed again, around centre E with diame-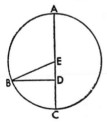ter AEC; apogee at A, perigee at C; and the sun at D. Now it has been shown that the greatest difference between regular and apparent movement occurs at the position which with respect to the apparent movement is midway between the apsides. For that reason let BD be erected perpendicular to AEC, and let it cut the circumference in point B, and let BE

be joined. Therefore, since in right triangle *BDE* two sides have been given, namely *BE* which is the radius of the circle and *DE* the distance of the sun from the centre; the triangle will have its angles given. And angle *DBE* will be given, which is the difference between the angle *BEA* of regular movement and right angle *EDB* the angle of apparent movement.

But as *DE* is made greater or less, the whole species of the triangle changes. Thus, before Ptolemy

$$\text{angle } B = 2°23';$$

in the time of al-Battani and Arzachel

$$\text{angle } B = 1°59';$$

but at present

$$\text{angle } B = 1°51'.$$

And for Ptolemy

$$\text{arc } AB = 92°23',$$

which is intercepted by angle *AEB*, and

$$\text{arc } BC = 87°37'.$$

For al-Battani

$$\text{arc } AB = 91°59'$$

and

$$\text{arc } BC = 88°1'.$$

And at present

$$\text{arc } AB = 91°51'$$

and

$$\text{arc } BC = 88°9'.$$

Whence too the remaining differences are manifest. For let any other arc *AB* be taken, as in the following figure: and let angle *AEB* be given, and the interior angle *BED*, and the two sides *BE* and *ED*. By the calculus of plane triangles there will be given [89ª] angle *EBD*, the additosubtraction, the difference between the regular and the apparent movement. And it is necessary for these differences to change on account of the change of side *ED*, as has already been said.

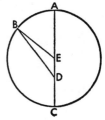

18. ON THE EXAMINATION OF THE REGULAR MOVEMENT IN LONGITUDE

These things have been set forth concerning the annual irregularity of the sun, but not by means of the simple difference so appearing but by means of the difference still mingled with that which the length of time has disclosed.

We shall distinguish them from one another later on. Meanwhile, the mean and regular movement of the centre of the Earth will be given in numbers which will be the more certain the more that movement is separated from any differences of irregularity and the more it extends in time. Now that will be established in this way.

We have taken that autumn equinox which was observed by Hipparchus at Alexandria in the 32nd year of the third period of Callippus—which, as was said above, was the 177th year after the death of Alexander—at midnight after the third intercalary day, which the fourth day followed. But according as Alexandria is approximately 1 hour to the east of Cracow in longitude, it was approximately 1 hour before midnight. Therefore according to the calculation handed on above the position of the autumn equinox in the sphere of the fixed stars was 176°10' from the head of Aries and that was the apparent position of the sun. It was $114^1/_2°$ distant from the highest apsis.

In accordance with this model let there be traced around centre D the circle ABC which the centre of the Earth describes. Let ADC be the diameter; and let the sun be situated on the diameter at point E; the apogee in A; and the perigee in C. But let B be the point where the sun appears in the autumn equinox, and let the straight lines BD and BE be joined.

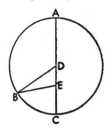

Since

$$\text{angle } DEB = 114^1/_2°,$$

and is seen to measure the distance of the sun from the apogee; and

$$\text{side } DE = 414,$$

$$\text{where } BD = 10,000;$$

therefore, by the fourth theorem on plane triangles, triangle BDE has its sides and angles given. And

$$\text{angle } BDE = \text{angle } BDA = \text{angle } BED = 2°10'.$$

[89ᵇ] But

$$\text{angle } BED = 114°30'.$$

Hence

$$\text{angle } BDA = 116°40';$$

and the mean or regular position of the sun is 178°20' from the head of the Ram in the sphere of the fixed stars.

With this we have compared the autumn equinox observed by us in Frauenburg under the same meridian of Cracow in the year of Our Lord 1515, on the 18th day before the Kalends of October, in the 1840th year since the death of Alexander, on the 6th day of Phaophi the second month by the Egyptian calendar, half an hour after sunrise. At this time the position of the autumn equinox by calculation and observation was 152°45' in the sphere of the fixed stars and was 83°29' distant from the highest apsis in accordance with the preceding demonstration.

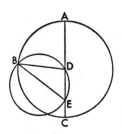

Now let

angle *BEA* = 83°20',

where 2 rt. angles = 180°;

and two sides of the triangle are given:

BD = 10,000

and

DE = 323.

By the fourth theorem on plane triangles

angle *DBE* ≑ 1°50'.

For, if a circle circumscribes triangle *BDE*, then, as on the circumference,

angle *BED* = 166°40',

where 2 rt. angles = 360°.

And

ch. *BD* = 19,864

where dmt. = 20,000.

And as

BD : *DE* is given,

ch. *DE* ≑ 640,

and

DE = ch. *DBE*

and, as on circumference,

angle *DBE* = 1°50';

but, as at centre,

angle *DBE* = 3°40'.

And this was the additosubtraction and difference between the regular and the apparent movement. And

angle *BDA* = angle *DBE* + angle *BED* = 1°50' + 83°20' = 85°10',

the distance of the regular movement from the apogee, and hence the mean position of the sun is 154°35' in the sphere of the fixed stars.

Therefore in the time between both observations there are 1662 Egyptian years 37 days 18 minutes (of a day) 45 seconds. And the mean and regular movement over and above the whole revolutions—of which there were 1660—is approximately 336°15', which is consonant with the number which we set out in the table of regular movements.

19. ON DETERMINING THE POSITIONS OF THE REGULAR MOVEMENT OF THE SUN AT THE BEGINNINGS (OF YEARS)

[90ª] Accordingly in the flow of time between the death of Alexander the Great and the observation made by Hipparchus there were 176 years 362 days 27$^{1}/_{2}$ minutes, in which the mean movement was 312°43', according to calculation. When these degrees are subtracted from the sum of the 178°20' of Hipparchus' observation and from the 360° of the circle, there will remain, for noon of the first day of Thoth the first month of the Egyptians at the beginning of the years named after the death of Alexander, a position of 225°37' beneath the meridian of Cracow and of Frauenburg, the place of our observation.

From this to the beginning of the Roman years of Julius Caesar in 278 years 118$^1/_2$ days the mean movement is 46°27' over and above the complete revolutions. The addition of these degrees to the degrees of the position of Alexander gives 272°4' as Caesar's position at midnight before the Kalends of January, from which the Romans are accustomed to take the beginning of their years and days.

Then in 45 years 12 days, or in 323 years 130$^1/_2$ days from the death of Alexander the Great, there arises the position of Christ at 272°31'.

And since Christ was born in the third year of the 194th Olympiad, the calculations which give 775 years and 12$^1/_2$ days from the beginning of the year of the first Olympiad to midnight before the Kalends of January similarly give 96°16' as the position of the first Olympiad at noon of the first day of the month Hekatombaion, the anniversary of which day is now the Kalends of July according to the Roman calendar.

In this way the beginnings of the simple movement of the sun are determined with respect to the sphere of the fixed stars. Moreover, the positions of the composite movement are given by the addition of the precession of the equinoxes and similarly to the others: the Olympic position at 90°59'; the position of Alexander at 226°38'; that of Caesar at 276°59'; and that of Christ at 278°2'. All these things, as we said, are taken with respect to the Cracow meridian.

20. ON THE SECOND AND TWOFOLD IRREGULARITY WHICH OCCURS IN THE CASE OF THE SUN ON ACCOUNT OF THE CHANGE OF THE APSIDES

[90^b] But there is now a greater difficulty in connexion with the inconstancy of the apsis of the sun, since, although Ptolemy thought it to be fixed, others have thought it to follow the movement of the starry sphere, according as they judged that the fixed stars moved too. Arzachel opined that this movement also was irregular, that is to say, as happening to retrograde—from the token that, although, as was said, al-Battani had found the apogee 7°44' to the west of the solstice (for previously during the 740 years after Ptolemy it had progressed approximately 17°), it seemed to Arzachel 193 years later to have retrograded approximately 4$^1/_2$°. And accordingly he thought there was some other movement made by the centre of the annual orbital circle in a small circle, in accordance with which (movement) the apogee was deflected back and forth and the centre of the circle (of the year) was at unequal distances from the centre of the world. That was a good enough device, but it was not accordingly accepted, because upon a universal comparison it is not consonant with the rest; that is to say, if the succession in the order of movement is considered: namely, that at some time before Ptolemy the movement came to a standstill, that during 740 years or thereabouts it traversed 17°, that in the 200 years thereafter it retrograded 4° or 5°, that in the time remaining down to us it progressed, and that no other retrogradation was perceived during the total time, and no more standstills, though they necessarily intervene in the case of contrary movements back and forth. And this can by no means be understood as occurring in uniform and circular movement. Wherefore it is believed by many that some error had crept into their observations. But each mathematician is alike in his care and industry, so that it is doubtful which one we should follow in preference to the other. At all events, I confess that nowhere is there greater difficulty than in determining the apogee of the sun, where we ratiocinate with very small and hardly perceptible magnitudes, since in the neighbourhood of the perigee and apogee (a movement of) 1° effects only a variation of approximately 2' in the additosubtraction, but in the neighbourhood of the mean apsides (a movement of) 1' effects 5° or 6° (in the additosubtraction); and so a slight [91^a] error can propagate itself greatly. Hence in placing the apogee at $6^2/_3$° of Cancer, we were not content to rely upon the instruments of the horoscope, unless the eclipses of the sun and moon gave us more certainty, since if any error lay concealed in our observations, the eclipses would uncover it without fail. Therefore, in accordance with most likelihood, we can apply our intelligence to conceiving the movement as a whole: it is eastward, but irregular, since after that standstill between Hipparchus and Ptolemy the apogee has appeared to be in continuous, orderly, and increased progression down to our time, with the exception of the movement which occurred erroneously—it is believed—between al-Battani and Arzachel, as all the rest seems to be in harmony. For it seems to follow from the same ratio of circular movement that the additosubtraction (of the movement) of the sun similarly does not stop decreasing and that corrections are made for these two irregularities in conjunction with the first and simple anomaly of the obliquity of the ecliptic or something similar.

But in order for this to become more clear, let the circle *AB* around centre *C* be in the plane of the ecliptic. And let the diameter be *ACB*, and on *ACB* let *D* be the globe of the sun as it were at the centre of the world; and let another quite small circle *EF* be described around centre *C* in such a way as not to comprehend the sun. And let it be understood that the centre of the annual revolution of the Earth moves around this small circle with a rather slow progress. And since the small circle *EF* together with line *AD* has a rather slow movement eastward and the centre of the annual revolution has

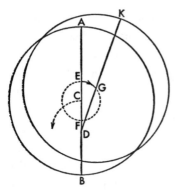

a rather slow movement westward along circle *EF*, sometimes the centre of the annual orbital circle will be found at its greatest distance which is *DE*, and sometimes at its least which is *DF*, and with a slower movement at the greatest distance and a faster movement at the least. And along the middle curves the small circle makes the distance between the centres increase and decrease with time, and it makes the highest apsis alternately precede and follow the apsis or apogee which is on line *ACD* as if in the middle position. In this way, if arc *EG* is taken and with *G* as centre a circle equal to *AB* is described, the then highest apsis will be on line *DGK* and *DG* will be a shorter distance than *DE*, by Euclid, III, 8.

And these things are demonstrated by means of a circle eccentric to an eccentric circle as above; and by means of the epicycle [91^b] on the epicycle as follows: Let circle *AB* be homocentric with the world and with the sun, and let *ABC* be the diameter, whereon the highest apsis is. And with *A* as centre let the epicycle *DE* be described; and again with *D* as centre, the epicycle *FG*, whereon the Earth revolves. And all in the same plane of the ecliptic. Let the movement of the first epicycle be eastward and approximately annual; and that of the second too, *i.e.*, *D*, be similar but westward. And let both have proportionately equal revolutions with respect to line *AC*. Moreover, let the centre of the Earth moving westward from *F* add a little movement to *D*.

From this it is clear that when the Earth is at *F*, it will make the apogee of the sun to be farthest away; and when at *G* it will make the apogee to be nearest; but in the mean arcs of epicycle *FG*, it will make the apogee precede or follow, increased or decreased, greater or less; and hence it will make the movement appear irregular, as has been demonstrated before of the epicycle and the eccentric circle.

Now let arc *AI* be taken. And with point *I* as centre let the epicyclical epicycle be taken again. And let *CI* be joined and extended in the straight line *CIK*.

angle *KID* = angle *ACI*

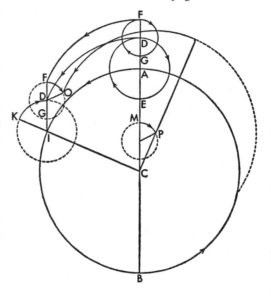

on account of the revolutions being proportionately equal. Therefore, as we demonstrated above, point D will describe around centre L an eccentric circle equal to homocentric circle AB, and with an eccentricity CL equal to DI; and F will describe an eccentric circle having an eccentricity CLM equal to IDF; and G similarly an eccentric circle having an eccentricity CN equal to IG. Meanwhile, if the centre of the Earth has by now measured [92ª] any arc FO on its second epicycle, point O will not describe an eccentric circle whose centre is on line AC but one whose centre is on a line parallel (to DO), such as LP. But if OI and CP are joined,

$$OI = CP,$$

but

$$OI < IF$$

and

$$CP < CM.$$

And

$$\text{angle } DIO = \text{angle } LCP,$$

by Euclid, I, 8. And that is the interval whereby the apogee of the sun on line CP will be seen to precede A.

From this moreover it is clear that the same thing occurs through the eccentric circle having an epicycle, since with the eccentric circle alone pre-existing which epicycle D describes around centre L, the centre of the Earth revolves through arc FO in accordance with the aforesaid conditions, i.e., (in a movement) less than the annual revolution. For it will describe, as before, another circle eccentric to the first, around centre P; and the same things will occur again. And since so many ways lead to the same number, I could not really say which one is right, except that the perpetual harmony of numbers and appearances compels us to believe that it is some one of them.

21. How Great the Second Difference in the Irregularity of the Sun Is

Therefore, since it has already been seen that the second irregularity follows after that first and simple anomaly of the obliquity of the ecliptic or its similitude, we shall have its fixed differences, if some error on the part of past observers does not stand in the way. For according to calculation we have a simple anomaly of approximately 165°39' for the year of Our Lord 1515, and also its beginning by a calculation backwards to approximately 64 years before the birth of Christ, from which time to us there has been a passage of 1580 years. Now the greatest eccentricity of that beginning has been found by us to be 414, whereof the radius is 10,000. But the eccentricity of our time, as was shown, is 323.

Now let *AB* be a straight line, and on it let *B* be the sun and centre of the world. Let *AB* be the greatest eccentricity and *BD* the least. Let a small circle be described whose diameter is *AD*, and let

$$\text{arc } AC = 165°39',$$

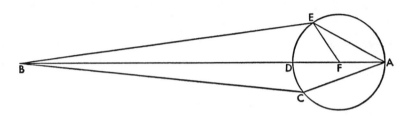

in proportion to the first simple anomaly. Since line *AB* has [92^b] been found at the beginning of the simple anomaly, *i.e.*, at *A*, and

$$AB = 414,$$

and now

$$\text{line } BC = 323,$$

therefore we shall now have triangle *ABC* with sides *AB* and *BC* given; and also one angle *CAD* given, because *CD* the remaining arc of the semicircle is given, *i.e.*,

$$\text{arc } CD = 14°21'.$$

Therefore, by what we have shown concerning plane triangles, there are given the remaining side *AC* and angle *ABC*, the difference between the mean and the irregular movement of the apogee; and inasmuch as line *AC* subtends the given arc, diameter *AD* of circle *ACD* will also be given. For since

$$\text{angle } CAD = 14°21',$$
$$CB = 2496$$

where diameter of circle circumscribing triangle = 20,000,

and since

$$BC : AB \text{ is given,}$$
$$AB = 3225 = \text{ch. } ACB = \text{ch. } 341°26'.$$

Hence by subtraction

$$\text{angle } CBD = 4°13',$$
$$\text{where 2 rt. angles} = 360°.$$

And

$$\text{ch. } CBD = AC = 735.$$

Therefore

$$AC \doteqdot 95,$$
$$\text{where } AB = 414.$$

182

And according as *AC* subtends the given arc, it will have a ratio to *AD* as to a diameter. Therefore

$$AD = 96,$$
$$\text{where } ADB = 414;$$

and, by subtraction,

$$DB = 321,$$

and that is the distance of the least eccentricity. But, as on the circumference

$$\text{angle } CBD = 4°13',$$

and as at the centre

$$\text{angle } CBD = 2°6\frac{1}{2}',$$

which is the additosubtraction to be subtracted from the regular movement of *AB* around centre *B*. Now let there be drawn the straight line *BE* touching the circle at point *E*; and with centre *F* taken, let *EF* be joined. Therefore, since in right triangle *BEF*,

$$\text{side } EF = 48,$$

and

$$\text{side } BDF = 369,$$
$$EF = 1300,$$
$$\text{where radius } FB = 10,000.$$

And

$$EF = \frac{1}{2} \text{ ch. 2 } EBF;$$

and

$$\text{angle } EBF = 7°28',$$
$$\text{where 4 rt. angles} = 360°;$$

and that is the greatest additosubtraction between the regular movement at *F* and the apparent at *E*.

Hence the remaining and particular differences can be discovered: for instance, if

$$\text{angle } AFE = 6°.$$

For we shall have the triangle with sides *EF* and *FB* and angle *EFB* given. Hence

$$\text{angle } EBF = 41',$$

which is the additosubtraction. [93ª] But if

$$\text{angle } AFE = 12°,$$
$$\text{add.} = 1°23'.$$

And if

$$\text{angle } AFE = 18°,$$
$$\text{add.} = 2°3';$$

and so for the rest in this way, as was said above in the case of the additosubtractions for the annual revolution.

22. How the Regular Movement of the Apogee of
the Sun and the Irregular Movement Are Unfolded

Therefore, since the time in which the greatest eccentricity coincided with the beginning of the first and simple anomaly was the third year of the 178th Olympiad but the 259th year of Alexander the Great by the Egyptian calendar, and on that account the simultaneously true and mean position of the apogee was at $5^1/_2°$ of Gemini, *i.e.*, $65^1/_2°$ from the spring equinox; and since the precession of the equinoxes—the true at that time coinciding with the mean—was 4°38': the subtraction of 4°38' from $65^1/_2°$ leaves 60°52' from the head of Aries in the sphere of the fixed stars as the position of the apogee.

Again in the second year of the 573rd Olympiad and in the 1515th year of Our Lord, the position of the apogee was found at $6^2/_3°$ of Cancer. But by calculation the precession of the spring equinox was $27^1/_4°$; and the subtraction of $27^1/_4°$ from 96°40' leaves 69°25'. Now it was shown that with a first anomaly of 165°39' existing at that time there was an additosubtraction of 2°7', by which the true locus preceded the mean. Wherefore it was clear that the mean locus of the apogee of the sun was 71°32'.

Therefore during the middle 1580 Egyptian years the mean and regular movement of the apogee was 10°41'. And when we have divided that by the number of the years, we shall have an annual rate of 24"20"'14"".

23. On the Correction of the Anomaly of the Sun
and the Determination of Its Prior Positions

[93^b] If we subtract these 24"20"'14"" from the simple annual movement, which was 359°44'49"7"'4"" there will remain an annual regular movement of anomaly of 359°44'24"46"'50"". Again, the distribution of 359°44'24"46"'50"" through the 365 days will give a daily rate of 59'8"7"'22"" in accord with what was set out above in the tables. Hence we shall have the positions at the established beginnings of years—starting at the 1st Olympiad. For it was shown that on the 18th day before the Kalends of October in the second year of the 573rd Olympiad at half an hour after sunrise the mean apogee of the sun was at 71°32', from which the sun had a distance of 82°58'. And from the first Olympiad there have been 2290 Egyptian years 281 days 46 minutes, during which the movement of anomaly—the whole cycles not being counted—was 42°33'. The subtraction of 42°33' from 82°58' leaves 40°25' as the position of anomaly for the first Olympiad.

And similarly, as above, the position for the Alexander years is 166°38'; for the Caesar years, 211°11'; and for the years of Our Lord, 211°19'.

24. TABLE OF THE DIFFERENCES BETWEEN REGULAR AND APPARENT MOVEMENT

But in order that those things which we have shown concerning the (additive and subtractive) differences between the regular and apparent movements of the sun may be better fitted up for use, we shall also set out a table of them, having sixty rows and six orders of columns.

For the two first columns of both semicircles—that is to say, of the ascending and the descending semicircles—will contain numbers increasing by 3°'s, as above in the case of the movements of the equinoxes.

In the third column will be inscribed the degrees of additosubtraction arising from the movement [94ᵃ] or anomaly of the solar apogee; and this additosubtraction ascends to the height of approximately $7^{1}/_{2}°$, according as it fits each row of degrees.

The fourth place is given over to the proportional minutes, which go up to 60'; and they are reckoned according to the differences between the greater and the lesser additosubtractions arising from the simple anomaly. For since the greatest of these differences is 32', the sixtieth part will be 32". Therefore in accordance with the magnitude of the difference, which we derive from the eccentricity by the mode described above, we put down a number up to 60 to correspond to the single items in the column of the 3°'s.

In the fifth column the single additosubtractions arising from the annual and first anomaly are set up in accordance with the least distance of the sun from the centre.

In the sixth and final column, the differences between these additosubtractions and the additosubtractions which occur at greatest eccentricity.[1] The table is as follows:

---

[1] The movements on the homocentric circle, on the first epicycle, and on the second epicycle we proportionately equal. Hence, from the first two columns of the table are to be taken the movement on the second epicycle, or arc *KJ*; and from the third column, the additosubtraction to be applied to the annual anomaly or movement of the first epicycle; this additosubtraction, angle *KFJ*, corrects the mean anomaly from *H* to *I*. (Proportional minutes corresponding to arc *KJ* are to be saved.) Then from the fifth column is to be taken the additosubtraction *GEI*, corresponding to angle *GFI*. But since the true position of the sun, is not at *I* but at *J*, the additosubtraction most be corrected for the difference between angle *FEI* and angle *FIJ*. The proportional minutes which have been saved enable one to adjust the final difference, angle *IEJ*, according as chord *FJ* varies in length between *FL* and *FK*: that is to say, the change in eccentricity according to the movement around circle *CO* may be considered as it were a variation in the length of the radius of the corrected epicycle *HK*.

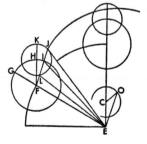

TABLE OF THE ADDITIONS-AND-SUBTRACTIONS OF THE MOVEMENT OF THE SUN

Common Numbers (Deg. Deg.)		Additions-and-subtractions arising from movement of the centre (Deg. Min.)		Proportional minutes	Additions-and-subtractions arising from eccentric orbital circle or first epicycle (Deg. Min.)		Differences (Min.)	Common Numbers (Deg. Deg.)		Additions-and-subtractions arising from movement of the centre (Deg. Min.)		Proportional minutes	Additions-and-subtractions arising from eccentric orbital circle or first epicycle (Deg. Min.)		Differences (Min.)
3	357	0	21	60	0	6	1	93	267	7	24	30	1	50	32
6	354	0	41	60	0	11	3	96	264	7	24	29	1	50	33
9	351	1	2	60	0	17	4	99	261	7	24	27	1	50	32
12	348	1	23	60	0	22	6	102	258	7	23	26	1	49	32
15	345	1	44	60	0	27	7	105	255	7	21	24	1	48	31
18	342	2	3	59	0	33	9	108	252	7	18	23	1	47	31
21	339	2	24	59	0	38	11	111	249	7	13	21	1	45	31
24	336	2	44	59	0	43	13	114	246	7	6	20	1	43	30
27	333	3	4	58	0	48	14	117	243	6	58	18	1	40	30
30	330	3	23	57	0	53	16	120	240	6	49	16	1	38	29
33	327	3	41	57	0	58	17	123	237	6	37	15	1	35	28
36	324	4	0	56	1	3	18	126	234	6	25	14	1	32	27
39	321	4	18	55	1	7	20	129	231	6	14	12	1	29	25
42	318	4	35	54	1	12	21	132	228	6	2	11	1	25	24
45	315	4	51	53	1	16	22	135	225	5	44	10	1	21	23
48	312	5	6	51	1	20	23	138	222	5	28	9	1	17	22
51	309	5	20	50	1	24	24	141	219	5	19	7	1	12	21
54	306	5	34	49	1	28	25	144	216	4	51	6	1	7	20
57	303	5	47	47	1	31	27	147	213	4	30	5	1	3	18
60	300	6	3	46	1	34	28	150	210	4	9	4	0	58	17
63	297	6	12	44	1	37	29	153	207	3	46	3	0	53	14
66	294	6	27	42	1	39	29	156	204	3	23	3	0	47	13
69	291	6	33	41	1	42	30	159	201	3	1	2	0	42	12
72	288	6	42	40	1	44	30	162	198	2	37	1	0	36	10
75	285	6	51	39	1	46	30	165	195	2	12	1	0	30	9
78	282	6	58	38	1	48	31	168	192	1	47	1	0	24	7
81	279	7	5	36	1	49	31	171	189	1	21	0	0	18	5
84	276	7	11	35	1	49	31	174	186	0	54	0	0	12	4
87	273	7	16	33	1	50	31	177	183	0	27	0	0	6	2
90	270	7	21	32	1	51	32	180	180	0	0	0	0	0	0

25. On the Calculation of the Apparent Movement
of the Sun

[95ᵇ] From that, I think, it is now sufficiently clear how the apparent position of the sun is calculated for any given time. For we must seek the true position of the spring equinox for that time or its precession together with its first and simple anomaly, as we have set forth above, and then the mean simple movement of the centre of the Earth—or you may call it the movement of the sun—and the annual anomaly, by means of the tables of regular movements; and they are added to their established beginnings. Accordingly you will take the number of the first simple anomaly found in the first or second column of the preceding table; and in the third column[1] you will find the corresponding additosubtraction for correcting the annual anomaly, and in the following column the proportional minutes; save the proportional minutes. Now add the additosubtraction to the annual anomaly, if the first (and simple anomaly) or its number contained in the first column—is less than a semicircle; otherwise subtract. For the remainder or aggregate will be the corrected anomaly of the sun; now by means of this take the additosubtraction arising from the annual (eccentric) orbital circle (or first epicycle)—which is found in the fifth column—and the difference in the following column. If this difference, when adjusted to the proportional minutes you have saved, amounts to something, it is always added to this additosubtraction, and the additosubtraction thus becomes corrected and is subtracted from the mean position of the sun, if the number of the annual anomaly is found in the first column or is less than a semicircle, but it is added, if the annual anomaly is greater or is found in one of the other columns of numbers. For that which in this way becomes the remainder or aggregate will determine the true position of the sun as measured from the head of the constellation of Aries, and if finally the true precession of the spring equinox is added to this (position of the sun), it will straightway show the distance of the sun from the equinox in degrees of the ecliptic among the twelve signs.

But if you wish to do that in another way, take the regular composite movement instead of the simple, and do the other things we spoke of, except that instead of the precession of the equinox, you add or subtract merely its additosubtraction, as the case demands. And so the rational explanation of the appearance of the sun by means of the mobility of the Earth is consonant with ancient and modern findings; and it is all the more [96ᵃ] presumed to hold for the future.

But furthermore, we are not ignorant of the fact that, if anyone thought that the centre of the annual revolutions were fixed as the centre of the world but that the sun moved in accordance with two movements similar and equal to those which we

[1] *i.e.*, since the movements on the first epicycle and on the second epicycle are proportionately equal to one another.

demonstrated in the case of the centre of the eccentric circle, everything will be manifest which was manifest before—the same numbers and the same demonstrations—since nothing else is changed in them except their situation, especially those which have to do with the sun. For then the movement of the centre of the Earth round the centre of the world would be absolute and simple, as the other two movements would be attributed to the sun itself. And on that account there will still remain some doubt as to which of these centres is the centre of the world, as we said ambiguously in the beginning that the centre of the world was at the sun or around the sun. But we shall say more about this question in our explanation of the five wandering stars; and we shall decide the issue to the extent that we are able, holding it enough, if we apply to the apparent movement of the sun calculations which have certitude and are not misleading.

26. ON THE ΝΥΧΘΗΜΕΡΟΝ, *i.e.*, THE DIFFERENCE OF THE NATURAL DAY

In connection with the sun there still remains something to be said about the inequality of the natural day. This time is comprehended by the space of twenty-four hours, which up to now we have used as the common and certain measure of the celestial movements. But some, like the Chaldees and the ancient Jews, define such a day as the time between two sunrises; others, like the Athenians, as that between two sunsets; or like the Romans, from midnight to midnight; or like the Egyptians, from noon to noon. Now it is clear that during this time the revolution proper to the terrestrial globe is completed together with that which is added by the annual revolution in accordance with the apparent movement of the sun.[1] The apparent irregular course of the sun in especial shows that this addition is unequal, as does the fact that the natural day takes place with respect to the poles of the equator, but the year with respect to the ecliptic. Wherefore that apparent time cannot be the common and certain measure of movement, since day does not accord with day in every respect; and so it was necessary to choose among them some mean and equal day, by which it would be possible [96ᵇ]

[1]In Ptolemy the daily revolution and the annual movement were in opposite directions, and thus the solar day was slightly longer than the sidereal day. Here the daily revolution of the Earth and its annual movement are both of them in the same direction, *i.e.*, eastward, and the solar day remains longer than the sidereal day on account of the third movement of the Earth, *i.e.*, the declination of the pole of the Earth, which is approximately equal to the annual revolution but in the opposite direction.

Let *A* be the sun, *CF* and *DEF* the Earth with centre *B* and *G*. And let *FBC* and *FGD* be the same meridian line. Let the centre of the Earth move from *B* to *G* during the space of 24 equatorial hours. As the movement of declination keeps the axis of the Earth parallel to itself, so too the meridian line *FBC* or *FGD* will be parallel to itself at the end of one daily revolution, but it will not be one with *GEA*, the line from the centre of the Earth to the centre of the sun, until the Earth has further revolved through arc *DE*. That is to say, the solar day is equal to the 360° of sidereal day *DEFD* plus arc *DE*.

to measure regularity of movement without trouble. Therefore since in the circle of the total year there are 365 revolutions around the poles of the Earth, to which there accretes approximately one whole supernumerary revolution on account of the daily addition made by the apparent progress of the sun: consequently one 365th part of that would fill out the natural day upon an equal basis.

Wherefore we must define and separate the equal day from the apparent and irregular. Accordingly we call that the equal day which comprehends the whole revolution of the equator and over and above that the portion which the sun is seen to traverse with regular movement during that time; but the unequal and apparent day that which comprehends the 360 "times"[1] of one revolution of the equator and in addition that which ascends in the horizon or meridian together with the apparent progress of the sun. Although the difference between these days is very slight and not immediately perceptible, nevertheless it becomes evident after the passage of a certain number of days.

There are two causes for this: the irregularity of the apparent movement of the sun and the unequal ascension of the oblique ecliptic. The first cause, which exists by reason of the irregular apparent movement of the sun, has already been explained, since in the case of the semicircle in which the highest apsis holds the midpoint there is a deficiency of $4^3/_4$ "times" with respect to the ecliptic, according to Ptolemy, and in the case of the other semicircle, in which the lowest apsis is, there is a similar excess of the same amount. Accordingly the total excess of one semicircle over the other was $9^1/_2$ "times."

But in the case of the other cause—which has to do with the rising and setting—the greatest difference occurs between the semicircles comprehending each solstice. This is the difference which exists between the shortest and the longest day and which is most variable, as being particular to each region. The difference which is measured from noon or midnight is comprehended by four termini everywhere, since from 16° of Taurus to 14° of Leo, 88° (of the ecliptic) cross the meridian together with approximately 93 "times"; and from 14° of Leo to 16° of Scorpio, 92° (of the ecliptic) and 87 "times" pass over the meridian, so that in the latter case there is a deficiency of 5 "times" and in the former case an excess of 5 "times." And so the sum of the days in the first segment exceeds those in the second by ten "times"—which make two thirds of one hour; and the same thing takes place conversely in the other semicircle within the remaining termini set diametrically opposite to these. Now the mathematicians chose [97ᵃ] to take the natural day from noon or midnight rather than from sunrise or sunset. For the difference which is taken from the horizon is more manifold; for it extends to a certain number of hours, and moreover it is not everywhere the same but varies manifoldly according to the obliquity of the sphere. But the one which pertains to the meridian is everywhere the same and is more simple. Therefore the total difference, which is constituted by the aforesaid causes: the apparent

---

[1]The unit parts of the equator we called "times" instead of degrees.

irregular progress of the sun and the irregular passage over the meridian, in the time before Ptolemy, took its beginning of decrease at the midpoint of Aquarius and, increasing from the beginning of Scorpio, added up to $8^1/_3$ "times"; and now decreasing from 20° of Aquarius or thereabouts to 10° of Scorpio and increasing from 10° of Scorpio to 20° of Aquarius, it has contracted to 7 "times" 48'. For these things too are changed on account of the inconstancy of the perigee and the eccentricity with the passage of time. Finally, moreover, if the greatest difference in the precession of the equinoxes is taken into account, the total difference of the natural days can extend itself to above 10 "times" for a period of years. In this the third cause of the inequality of days was hidden up to now, because the revolution of the equator was found regular in respect to the mean and regular equinox but not in respect to the apparent equinoxes, which—as is clear enough—are not wholly regular. Therefore the doubling of the 10 "times" makes $1^1/_3$ hours, by which sometimes the longer days can exceed the shorter.

These things can perhaps be neglected this side of manifest error in connexion with the annual progress of the sun and rather slow movement of the fixed stars; but on account of the speed of the moon—by reason of which an inexactitude of $^5/_6$° in the movement of the sun can cause error—they are by no means to be neglected. Accordingly, the method of reducing the irregular and apparent time—wherein all differences agree—to the equal time, is as follows.

For any period of time proposed there must be sought in each limit of the time—I mean in the beginning and the end—the mean position of the sun with respect to the mean equinox according to its regular movement which we called composite, and also the true apparent position with respect to the true equinox; and we must consider how many "times" the right ascensions [97b] at midday or midnight have amounted to, or how many "times" intervened between the first true position and the second true position. For if they are equal to the degrees between the two mean positions, then the apparent time assumed will be equal to the mean time. But if the "times" exceed, the excess should be added to the given time; and if they are deficient, the deficiency should be subtracted from the apparent time. For if we take the sums and remainders, we shall have the time reduced to equality by taking for one "time" four minutes of an hour or ten seconds of a minute of a day. But if the equal time is given, and you want to know how much apparent time corresponds to it, you will do the contrary.

Now for the first Olympiad we have the mean position of the sun at 90°59' in relation to the mean spring equinox, on noon of the first day of Hekatombaion, the first month by the Athenian calendar, and at 0°36' of Cancer in relation to the apparent equinox. But for the years of Our Lord we have the mean movement of the sun at 8°2' of Capricorn and the true movement at 8°48' of the same. Therefore 178 "times" 54' ascend in the right sphere from 0°36' of Cancer to 8°48' of Capricorn, and they exceed the distance of the mean positions by 1 "times" 51', which make 7 minutes of an hour. And so for the rest, by means of which the course of the moon can be examined most accurately: we shall speak of that in the following book.

BOOK FOUR

[98ᵃ] Since in the preceding book, to the extent that our mediocrity was able, we explained the appearances due to the movement of the Earth around the sun, and we proposed by that same means to determine the movements of all the planets; the circular movement of the moon interrupts us now and does so of necessity because through her in particular, who shares in both night and day, the positions of the stars are apprehended and examined; then, because she alone of all the planets refers her revolutions however irregular directly to the centre of the earth and is most closely akin to the earth. And on that account, in so far as she is considered in herself, she does not indicate anything about the mobility of the Earth—except perhaps in the case of the daily movement; and for that reason the ancients believed that the Earth was the centre of the world and the centre common to all revolutions. In our explanation of the circular movement of the moon we do not differ from the ancients as regards the opinion that it takes place around the Earth. But we shall bring forward certain things which are different from what we received from our elders and are more consonant; by means of them we shall try to set up the movement of the moon with more certitude, in so far as that is possible.

1. The Hypotheses of the Circles of the Moon According to the Opinion of the Ancients

Accordingly the movement of the moon has the following property: it does not follow the ecliptic but follows an incline proper to itself, which bisects the ecliptic and is in turn bisected by it, and from this line of intersection the moon crosses over into both latitudes. These facts are as firmly established as the solstices in the annual movement of the sun. As the year belongs to the sun, so the month belongs to the moon. Now the middle positions at the sections are called (by some) ecliptic; by others, nodes—and the conjunctions and oppositions of the sun and moon occurring at those positions are called ecliptic. [98ᵇ] For there are not any other points common to both circles except these in which the eclipses of the sun and moon can take place. For in other places the divagation of the moon keeps the sun and moon from opposing one another with their lights; but, as they pass by, they do not block one another. Moreover, the orbital circle of the moon with its four "hinges" or cardinal points revolves obliquely around the centre of the Earth in a regular movement of approximately 3' per day, and it completes its revolution in 19 years. Accordingly the moon is perceived always to move eastward in this orbital circle and in its plane, but sometimes with least velocity and at other times with greatest velocity. For it is slower, the higher up it is; and faster, the nearer to Earth; and this fact can be apprehended more easily in the case of the moon than in that of any other planet on account of the nearness of the moon.

Accordingly the ancients understood that change in velocity to occur on account of an epicycle; in running around this epicycle the moon, when in the upper semicircle, subtracts from the regular movement, but when in the lower semicircle, it adds the same amount to it. Besides, it has been demonstrated that those things which take place through an epicycle can take place through an eccentric circle. But the ancients chose the epicycle because the moon seemed to admit to a twofold irregularity. For when it was at the highest or the lowest apsis of the epicycle, there was no apparent difference from regular movement. But around the point of contact of the epicycle and the greater circle there was a variable difference, for the difference was far greater when the half moon was waxing or waning than when there was a full moon; and this in a fixed and orderly succession. Wherefore they thought that the circle in which the epicycle moved was not homocentric with the Earth; but that there was an eccentric circle carrying an epicycle in which the moon was moved in accordance with the law that in all mean oppositions and conjunctions of the sun and moon the epicycle should be at the apogee of the eccentric circle but in the mean quadrants of the (synodic) circle[1] at the perigee of the eccentric circle. Therefore they imagined two equal and mutually opposing movements around the centre of the Earth—namely, that of the epicycle eastward and that of the centre of the eccentric circle and its apsides westward, with the line of the mean position of the sun always half-way between them. And in this way the epicycle traverses the eccentric circle twice a month.

And in order that these things may be brought before our eyes, let *ABCD* be the oblique lunar circle homocentric with the Earth. Let it be quadrisected by the diameters *AEC* and *BED*; and let *E* be the centre of the Earth. Now on line *AC* there will be the mean conjunction of the sun and moon and at the same position and time the apogee of the eccentric circle—whose centre is *F*—and the centre [99ª] of the epicycle *MN*. Now let the apogee of the eccentric circle be moved as far westward as the epicycle eastward, and let them both move regularly around *E* in regular and monthly revolutions as measured by the mean conjunctions or oppositions of the sun. And let line *AEC* of the mean position of the sun be always half way between them; and furthermore let the moon move westward from the apogee of the

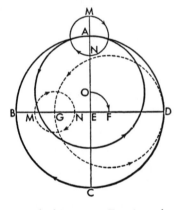

epicycle. For astronomers think that the appearances agree with this set-up. For since the epicycle in the time of half a month moves the distance of a semicircle away from the sun but completes a full revolution from the apogee of the eccentric circle; as a consequence,

---

[1]The synodic circle is the lunar cycle of revolution with respect to the sun.

at the midpoint of this time—when the moon is half full—the moon and the apogee are in opposition to one another along diameter *BD*; and the epicycle is at the perigee of the eccentric circle, as at point *G* where—having become nearer to the Earth—it makes greater differences of irregularity. For when equal magnitudes are set out at unequal intervals, the one which is nearer to the eye appears the greater. Accordingly the differences will be least when the epicycle is at *A*, but greatest when it is at *G*, since the diameter *MN* of the epicycle will be to line *AE* in its least ratio but will be to *GE* in a greater ratio than to all the other lines which are found in the rest of the positions, since *GE* is the shortest line and *AE* or its equal *DE* the longest of all those lines which can be extended from the centre of the Earth to the eccentric circle.

2. On the Inadequacy of Those Assumptions

Our predecessors assumed that such a composition of circles was consonant with lunar appearances. But if we consider the thing itself rather carefully, we shall not find this hypothesis very fitting or adequate, as we can prove by reason and sense. For while they admit that the movement of the centre of the epicycle is regular around the centre of the Earth, they must also admit that the movement is irregular on its own eccentric circle which it describes. If—for example—it is assumed that

angle *AEB* = angle *AED* = 45°,

so that by addition

angle *BED* = 90°;

and if the centre of the epicycle is taken in *G* [99b] and *GF* is joined; it is clear that

angle *GFD* > angle *GEF*,

the exterior than the interior and opposite angle. Wherefore the dissimilar arcs *DAB* and *DG* are both described during one period of time, so that when

arc *DAB* = 90°;

arc *DG* > 90°,

and arc *DG* has been described by the centre of the epicycle during this same time. But it is clear at half moon

arc *DAB* = arc *DG* = 180° :

therefore the movement of the epicycle on the eccentric circle which it describes is not regular.

But if this is so, what shall we reply to the axiom: *The movement of the heavenly bodies is regular except for seeming irregular with respect to appearances;* if the apparent regular movement of the epicycle is really irregular and takes place utterly contrary to the principle set up and assumed? But if you say that the epicycle moves regularly around the centre of the Earth and that that takes care sufficiently of the regularity, then what sort of regularity will that be which occurs in a circle foreign to the epicycle, in which its movement does not exist, and not in its own eccentric circle?

We also are amazed at the fact that they mean the regularity of the moon in its epicycle to be understood not in relation to the centre of the Earth, namely, in respect to line *EGM*, to which the regularity having to do with the centre of the epicycle should rightly be referred, but in relation to some other different point, which has the Earth midway between it and the centre of the eccentric circle, and that line *IGH* is, as it were, the index of the regularity of the moon in the epicycle. And that shows well enough that this movement is really irregular. For the appearances which in part follow upon this hypothesis force this admission. And now that the moon traverses its own epicycle irregularly, we may mark what the line of reasoning would be like if we should try to confirm the irregularity of apparent movement by means of real irregularities. For what else shall we be doing except giving a hold to those who detract from this art?

Furthermore, experience and sense-perception teach us that the parallaxes of the moon are not consonant with those which the ratio of the circles promises. For the parallaxes, which are called commutations, take place on account of the magnitude of the Earth being evident in the neighbourhood of the moon. For since the straight lines which are extended from the centre of the Earth and its surface do not appear parallel but [100ᵃ] in accord with a manifest inclination cut one another in the body of the moon, they are necessarily able to make for irregularity in the apparent movement of the moon, so that the moon is seen in a different position by those viewing it obliquely along the convexity of the Earth and by those who behold the moon from the centre or vertex (of the Earth). Accordingly such parallaxes vary in proportion to the distance of the moon from the Earth. For by the consensus of all the mathematicians the greatest distance is $64^1/_6$ units whereof the radius of the Earth is one unit; but in accordance with the commensurability of these things the least distance should be 33ᴾ33', so that the moon would move towards us through approximately half the total distance— and by the ensuing proportion it was necessary for the parallaxes at greatest and least distance to differ from one another in the ratio of the squares.[1] But we see that those parallaxes, which occur at the time of the half moon waxing or waning, even in the perigee of the epicycle differ slightly or not at all from those which occur at the eclipses of the sun and moon, as we shall show satisfactorily in the proper place.

---

[1] Literally, in duplicate ratio.

But the body itself of the moon makes perfectly clear that error, because for the same reason it would appear twice as large and twice as small in its diameter. But just as circles are in the ratio of the squares[1] of their diameters, the moon should seem almost four times greater in its quadratures when nearest the earth than when opposite the sun, if it were a full moon shining; but since a half moon is shining, nevertheless it should shine with twice the area of light as a full moon there—although the contrary of this is self-evident. If someone who is not content with simple sight wishes to make an experiment with the dioptra of Hipparchus or some other instruments by which the diameter of the moon may be determined, he will find that the diameter does not vary except in so far as the epicycle without the eccentric circle demands. For that reason Menelaus and Timochares in investigating the fixed stars by means of the positions of the moon did not hesitate to use the same lunar diameter always as $^1/_2$°, which the moon was seen to occupy most of the time.

3. ANOTHER THEORY OF THE MOVEMENT OF THE MOON

In this way it is perfectly clear that it is not an eccentricity which makes the epicycle appear greater and smaller, but some other relation of circles. [100ᵇ] For let *AB* be the epicycle which

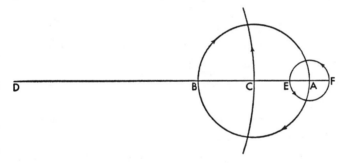

we shall call first and greater; and let *C* be its centre. Let *D* be the centre of the Earth, and from *D* let the straight line *DC* be extended to the highest apsis of the epicycle; and with *A* as centre let another small epicycle *EF* also be described—and all this in the same plane of the oblique circle of the moon. Now let *C* be moved eastward but *A* westward; and again let the moon be moved eastward from *F* the upper part of *EF*. And let such an order be kept that when line *DE* is one with line of the mean position of the sun, the moon is always nearest to centre *C*, *i.e.*, is in point *E*, but in the quadratures is farthest away at *F*. I say that the lunar appearances agree with this set-up.

For it follows that twice a month the moon runs around epicycle *EF*, during which time *C* makes one revolution with respect to the mean position of the sun. And the new and the full moon will be seen to cause the least circle, namely, that whereof the radius is *CE*; but the moon in its quadratures will cause the greatest circle with radius *CF*; and thus again in the conjunctions and oppositions it will make lesser differences between the regular and the apparent movement, but in the quadratures greater differences by

---

[1]Literally, in duplicate ratio.

means of similar but unequal arcs around centre *E*. And since the centre of the epicycle is always in a circle homocentric with the Earth, it will not exhibit such diverse parallaxes but parallaxes in conformity with the epicycle. And the reason will be evident why the body of the moon is seen somehow similar to itself; and all the other things which are perceived in the movement of the moon will come about in this way.

We shall demonstrate them successively by means of our hypothesis, although the same things can take place through eccentric circles, as in the case of the sun—the due proportion being kept. Now we shall take our start from the regular movements, as we did above, without which the irregular movement cannot be separated out. But here there is no small difficulty on account of the parallaxes, which we mentioned; and for that reason the position (of the moon) is not observable by means of astrolabes and other such instruments. But the kindness of nature makes provision for human longing even in this respect, so that the position (of the moon) is more surely determinable through its eclipses than by means of instruments and without any suspicion of error. [101ª] For since the other parts of the world are pure and are filled with the light of day, it stands to reason that night is nothing except the shadow of the Earth, which has the figure of a cone and ends in a point. Falling upon this cone, the moon is dimmed; and, when placed at the midpoint of its darkness, is understood to have arrived without any doubt at the position opposite the sun. But the eclipses of the sun, which take place when the moon moves in front of it, do not offer such a certain determination of the position of the moon. For it happens that the conjunction of the sun and moon is seen by us at some given time, although in relation to the centre of the Earth the conjunction has passed or has not yet taken place, because of the aforesaid parrallax. And accordingly in different parts of the Earth we view the same eclipse of the sun as unequal in magnitude and duration and not similar in all respects. But in the case of the eclipses of the moon no such hindrance occurs, since the Earth transmits the axis of that blotting shadow through its centre from the sun; and for that reason the eclipses of the moon are best fitted for determining the course of the moon with the utmost certainty.

4. ON THE REVOLUTIONS OF THE MOON AND ITS PARTICULAR MOVEMENTS

Among the ancients who cared to hand these things down to posterity by means of numbers was found Meton the Anthenian, who flourished around the 37th Olympiad. He reported that in 19 solar years there were 235 full months, whence that great ἐννεαδεκάτερις year, *i.e.*, year of nineteen years, was called the Metontic year. That number was so suitable that it was set up publicly in the market place at Athens and other famous cities, and even down to the present it has remained in common use, because they think that through it the beginnings and ends of the months are established in a sure order and that through it also the solar year of $365^1/_4$ days is commensurable with the months. Hence the Callippic period of 76 years, in which there is an intercalation of 19 days and which they call the Callippic year.

But Hipparchus discovered through careful study that in 304 years there was an excess of a total day, and that (the Callippic year) was verifiable only when the solar year was $^1/_{300}$ of a day smaller. And so by some men that year was called the great year of Hipparchus, [101^b] in which there were 3760 full months. These years are called more simply and crassly, so to speak, Minerva's, when the recurrences of anomaly and of latitude are also sought, for the sake of which that same Hipparchus was making further investigations. For by comparing the readings which he took in making careful observation of the lunar eclipses with those which he had got from the Chaldees, he determined the time in which the revolutions of the months and of the anomaly recurred simultaneously to be 345 Egyptian years 82 days and 1 hour; and during that time there were 4267 full months and 4573 cycles of anomaly. Therefore when the number of months has been reduced to days and there are 126,007 days 1 hour, one month is found equal to 29 days 31 minutes 50 seconds 8 thirds 9 fourths 20 fifths. According to that ratio the movement during any time is manifest. For the division of the 360° of revolution of one month by the number of days in a month produces a daily movement of the moon in relation to the sum of 12°11'26"41"'20""18"""'. The multiplication of that by 365 makes—in addition to the 12 revolutions—an annual movement of 129°37'21"28"'29"".

Furthermore, since the 4267 months and 4573 cycles of anomaly are given in numbers which are composite with respect to one another, that is, as being numbered by the common measure of 17, the ratio of 4267 months to 4573 cycles of anomaly will in least terms be the ratio of 251 to 269; and by Euclid, X, 15, we shall have the proportion of the revolution of the moon to the movement of anomaly in that ratio. Accordingly, when we have multiplied the (annual) movement of the moon by 269 and divided the product by 251, the quotient will be annual movement of anomaly of 13 full revolutions and 88°43'8"40"'20"" and hence a daily movement of 13°3'53"56"'29"".

But the revolution in latitude has another ratio. For it does not agree with the prescribed time in which the anomaly has recurred; but we understand that the latitude of the moon has returned only at that time when a later eclipse of the moon is in every respect similar and equal to an earlier, so that both obscurations are in the same part of the moon and are equal, *i.e.*, in magnitude and duration. And this happens when the distances of the moon from the highest or the lowest apsis are equal. For then the moon is understood to have traversed equal shadows in equal time. [102ᵃ] Now according to Hipparchus such a returning occurs once in 5458 months, to which there correspond 5923 revolutions of latitude. And by that ratio the particular movements of latitude in years and days will be established, as in the case of the others. For when we have multiplied the movement of the moon away from the sun by 5923 months and divided the product by 5458, we shall have an annual movement of the moon in latitude of 13 revolutions 148°42'46"49'"3"" and a daily movement of 13°13'45"39'"40"".

Hipparchus gave this as the rate of the regular movements of the moon, and no one before him had made a closer approximation. Nevertheless the succeeding ages did not show these movements absolved by all the same numbers. For Ptolemy found the same mean movement away from the sun as did Hipparchus, but an annual movement of anomaly deficient with respect to the former in 1"11'"39"" and an annual movement of latitude with an excess of 53'"41"". But now after the passage of many ages since Hipparchus we also found a mean annual movement deficient in 1" 2'"49"" and a movement of anomaly deficient in only 24'"49"". Moreover, there is an excess of 1"1'"44"" in the movement in latitude. And so the regular movement of the moon, whereby it differs from the terrestrial movement, will be an annual movement of 129°37'22"32'"40"", a movement of anomaly of 88°43'9"5'"9"" and a movement in latitude of 148°42'45"17'"21"".

MOVEMENT OF THE MOON IN YEARS AND PERIODS OF SIXTY YEARS

Egyptian Years	Movement 60°	°	′	″	‴		Egyptian Years	Movement 60°	°	′	″	‴
1	2	9	37	22	36		31	0	58	18	40	48
2	4	19	14	45	12		32	3	7	56	3	25
3	0	28	52	7	49		33	5	17	33	26	1
4	2	38	29	30	25		34	1	27	10	48	38
5	4	48	6	53	2		35	3	36	48	11	14
6	0	57	44	15	38		36	5	46	25	33	51
7	3	7	21	38	14		37	1	56	2	56	27
8	5	16	59	0	51		38	4	15	40	19	3
9	0	26	36	23	27		39	0	25	17	41	40
10	3	36	13	46	4		40	2	34	55	4	16
11	5	45	51	8	40		41	4	44	32	26	53
12	1	55	28	31	17		42	0	53	9	49	29
13	4	5	5	53	53		43	2	3	47	12	5
14	0	14	43	16	29		44	5	13	24	34	42
15	2	24	20	39	6		45	1	22	1	57	18
16	4	33	58	1	42		46	3	32	39	19	55
17	0	43	35	24	19		47	5	41	16	42	31
18	2	53	12	46	55		48	1	51	54	5	8
19	5	2	50	9	31		49	4	1	31	27	44
20	1	12	27	32	8		50	0	10	8	50	20
21	3	22	4	54	44		51	2	20	46	12	57
22	5	31	42	17	21		52	4	30	23	35	33
23	1	41	19	39	57		53	0	39	0	58	10
24	3	50	57	2	34		54	2	49	38	20	46
25	0	0	34	25	10		55	4	58	15	43	22
26	2	10	11	47	46		56	0	8	53	5	59
27	4	19	49	10	23		57	3	18	30	28	35
28	0	29	26	32	59		58	5	27	7	51	12
29	2	39	3	55	36		59	1	37	45	13	48
30	4	48	41	18	12		60	3	47	22	36	25

Position of the Birth of Christ—209°58′

MOVEMENT OF THE MOON IN DAYS AND PERIODS OF SIXTY DAYS

Days	Movement 60°	°	′	″	‴		Days	Movement 60°	°	′	″	‴
1	0	12	11	26	41		31	6	17	54	47	26
2	0	24	22	53	23		32	6	30	6	14	8
3	0	36	34	20	4		33	6	42	17	40	49
4	0	48	45	46	46		34	6	54	29	7	31
5	1	0	57	13	27		35	7	6	40	34	12
6	1	13	8	40	9		36	7	18	52	0	54
7	1	25	20	6	50		37	7	31	3	27	35
8	1	37	31	33	32		38	7	43	14	54	17
9	1	49	43	0	13		39	7	55	26	20	58
10	2	1	54	26	55		40	8	7	37	47	40
11	2	14	5	53	36		41	8	19	49	14	21
12	2	26	17	20	18		42	8	32	0	41	3
13	2	38	28	47	0		43	8	44	12	7	44
14	2	50	40	13	41		44	8	56	23	34	26
15	3	2	51	40	22		45	9	8	35	1	7
16	3	15	3	7	4		46	9	20	46	27	49
17	3	27	14	33	45		47	9	32	57	54	30
18	3	39	26	0	27		48	9	45	9	21	12
19	3	51	37	27	8		49	9	57	20	47	53
20	4	3	48	53	50		50	10	9	32	14	35
21	4	16	0	20	31		51	10	21	43	41	16
22	4	28	11	47	13		52	10	33	55	7	58
23	4	40	23	13	54		53	10	46	6	34	40
24	4	52	34	40	36		54	10	58	18	1	21
25	5	4	46	7	17		55	11	10	29	28	2
26	5	16	57	33	59		56	11	22	40	54	43
27	5	29	9	0	40		57	11	34	52	21	25
28	5	41	20	27	22		58	11	47	3	48	7
29	5	53	31	54	3		59	11	59	15	14	48
30	6	5	43	20	45		60	12	11	26	41	31

Position of the Birth of Christ—209°58′

MOVEMENT OF ANOMALY OF THE MOON IN YEARS AND PERIODS OF SIXTY YEARS

Egyptian Years	Movement					Egyptian Years	Movement				
	60°	°	′	″	‴		60°	°	′	″	‴
1	1	28	43	9	7	31	3	50	17	42	44
2	2	57	26	18	14	32	5	19	0	51	52
3	4	26	9	27	21	33	0	47	43	0	59
4	5	54	52	36	29	34	2	16	27	10	6
5	1	23	35	45	36	35	3	45	10	19	13
6	2	52	18	54	43	36	5	13	53	28	21
7	4	21	2	3	59	37	0	42	36	37	28
8	5	49	45	12	58	38	2	11	19	46	35
9	1	18	28	22	5	39	3	40	2	55	42
10	2	47	11	31	12	40	5	8	46	4	50
11	4	15	54	40	19	41	0	37	29	13	57
12	5	44	37	49	27	42	2	6	12	23	4
13	1	13	20	58	34	43	3	34	55	32	11
14	2	42	4	7	41	44	5	3	38	41	19
15	4	10	47	16	48	45	0	32	21	50	26
16	5	39	30	25	56	46	2	1	4	59	33
17	1	8	13	35	3	47	3	29	48	8	40
18	2	36	56	44	10	48	4	58	31	17	48
19	4	5	39	53	17	49	0	27	14	26	55
20	5	34	23	2	25	50	1	55	57	36	2
21	1	3	6	11	32	51	3	24	40	45	9
22	2	31	49	20	39	52	4	53	23	54	17
23	4	0	32	29	46	53	0	22	7	3	24
24	5	29	15	38	54	54	1	50	50	12	31
25	0	57	58	48	1	55	3	19	33	21	38
26	2	26	41	57	8	56	4	48	16	30	46
27	3	55	25	6	15	57	0	16	59	39	53
28	5	24	8	15	23	58	1	45	42	49	0
29	0	52	51	24	30	59	3	14	25	58	7
30	2	21	34	33	37	60	4	43	9	7	15

Position of the Birth of Christ—207° 7′

MOVEMENT OF LUNAR ANOMALY IN PERIODS OF SIXTY DAYS

Days	Movement					Days	Movement				
	60°	°	′	″	‴		60°	°	′	″	‴
1	0	13	3	53	56	31	6	45	0	52	11
2	0	26	7	47	53	32	6	58	4	46	8
3	0	39	11	41	49	33	7	11	8	40	4
4	0	52	15	35	46	34	7	24	12	34	1
5	1	5	19	29	42	35	7	37	16	27	57
6	1	18	23	23	39	36	7	50	20	21	54
7	1	31	27	17	35	37	8	3	24	15	50
8	1	44	31	11	32	38	8	16	28	9	47
9	1	57	35	5	28	39	8	29	32	3	43
10	2	10	38	59	25	40	8	42	35	57	40
11	2	23	42	53	21	41	8	55	39	51	36
12	2	36	46	47	18	42	9	8	43	45	33
13	2	49	50	41	14	43	9	21	47	39	29
14	3	2	54	35	11	44	9	34	51	33	26
15	3	15	58	29	7	45	9	47	55	27	22
16	3	29	2	23	4	46	10	0	59	21	19
17	3	42	6	17	0	47	10	14	3	15	15
18	3	55	10	10	57	48	10	27	7	9	12
19	4	8	14	4	53	49	10	40	11	3	8
20	4	21	17	58	50	50	10	53	14	57	5
21	4	34	21	52	46	51	11	6	18	51	1
22	4	47	25	46	43	52	11	19	22	44	58
23	5	0	29	40	39	53	11	32	26	38	54
24	5	13	33	34	36	54	11	45	30	32	51
25	5	26	37	28	32	55	11	58	34	26	47
26	5	39	41	22	29	56	12	11	38	20	44
27	5	52	45	16	25	57	12	24	42	14	40
28	6	5	49	10	22	58	12	37	46	8	37
29	6	18	53	4	18	59	12	50	50	2	33
30	6	31	56	58	15	60	13	3	53	56	30

Position of the Birth of Christ—207° 7′

LUNAR MOVEMENT IN LATITUDE IN YEARS AND PERIODS OF SIXTY YEARS

Egyptian Years	Movement 60°	°	′	″	‴
1	2	28	42	45	17
2	4	57	25	30	34
3	1	26	8	15	52
4	3	54	51	1	9
5	0	23	33	46	26
6	2	52	16	31	44
7	5	20	59	17	1
8	1	49	42	2	18
9	4	18	24	47	36
10	0	47	7	32	53
11	3	15	50	18	10
12	5	44	33	3	28
13	2	13	15	48	45
14	4	41	58	34	2
15	1	10	41	19	20
16	3	39	24	4	37
17	0	8	6	49	54
18	2	36	49	35	12
19	5	5	32	20	29
20	1	34	15	5	46
21	4	2	57	51	4
22	0	31	40	36	21
23	3	0	23	21	38
24	5	29	6	6	56
25	1	57	48	52	13
26	4	26	31	37	30
27	0	55	14	22	48
28	3	23	57	8	5
29	5	52	39	53	22
30	2	21	21	38	40
31	4	50	5	23	57
32	1	18	48	9	14
33	3	47	30	54	32
34	0	16	13	39	48
35	2	44	56	25	6
36	5	13	39	10	24
37	1	42	21	55	41
38	4	11	4	40	58
39	0	39	47	26	16
40	3	8	30	11	33
41	5	37	12	56	50
42	2	5	55	42	8
43	4	34	38	27	25
44	1	3	21	12	42
45	3	32	3	58	0
46	0	0	46	43	17
47	2	29	29	28	34
48	4	58	12	13	52
49	1	26	54	59	8
50	3	55	37	44	26
51	0	24	20	29	44
52	2	53	3	15	1
53	5	21	46	0	18
54	1	50	28	45	36
55	4	19	11	30	53
56	0	47	54	16	10
57	3	16	37	1	28
58	5	45	19	46	45
59	2	14	2	32	2
60	4	42	45	17	21

Position of the Birth of Christ—129 °45′

MOVEMENT IN LATITUDE OF THE MOON IN DAYS AND PERIODS OF SIXTY DAYS

Days	Movement 60°	°	′	″	‴
1	0	13	13	45	39
2	0	26	27	31	18
3	0	39	41	16	58
4	0	52	55	2	37
5	1	6	8	48	16
6	1	19	22	33	56
7	1	32	36	19	35
8	1	45	50	5	14
9	1	59	3	50	54
10	2	12	17	36	33
11	2	25	31	22	13
12	2	38	45	7	52
13	2	51	58	53	31
14	3	5	12	39	11
15	3	18	26	24	50
16	3	31	40	10	29
17	3	44	53	56	9
18	3	58	7	41	48
19	4	11	21	27	28
20	4	24	35	13	7
21	4	37	48	58	46
22	4	51	2	44	26
23	5	4	16	30	5
24	5	17	30	15	44
25	5	30	44	1	24
26	5	43	57	47	3
27	5	57	11	32	43
28	6	10	25	18	22
29	6	23	39	4	1
30	6	36	52	49	41
31	6	50	6	35	20
32	7	3	20	20	59
33	7	16	34	6	39
34	7	29	47	52	18
35	7	43	1	37	58
36	7	56	15	23	37
37	8	9	29	9	16
38	8	22	42	54	56
39	8	35	56	40	35
40	8	49	10	26	14
41	9	2	24	11	54
42	9	15	37	57	33
43	9	28	51	43	13
44	9	42	5	28	52
45	9	55	19	14	31
46	10	8	33	0	11
47	10	21	46	45	50
48	10	35	0	31	29
49	10	48	14	17	9
50	11	1	28	2	48
51	11	14	41	48	28
52	11	27	55	34	7
53	11	41	9	19	46
54	11	54	23	5	26
55	12	7	36	51	5
56	12	20	50	36	44
57	12	34	4	22	24
58	12	47	18	8	3
59	13	0	31	53	43
60	13	13	45	39	22

Position of the Birth of Christ—129 °45′

5. DEMONSTRATION OF THE FIRST IRREGULARITY OF THE MOON WHICH OCCURS AT THE NEW AND AT THE FULL MOON

[105ᵇ] We have set out the regular movements of the moon, according as they can be known by us at present. Now we must approach the ratio of irregularity which we shall demonstrate by way of the epicycle, and first the irregularity which occurs in the conjunction and oppositions with the sun, in connexion with which the ancient mathematicians exercised their amazing genius in triads of lunar eclipses. We shall also follow the road thus prepared for us by them, and we shall take three eclipses carefully observed by Ptolemy and compare them with three others noted with no less care, in order to examine the regular movements already set out, to see if they have been set out correctly. In explaining them we shall in imitation of the ancients employ as regular the mean movement of the sun and moon away from the position of the spring equinox, since the variation which occurs on account of the irregular precession of equinoxes is not perceptible in such a short time or even in ten years.

Accordingly, Ptolemy took as first the eclipse occurring in the 17th year of Hadrian's reign, after the close of the 20th day of the month Pauni by the Egyptian calendar; and it was the year of Our Lord 133 on the 6th day of May or the day before the Nones. There was a total eclipse, the midtime of which was a quarter of an equal hour before midnight at Alexandria; but at Frauenburg or Cracow it was an hour and a quarter before the midnight which the seventh day followed; and the sun was at $12^1/_4°$ of Taurus, but according to the mean movement at 12°21' of Taurus.

He says that the second occurred in the 19th year of Hadrian, when two days of Chiach—the fourth Egyptian month had passed: that was in the year of Our Lord 134, 13 days before the Kalends of November. There was an eclipse from the north covering ten twelfths of its diameter. The midtime was one equatorial hour before midnight at Alexandria, but two hours before midnight at Cracow; and the sun was at $25^1/_6°$ of Libra but by its mean movement at 26°43' of the same.

The third eclipse occurred in the 20th year of Hadrian, when 19 days of Pharmuthi—the eighth Egyptian month—had passed; in the year of Our Lord [106ᵃ] 135, when the 6th day of March had passed. The moon was again eclipsed in the north to the extent of half its diameter. The midtime was four equatorial hours past midnight at Alexandria, but at Cracow it was three hours after midnight, that morning being the Nones of March. At that time the sun was at $14^1/_2°$ of Pisces, but by its mean movement at 11°44' of Pisces.

Now it is clear that in the middle space of time between the first and the second eclipse the moon traversed as much space as the sun in its apparent movement—not counting the full circles—*i.e.*, 161°55'; and between the second and the third eclipse, 138°55'. Now in the first interval there were 1 year 166 days $23^3/_4$ equal hours according to apparent time, but by corrected time $23^5/_8$ hours; but in the second interval 1 year 137 days 5 hours simply, but $5^1/_2$ hours correctly.

And during the first interval the regular movement of the sun and the moon measured as one—not counting the circles—was 169°37', and there was a movement of anomaly of 110°21'; in the second interval the similarly regular movement of the sun and the moon was 137°34' and there was a movement of anomaly of 81°36'. Therefore it is clear that during the first interval the 110°21' of the epicycle subtract 7°42' from the mean movement of the moon; and during the second interval the 81°36' of the epicycle add 1°21'.

With these things thus before us, let there be described the lunar epicycle *ABC*, in which the first eclipse of the moon is at *A*, the second at *B*, and the remaining one at

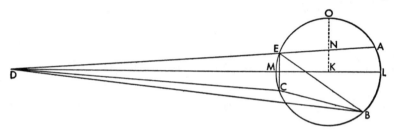

C, and in the order as above let the transit of the moon be understood as occurring westward. And let

arc *AB* = 110°21',

hence

– add. *AB* = 7°42',

as we said; and let

arc *BC* = 81°36',

hence

+ add. *BDC* = 1°21'.

And, as the remainder of the circle,

arc *CA* = 168°3'

and it adds the remainder of the additosubtraction, *i.e.*,

+ add. *CDA* = 6°21'.

Since on the ecliptic

arc *AB* = 7°42',

therefore

angle *ADB* = 7°42',

where 2 rt. angles = 180°.

But

angle *ADB* = 15°24',
[106b] where 2 rt. angles = 360°.

And, as on the circumference and as an exterior angle of triangle *BDE*,

angle *AEB* = 110°21' :

therefore

angle *EBD* = 94°57'.

But the sides of triangles whose angles are given are themselves given:

DE = 147,396

and

BE = 26,798,

where diameter of circle circumscribing triangle = 200,000.

Again, since on the ecliptic

arc *AEC* = 6°21',
angle *EDC* = 6°21',
where 2 rt. angles = 180°.

But

angle *EDC* = 12°42',
where 2 rt. angles = 360°.

And

angle *AEC* = 191°57'.

And

angle *ECD* = angle *AEC* – angle *CDE* = 179°15'.

Therefore the sides are given:

DE = 199,996

and

CE = 22,120,

where the diameter of the circle circumscribing triangle = 200,000.

But

CE = 16,302

and

BE = 26,798,

where *DE* = 147,396.

Again, since in triangle *BEC*

side *BE* is given,
side *EC* is given,

and

$$\text{angle } CEB = 81°36'$$
and hence
$$\text{arc } BC = 81°36' :$$
therefore, by the proofs concerning plane triangles,
$$\text{side } BC = 17,960.$$
But since the diameter of the epicycle = 200,000,
and
$$\text{arc } BC = 81°36',$$
$$\text{chord } BC = 130,684.$$
And in accordance with the ratio given
$$ED = 1,072,684$$
and
$$CE = 118,637,$$
and
$$\text{arc } CB = 72°46'10".$$
But, by construction,
$$\text{arc } CEA = 168°3'.$$
Therefore, by subtraction,
$$\text{arc } EA = 95°16'50"$$
and
$$\text{chord } EA = 147,786.$$
Hence by addition
$$\text{line } AED = 1,220,470.$$
But since segment EA is less than a semicircle, the centre of the epicycle will not be in [107^a] it but in the remainder $ABCE$. Therefore let K be the centre, and let $DMKL$ be drawn through both apsides, and let L be the highest apsis and M the lowest, Now, by Euclid, III, 36, it is clear that
$$\text{rect. } AD, DE = \text{rect. } LD, DM.$$
Now since LM, the diameter of the circle—to which DM is added in a straight line—is bisected at K, then
$$\text{rect. } LD, DM + \text{sq. } KM = \text{sq. } DK.$$
Therefore
$$DK = 1,148,556$$
$$\text{where } KL = 100,000;$$
and on that account,
$$LK = 8,706$$
$$\text{where } DKL = 100,000$$

and *LK* is the radius of the epicycle. Having done that, draw *KNO* perpendicular to *AD*. Since *KD*, *DE*, and *EA* have their ratios to one another given in the parts whereof *LK* = 100,000, and since

$$NE = {}^{1}/_{2} AE = 73,893 :$$

therefore, by addition,

$$DEN = 1,146,577.$$

But in triangle *DKN*

side *DK* is given,

side *ND* is given,

and

angle *N* = 90°;

on that account, at the centre,

angle *NKD* = 86°38$^{1}/_{2}$'

and

arc *MEO* = 86°38$^{1}/_{2}$'.

Hence,

arc *LAO* = 180° – arc *NEO* = 93°21$^{1}/_{2}$'.

Now

arc *OA* = $^{1}/_{2}$ arc *AOE* = 47°38$^{1}/_{2}$';

and

arc *LA* = arc *LAO* – arc *OA* = 45°43',

which is the distance—or position of anomaly—of the moon from the highest apsis of the epicycle at the first eclipse. But

arc *AB* = 110°21'.

Accordingly, by subtraction,

arc *LB* = 64°38',

which is the anomaly at the second eclipse. And by addition

arc *LBC* = 146°14',

where the third eclipse falls. Now it will also be clear that since

angle *DKN* = 86°38$^{1}/_{2}$',

where 4 rt. angles = 360°,

angle *KDN* = 90° – angle *DKN* = 3°21$^{1}/_{2}$';

and that is the additosubtraction which the anomaly adds at the first eclipse. Now

angle *ADB* = 7°42';

therefore, by subtraction,

angle *LDB* = 4°20$^{1}/_{2}$',

which arc *LB* subtracts from the regular movement of the moon at the second eclipse. And since

[107^b] angle *BDC* = 1°21',

and therefore, by subtraction,

angle *CDM* = 2°49',

the subtractive additosubtraction caused by arc *LBC* at the third eclipse; therefore the mean position of the moon, *i.e.*, of centre *K*, at the first eclipse was 9°53' of Scorpio, because its apparent position was at 13°15' of Scorpio; and that was the number of degrees of the sun diametrically opposite in Taurus. And thus the mean movement of the moon at the second eclipse was at 29$\frac{1}{2}$° of Aries; and in the third eclipse, at 17°4' of Virgo. Moreover, the regular distances of the moon from the sun were 177°33' for the first eclipse, 182°47' for the second, 185°20' for the last. So Ptolemy.

Following his example, let us now proceed to a third trinity of eclipses of the moon, which were painstakingly observed by us. The first was in the year of Our Lord 1511, after October 6th had passed. The moon began to be eclipsed 1$\frac{1}{8}$ equal hours before midnight, and was completely restored 2$\frac{1}{3}$ hours after midnight, and in this way the middle of the eclipse was at $\frac{7}{12}$ hours after midnight—the morning following being the Nones of October, the 7th. There was a total eclipse, while the sun was in 22°25' of Libra but by regular movement at 24°13' of Libra.

We observed the second eclipse in the year of Our Lord 1522, in the month of September, after the lapse of five days. The eclipse was total, and began at $\frac{2}{5}$ equal hours before midnight, but its midpoint occurred 1$\frac{1}{3}$ hours after midnight, which the 6th day followed—the 8th day before the Ides of September. The sun was in 22$\frac{1}{5}$° of Virgo but, according to its regular movement, in 23°59' of Virgo.

We observed the third in the year of Our Lord 1523, at the close of August 25th. It began 2$\frac{4}{5}$ hours after midnight, was a total eclipse, and the midtime was 4$\frac{5}{12}$ hours after the midnight prior to the 7th day before the Kalends of September. The sun was in 11°21' of Virgo but according to its mean movement at 13°2' of Virgo.

And here it is also manifest that the distance between the true positions of the sun and the moon from the first eclipse to the second was 329°47', [108^a] but from the second to the third it was 349°9'. Now the time from the first eclipse to the second was 10 equal years 337 days $\frac{3}{4}$ hours according to apparent time, but by corrected equal time $\frac{4}{5}$ hours. From the second to the third there were 354 days 3 hours 5 minutes; but according to equal time 3 hours 9 minutes.

During the first interval the mean movement of the sun and the moon measured as one—not counting the complete circles—amounted to 334°47', and the movement of anomaly to 250°36', subtracting approximately 5° from the regular movement; in the second interval the mean movement of the sun and moon was 346°10'; and the movement of anomaly was 306°43', adding 2°59' to the mean movement.

Now let *ABC* be the epicycle, and let 21 be the position of the moon at the middle of the first eclipse, *B* at the second, *C* at the third. And let the movement of the

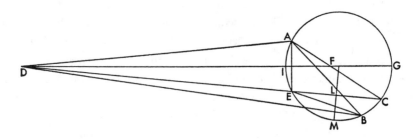

epicycle be understood as proceeding from *C* to *B* and from *B* to *A*, *i.e.*, from above, westward, and from below, eastward. And

arc *ACB* = 250°36',

and, as we said, it subtracts 5° from the mean movement during the first interval of time. But

arc *BAC* = 306°43',

which adds 2°59' to the mean movement of the moon; and accordingly by subtraction the remainder

arc *AC* = 197°19',

which subtracts the remaining 2°1'. But since arc *AC* is greater than a semicircle and is subtractive, then it must contain the highest apsis. For the highest apsis cannot be in area *BA* or *CBA*, which are additive and each less than a semicircle; but the lesser movement is placed by the apogee. Therefore let *D* be taken opposite as the centre of the Earth; and let *AD*, *DB*, *DEC*, *AB*, *AE*, and *EB* be joined.

Now since in triangle *DBE*

exterior angle *CEB* = 53°17',

because angle *CEB* intercepts arc *CB*, and

arc *CB* = 360° − arc *BAC*;

and since, as at the centre,

angle *BDE* = 2°59',

but, as at the circumference,

angle *BDE* = 5°58';

and since, therefore, by subtraction,

angle *EBD* = 47°19';

wherefore

side *BE* = 1042

and

side *DE* = 8024

= [108^b] where radius of circle circumscribing the triangle = 10,000.
Similarly, as standing on arc *AC* of the circumference,

angle *AEC* = 197°19',

and, as at the centre,

angle *ADC* = 2°1',

but, as on the circumference,

angle *ADC* = 4°2';

therefore, by subtraction,

angle *DAE* = 193°17',

where 2 rt. angles = 360°.

Therefore the sides are also given in the parts whereof the radius of the circle circumscribing triangle *ADE* = 10,000:

AE = 702

and

DE = 19,865 :

but whereas

DE = 8,024,
AE = 283

and

BE = 1042.

Therefore once more we shall have triangle *ABE*, wherein

side *AE* is given,
side *EB* is given,

and

angle *AEB* = 250°36',

where 2 rt. angles = 360°.

Accordingly by what we have shown concerning plane triangles

AB = 1,227

where EB = 1,042.

Accordingly in this way we have got hold of the ratios of the three lines *AB*, *EB*, and *ED*; and hence they will become manifest in terms of the parts whereof the radius of the epicycle = 10,000:

ch. AB = 16,323,
ED = 106,751,

and

ch. EB = 13,853.

Whence also

arc EB = 87°41';

209

and
$$\text{arc } EBC = \text{arc } EB + \text{arc } BC = 140°58';$$
and
$$\text{ch. } CE = 18,851,$$
and, by addition,
$$CED = 125,602.$$

Now let the centre of the epicycle be set forth: it necessarily falls in segment *EAC* as being greater than a semicircle. And let *F* be the centre; and let *DIFG* be extended in a straight line through both apsides, *I* the lowest and *G* the highest. Again it is clear that
$$\text{rect. } CD, DE = \text{rect. } GD, DI.$$
But
$$\text{rect. } GD, DI + \text{sq. } FI = \text{sq. } DF.$$
Therefore
$$DIF = 116,226,$$
$$\text{where } FG = 10,000.$$

Accordingly
$$FG = 8,604,$$
$$\text{where } DF = 100,000,$$
—which agrees with what we find reported by most of our predecessors after Ptolemy's time.

[109ª] Now from centre *F* let *FL* be drawn at right angles to *EC* and extended in the straight line *FLM*. It will bisect *CE* at point *I*. Now since
$$\text{line } ED = 106,751$$
and
$$1/_2 CE = LF = 9,426;$$
therefore, by addition,
$$DEL = 116,177$$
$$\text{where } FG = 10,000$$
$$\text{and where } DF = 116,226.$$

Therefore, in triangle *DFL*
$$\text{side } DF \text{ is given,}$$
$$\text{side } DL \text{ is given,}$$
$$\text{angle } DFL = 88°21',$$
and, by subtraction,
$$\text{angle } FDL = 1°39';$$
and similarly
$$\text{arc } IEM = 88°21'$$

and
$$\text{arc } MC = {}^1/_2 \text{ arc } EBC = 70°29';$$
hence, by addition,
$$\text{arc } IMC = 158°50',$$
and
$$\text{arc } GC = 180° - \text{arc } IMC = 21°10'.$$
And this was the distance of the moon from the apogee of the epicycle, or the position of anomaly at the third eclipse. And at the second eclipse
$$\text{arc } GCB = 74°27';$$
and at the first eclipse
$$\text{arc } GBA = 183°51'.$$
Again at the third eclipse, and as at the centre,
$$\text{angle } IDE = 1°39',$$
which is the subtractive additosubtraction. And at the second eclipse
$$\text{angle } IDB = 4°38',$$
which is still a subtractive addition-and-subtraction, because
$$\text{angle } IDB = \text{angle } GDC + \text{angle } CDB = 1°39' + 2°59'.$$
And accordingly
$$\text{angle } ADI = \text{angle } ADB - \text{angle } IDB = 5° - 4°38' = 22'$$
which are added to the regular movement at the first eclipse.

For that reason the position of regular movement of the moon in the first eclipse was 22°3' of Aries, but the position of the apparent movement was at 22°25'; and the sun was opposite, at the same number of degrees of Libra. In this way too the mean position of the moon in the second eclipse was at 26°50' of Pisces, but in the third eclipse, at 13° of Pisces, and the mean lunar movement by which it is separated from the annual movement of the Earth, was 177°50' at the first eclipse; at the second eclipse, 182°51'; and at the third eclipse, 179°58'.

6. CONFIRMATION OF WHAT HAS BEEN SET OUT CONCERNING THE MOON'S MOVEMENTS OF ANOMALY IN LONGITUDE

Moreover, by means of these things which are set out concerning the eclipses of the moon, it will be possible to test whether we have set out the regular movements of the moon correctly. For it was shown that in the second of the two eclipses the distance of the moon from the sun was 182°47', and (the movement) of anomaly was 64°38'; [109b] but in the second of those eclipses occurring in our time the movement of the moon away from the sun was 182°51' but (the movement) of anomaly was 74°27'. It is clear that in the intervening time there were 17,166 full months and as it were a movement of 4' and a movement of anomaly—not counting the whole cycles—of 9°49'. Now the time which intervenes between the 19th year of Hadrian on the 2nd day of the Egyptian month Chiach 2 hours before midnight, followed by the 3rd day of the month, and the 1522nd year of Our Lord on September 5th, $1^1/_3$ hours after midnight amounts to 1388 Egyptian years 302 days $3^1/_3$ hours by apparent time; and when corrected, 3 hours 34 minutes after midnight.

And in that time after the 17,165 complete revolutions of equal months there was according to Ptolemy and Hipparchus a movement away from the sun of 359°38'. And according to Hipparchus the movement of anomaly was 9°39', but according to Ptolemy 9°11'. Accordingly the lunar movement away from the sun calculated by Hipparchus and Ptolemy is deficient in 26', and the movement of anomaly of Ptolemy and of Hipparchus is deficient in 38'. These minutes swell our movements, and are consonant with the numbers which we have set out.

7. ON THE POSITIONS OF THE MOON IN LONGITUDE AND ANOMALY

Now we shall speak of these things, as above; and here we are to determine positions for the established beginnings of calendar years of the Olympiads, of the years of Alexander, Caesar, and Our Lord, and any additional one desired. Therefore if we consider the second of the three ancient eclipses—the one which occurred in the 19th year of Hadrian, on the 2nd day of the Egyptian month Chiach, one equatorial hour before midnight at Alexandria but for us under the Cracow meridian at 2 hours before midnight—we shall find from the beginning of the years of Our Lord to this movement 133 Egyptian years 325 days 22 hours simply, but 21 hours 37 minutes correctly. During this time the movement of the moon according to our calculation was 332°49' and (the movement) of anomaly was 217°32'. [110ª] And when they have been subtracted from the findings for the eclipse, each from its own kind, there remain 209°58' as the mean position of the moon away from the sun, and a position of anomaly of 207°7' at the beginning of the years of Our Lord at midnight before the Kalends of January.

Again (from the 1st Olympiad) to the beginning of the years of Our Lord, there are 193 Olympiads 2 years $194^1/_2$ days, which make 775 Egyptian years $12^1/_2$ days, but by corrected time 12 hours 11 minutes. Similarly from the death of Alexander to the birth of Christ, they compute 323 Egyptian years $130^1/_2$ days by apparent time, but by corrected time 12 hours 16 minutes. And from Caesar to Christ there are 45 Egyptian years 12 days, in which the ratios of equal and apparent time agree.

Accordingly when we have deducted the movements corresponding to the intervals of time from the positions at the birth of Christ, by subtracting single items from single items, we shall have for noon of the 1st day of the month Hekatombaion of the

1st Olympiad a regular lunar distance from the sun of 39°43' and a distance of anomaly of 46°20'.

At the beginning of the years of Alexander at noon on the first day of the month Thoth the moon was 310°44' distant from the sun, and the movement of anomaly was 85°41'.

And at the beginning of the years of Julius Caesar at midnight before the Kalends of January the moon was 350°39' distant from the sun, and the movement of anomaly was 17°58'. All this with reference to the Cracow meridian, since Gynopolis—commonly called Frauenburg—where we took our observations at the mouth of the Vistula, lies under this meridian, as the eclipses of the sun and moon observed in both places at the same time teach us; and Dyrrhachium in Macedonia—which was called Epidamnum in antiquity—is also under this meridian.

8. On the Second Irregularity of the Moon and What Ratio the First Epicycle Has to the Second

Accordingly, in this way the regular movement of the moon together with its first irregularity has been demonstrated. Now we must inquire into what ratio the first epicycle has to the second and both of them to the distance of the centre of the Earth. But, as we said, the greatest difference (between regular and apparent movement) is found in the mean quadratures when the half moon is waxing or waning, and that difference is $7^2/_3$°, [110^b] as even the observations of the ancients record. For they were making observations of the time in which the half moon had nearly reached the mean distance of the epicycle and was in the neighbourhood of the tangent from the centre of the Earth—and that is easily perceptible by means of the calculus set forth above. And as the moon was then at about 90° of the ecliptic measured from its rising or setting, they were aware of the error which the parallax could bring into the movement of longitude. For at that time the circle through the vertex of the horizon divides the ecliptic at right angles and does not admit any parallax in longitude but the parallax falls wholly in latitude. Then by means of the astrolabe they determined the position of the moon in relation to the sun. When they made their comparison, the moon was found to differ from its regular movement by $7^2/_3$°, as we said, instead of by 5°.

Now let epicycle AB be described; and let its centre be C. Let the centre of the Earth be D, and from D let the straight line DBCA be extended. Let A be the apogee of the epicycle, B the perigee; and let DE be drawn tangent to the epicycle, and let CE be joined. Accordingly since the greatest additosubtraction is at the tangent and in this case is 7°40', and hence

$$\text{angle } BDE = 7°40',$$

and

$$\text{angle } CED = 90°,$$

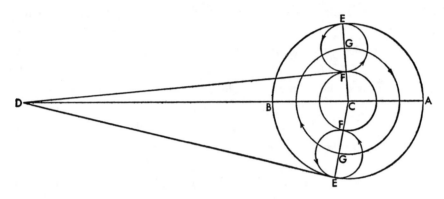

as being at the point of tangency of circle *AB*:
wherefore

$$CE = 1334,$$

where radius *CD* = 10,000.

But at the full moon it was much shorter, that is,

$$CE \fallingdotseq 860.$$

Let *CE* be such again, and let

$$CF = 860.$$

Point *F*, which the new moon and the full moon occupy, will be circumcurrent around the same centre; and accordingly, by subtraction,

$$FE = 474$$

and is the diameter of the second epicycle. Let *FE* be bisected at centre *G*, and by addition

$$CFG = 1097$$

and will be the radius of the circle which the centre of the second epicycle describes. And so it is established that

$$CG : GE = 1097 : 237,$$

where *CD* = 10,000.

9. ON THE REMAINING DIFFERENCE, BY REASON OF WHICH THE MOON SEEMS TO MOVE IRREGULARLY AWAY FROM THE HIGHEST APSIS OF ITS EPICYCLE

[111ª] By this induction it is given to understand how the moon is moved irregularly in its first epicycle, and that its greatest difference occurs when the half moon is horn-shaped or gibbous. Once more, let *AB* be that first epicycle, which the centre of the second epicycle describes through its mean movement; let *C* be the centre, *A* the highest apsis, and *B* the lowest. Let point *E* be taken anywhere in the circumference, and let *CE* be joined. Now let

$$CE : EF = 1097 : 237.$$

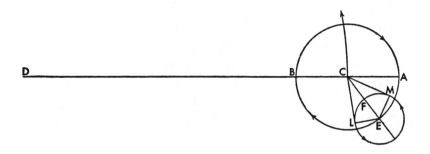

And with radius *EF* let the second epicycle be described around centre *E*. And let the straight lines *CL* and *CM* be drawn tangent to it on both sides. Let the movement of the small epicycle be from *A* to *E*, *i.e.*, from above, westward; and let the movement of the moon be from *F* to *L*, still westward. Accordingly it is clear that, since movement *AE* is regular, the second epicycle by virtue of its motion *FL* adds arc *EL* to the regular movement and by virtue of *MF* subtracts from the regular movement. But since in triangle *CEL*

$$\text{angle } L = 90°,$$

and

$$EL = 237,$$
$$\text{where } CE = 1,097.$$

Therefore

$$EL = 2,160,$$
$$\text{where } CE = 10,000.$$

And by the table

$$EL = {}^{1}/_{2} \text{ ch. 2 } ECL.$$

And

$$\text{angle } ECL = \text{angle } MCF,$$

since the triangles are similar and equal. And that is the greatest difference by which the moon varies in its movement from the highest apsis of the first epicycle. It occurs when the moon by its mean movement is 38°46' distant on either side of the line of mean movement of the Earth. And so it is perfectly clear that these greatest addito-subtractions occur at a mean distance of 38°46' between the sun and moon, and at the same distance on either side of the mean opposition.

10. HOW THE APPARENT MOVEMENT OF THE MOON IS DEMONSTRATED FROM THE REGULAR MOVEMENTS

[111ᵇ] Having seen all that first, we now wish to show by means of diagrams how the regular and apparent movements of the moon are separated out from those regular lunar movements which were set before us, taking our example from among the observations of Hipparchus, so that in this way our teaching may at the same time be confirmed experimentally. Accordingly in the 197th year from the death of Alexander, on the 17th day of Pauni, which is the 10th month in the Egyptian calendar, when $9^{1}/_{3}$ hours of the day had passed at Rhodes, Hipparchus by an observation of the sun and moon through an astrolabe found that they were 48°6' distant from one another, and that the moon was to the east of the sun. And since he judged that the position of the sun was in $10^{9}/_{10}$° of Cancer, as a consequence the moon was at 29° of Leo. At that time 29° of Scorpio was rising, and 10° of Virgo was in the middle of the heavens over Rhodes, which has an elevation of the north pole of 36°. By this argument it is clear that the moon, which was situated at 90° of the ecliptic from the meridian had at that time admitted no parallax in vision or else one imperceptible in longitude. But since this observation was made $3^{1}/_{3}$ hours after midday of the 17th—which corresponds at Rhodes to four equatorial hours—at Cracow it was $3^{1}/_{6}$ equatorial hours after midday in accordance with the distance which makes Rhodes a sixth of an hour nearer to us than Alexandria. Accordingly from the death of Alexander there were 196 years, 286 days, $3^{1}/_{6}$ hours simply, but $3^{1}/_{3}$ hours by equal time. At that time the sun by its mean movement had arrived at 12°3' of Cancer, but by its apparent movement at 10°40' of Cancer, whence the moon appeared in truth to be at 28°37' of Leo. But the regular movement of the moon according to the monthly revolution was at 45°9', and the movement of anomaly was 333° away from the highest apsis by our calculations.

With this example before us let us describe the first epicycle *AB*. Let *C* be its centre. [112ª] Let *ACB* be its diameter, and let *ACB* be extended as *ABD* in a straight line to the centre of the Earth. And in the epicycle, let

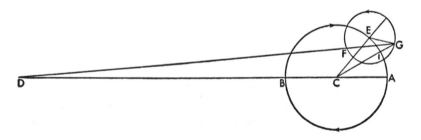

arc *ABE* = 333°.

Let *CE* be joined and again cut in *F*, so that

EF = 237,

where *EC* = 1,097.

And with *E* as centre and *EF* as radius, let *FG* the epicycle of the epicycle be described. Let the moon be at point *G*; and let

arc *FG* = 90°18',

in the ratio of double to the regular movement away from the sun, which was 45°9'. And let CG, EG, and DG be joined. Accordingly, since in triangle CEG two sides are given:

$$CE = 1,097$$

and

$$EG = EF = 237;$$

and

$$\text{angle } GEC = 90°18';$$

therefore, by what we have shown concerning plane triangles

$$\text{side } CG = 1,123$$

and

$$\text{angle } ECG = 12°11'.$$

By this means there are determined arc EI and the additive additosubtraction caused by the anomaly; and, by addition,

$$\text{arc } ABEI = 345°11'.$$

And, by subtraction,

$$\text{angle } GCA = 14°49',$$

and is the true distance of the moon from the highest apsis of epicycle AB; and

$$\text{angle } BCG = 165°11'.$$

Wherefore in triangle GDC two sides are given also.

$$GC = 1,123,$$

$$\text{where } CD = 10,000;$$

and

$$\text{angle } GCD = 165°11'.$$

Hence

$$\text{angle } CDG = 1°29',$$

the additosubtraction which was added to the mean movement of the moon. Hence the true distance of the moon from the mean movement of the sun is 46°34', and its apparent position is at 28°37' of Leo, and is 47°57' distant from the true position of the sun. And there is a deficiency of 9' according to Hipparchus' observation.

But in order that no one on that account should suspect that either his investigation or ours is wrong—though the deficiency is very slight—nevertheless I shall show that neither he nor we committed any error but that this is the way things rightly are. For if we recollect that the lunar circle which the moon itself follows is oblique, we will admit that it produces some sort of longitudinal irregularity in the ecliptic, especially around the mean positions, which lie between the northern and the southern limits of latitude and the ecliptic sections, in approximately the same way as between the oblique [112b] ecliptic and the equator, as we expounded in connection with the inequality of the natural day. And so if we transfer the ratios to the orbital circle of the moon, which Ptolemy recorded as being inclined to the ecliptic, we find that at those positions the ratios cause a 7' difference in longitude in relation to the ecliptic—and twice that difference is 14'; and the difference increases and decreases proportionally, since when the sun and moon are a quadrant of a circle distant from one another, and if the limit of northern or southern latitude is at the midpoint between them, then the arc intercepted on the ecliptic will be 14' greater than a quadrant of the lunar circle; and conversely in the other quadrants, which the ecliptic sections halve, the circles through the poles of the ecliptic intercept an arc that much less than a quadrant. So in the present case. Since the moon was in the neighbourhood of the mean position between the southern limit of latitude and the ascending ecliptic section—which the moderns call the head of the Dragon—and the sun had already passed by the other descending section—which they call the tail; it is not surprising if when the moon's distance of 47°57' in its own orbital circle was referred to the ecliptic, it increased by at least 7', without the fact of the sun declining in the west causing any subtractive parallax of vision. We shall speak more clearly of all that in our explanation of the parallaxes.

And so the distance of the luminaries, which Hipparchus determined by his instrument as being 48°6', agrees with our calculation perfectly and as it were unanimously.

11. ON THE TABLE OF THE LUNAR ADDITIONS-AND-SUBTRACTIONS OR *AEQUATIONES*

Accordingly, I judge that the mode of determining the motions of the moon is understood generally from this example, since in triangle *CEG*, the two sides *GE* and *CE* always remain the same. But we determine the remaining side *GC* together with angle *ECG*—which is the additosubtraction to be used in correcting the anomaly—according to angle *GEC* which changes continually but which is given. Then in triangle *CDG*, since the two sides *DC* and *GC* together with angle *DCG* have been computed, in the same way angle *D* at the centre of the Earth, the angular difference between the true and the regular movement, becomes established.

So that these things may be at hand, [113ª] we shall set out a table of the additosubtractions, which will contain six columns. For after the two columns of common numbers of the circle, in the third column will come the additosubtractions which are caused by the small epicycle and vary the regular movement of the first epicycle in accordance with the bi-monthly revolution. Then, we shall leave the fourth column vacant for the time being, and fill up the fifth column first, in which we shall inscribe the additosubtractions caused by the first and greater epicycle which occur at the mean conjunctions and oppositions of the sun and moon, and the greatest is 4°56'. In the next to the last column will be placed the numbers whereby the additosubtractions which occur at half moon exceed the former additosubtractions, and the greatest of these excesses is 2°44'. But in order that the other excesses may be evaluated, the proportional minutes have been worked out, and this is the ratio of them. For we have taken 2°44' as 60 minutes in relation to any other excesses occurring at the point of tangency of the (small) epicycle (with the line from the centre of the Earth).

In this way, in the same example,

$$CG = 1123,$$

where $CD = 10,000$.

And that makes the greatest additosubtraction at the point of tangency of the (small) epicycle (with the line from the centre of the Earth) to be 6°29', which exceeds the first additosubtraction by 1°33'. But

$$2°44' : 1°33' = 60' : 34';$$

and so we have the ratio of the excess which occurs at the semicircle of the small epicycle to the excess corresponding to the given arc of 90°18'. Therefore we shall write down 34 minutes in that part of the table corresponding to 90°. In this way we shall find the minutes which are proportional to the arcs inscribed in the table; and we shall set them out in the fourth column.

Finally we have added the degrees of northern and southern latitude in the last column, and we shall speak of them below. For convenience and ease of operation advise us to put them in this order.

TABLE OF ADDITIONS-AND-SUBTRACTIONS OF THE MOON

Common Numbers		Additions-and-subtractions caused by small epicycle		Proportional Minutes	Additions-and-subtractions caused by great epicycle		Excesses		Degrees of Northern Latitude	
Deg.	Deg.	Deg.	Min.		Deg.	Min.	Deg.	Min.	Deg.	Min.
3	357	0	51	0	0	14	0	7	4	59
6	354	1	40	0	0	28	0	14	4	58
9	351	2	28	1	0	43	0	21	4	56
12	348	3	15	1	0	57	0	28	4	53
15	345	4	1	2	1	11	0	35	4	50
18	342	4	47	3	1	24	0	43	4	45
21	339	5	31	3	1	38	0	50	4	40
24	336	6	13	4	1	51	0	56	4	34
27	333	6	54	5	2	5	1	4	4	27
30	330	7	34	5	2	17	1	12	4	20
33	327	8	10	6	2	30	1	18	4	12
36	324	8	44	7	2	42	1	25	4	3
39	321	9	16	8	2	54	1	30	3	53
42	318	9	47	10	3	6	1	37	3	43
45	315	10	14	11	3	17	1	42	3	32
48	312	10	30	12	3	27	1	48	3	20
51	309	11	0	13	3	38	1	52	3	8
54	306	11	21	15	3	47	1	57	2	56
57	303	11	38	16	3	56	2	2	2	44
60	300	11	50	18	4	5	2	6	2	30
63	297	12	2	19	4	13	2	10	2	16
66	294	12	12	21	4	20	2	15	2	2
69	291	12	18	22	4	27	2	18	1	47
72	288	12	23	24	4	33	2	21	1	33
75	285	12	27	25	4	39	2	25	1	18
78	282	12	28	27	4	43	2	28	1	2
81	279	12	26	28	4	47	2	30	0	47
84	276	12	23	30	4	51	2	34	0	31
87	273	12	17	32	4	53	2	37	0	16
90	270	12	12	34	4	55	2	40	0	0
93	267	12	3	35	4	56	2	42	0	16
96	264	11	53	37	4	56	2	42	0	31
99	261	11	41	38	4	55	2	43	0	47
102	258	11	27	39	4	54	2	43	1	2
105	255	11	10	41	4	51	2	44	1	18
108	252	10	52	42	4	48	2	44	1	33
111	249	10	35	43	4	44	2	43	1	47
114	246	10	17	45	4	39	2	41	2	2
117	243	9	57	46	4	34	2	38	2	16
120	240	9	35	47	4	27	2	35	2	30
123	237	9	13	48	4	20	2	31	2	44
126	234	8	50	49	4	11	2	27	2	56
129	231	8	25	50	4	2	2	22	3	9
132	228	7	59	51	3	53	2	18	3	21
135	225	7	33	52	3	42	2	13	3	32
138	222	7	7	53	3	31	2	8	3	43
141	219	6	38	54	3	19	2	1	3	53
144	216	6	9	55	3	7	1	53	4	3
147	213	5	40	56	2	53	1	46	4	12
150	210	5	11	57	2	40	1	37	4	20
153	207	4	42	57	2	25	1	28	4	27
156	204	4	11	58	2	10	1	20	4	34
159	201	3	41	58	1	55	1	12	4	40
162	198	3	10	59	1	39	1	4	4	45
165	195	2	39	59	1	23	0	53	4	50
168	192	2	7	59	1	7	0	43	4	53
171	189	1	36	60	0	51	0	33	4	56
174	186	1	4	60	0	34	0	22	4	58
177	183	0	32	60	0	17	0	11	4	59
180	180	0	0	60	0	0	0	0	5	0

12. ON THE COMPUTATION OF THE COURSE OF THE MOON

[114^b] Accordingly the method of computing the apparent movement of the moon is clear from what has been shown and is as follows. We shall reduce to equal time the time for which we are seeking the position of the moon proposed to us. By means of the time we shall deduce the mean movements of longitude, anomaly, and latitude—which last we shall also define soon—as we did in the case of the sun, from the given beginning of the years of Our Lord, or from some other beginning, and we shall declare the positions of the single movements at the time set before us. Then we shall seek in the table twice the regular longitude of the moon or twice its angular distance from the sun and[1] the corresponding additosubtraction found in the third column; and we shall note the proportional minutes which are in the next column. Accordingly if the number with which we entered upon the table was found in the first column or is less than 180°, we shall add the additosubtraction to the lunar anomaly; but if it is greater than 180° or is in the second column, the additosubtraction will be subtracted from the anomaly; and we shall have the corrected anomaly of the moon and its true angular distance from the highest apsis.

And entering the table again with this (distance) we shall determine the corresponding additosubtraction in the fifth column and the excess which follows in the sixth column, which the second (the small) epicycle adds (to the additosubtraction), over and above the first epicycle. The proportional part of this excess taken in accordance with the ratio of the 60 minutes is always added to this additosubtraction. The sum is subtracted from the mean movement of longitude or latitude, if the corrected anomaly is less than 180° or a semicircle; and it is added, if the anomaly is greater. And in this way we shall have the true distance of the moon from the mean position of the sun and the corrected movement of latitude. Wherefore the true position of the moon will not be unknown, either its distance from the first star of Aries in the case of the simple movement of the sun or its distance from the spring equinox in the case of the composite movement or the addition of the precession. Finally by means of the corrected movement in latitude we shall have in the seventh and last place of the table the degrees of latitude which measure the distance of the moon from the ecliptic. That latitude will be northern at the time when the movement of latitude is found in the first part of the table, [115^a] *i.e.*, if it is less than 90° or greater than 270°; otherwise it will be following a southern latitude. And so the moon will be coming down from the north to 180°, and afterwards it will be going up from the southern limit, until it has completed the remaining parts of the circle. Thus the apparent course of the moon has somehow as many affairs around the centre of the Earth as the Earth has around the sun.

---

[1]Because the moon traverses the small epicycle twice during one synodic month, the time of one revolution with respect to the sun.

13. How the Movement of Lunar Latitude Is
Examined and Demonstrated

Now too we must give the ratio of the lunar movement in latitude, and it seems more difficult to discover, as it is complicated by more attendant circumstances. For, as we said before, if two eclipses of the moon were similar and equal in all respects, *i.e.*, with the parts eclipsed having the same position to the north or to the south and at the same ascending or descending ecliptic section: its distance from the Earth or from the highest apsis would be equal, since in this harmony the moon is understood to have completed its whole circles of latitude by true movement. For since the shadow of the Earth is conoid, and if a right cone is cut in a plane parallel with the base, the section is a circle which is smaller the greater the distance from the base and greater the shorter the distance from the base, and similarly equal at an equal distance. And so the moon at equal distances from the Earth traverses equal circles of shadow and presents to our vision equal disks of itself. Hence the moon, standing out with equal parts in the same direction according to an equal distance from the centre of the shadow makes us certain of equal latitudes, from which it necessarily follows that the moon has returned to its former position in latitude and is now distant from the same ecliptic node by an equal interval. But that is especially true if the position fulfils two of those conditions. For its approach to the Earth or withdrawal from it changes the total magnitude of the shadow, [115^b] but so slightly that it can hardly be grasped. Accordingly the greater the interval of time between both eclipses, the more definite can we have the movement in latitude of the moon, as was said in the case of the sun.

But since you rarely find two eclipses agreeing in these conditions—and up to now none have come our way—nevertheless we note there is another method which will give us the same result, since—the other conditions remaining—if the moon is eclipsed in different directions and at opposite sections, then it will signify that at the second eclipse the moon has arrived at a position diametrically opposite to the former and in addition to the whole circles has described a semicircle; and that will seem to be satisfactory for investigating the thing.

Accordingly we have found two eclipses fairly close in these respects: the first in the 7th year of Ptolemy Philometor, which was the 150th year of Alexander when—as Claud says—27 days of Phamenoth the 7th month of the Egyptians had passed, in the night which the 28th day followed. And the moon was eclipsed from the beginning of the 8th hour till the end of the 10th hour in Alexandrian nocturnal seasonal hours, to the extent of seven-twelfths of the lunar diameter, and it was eclipsed from the north around a descending section. Therefore the midtime of the eclipse was, he says, 2 seasonal hours after midnight, which make $2^1/_3$ equatorial hours, since the sun was at 6° of Taurus, but $1^1/_3$ hours after midnight at Cracow.

222

We have taken the second eclipse beneath the same Cracow meridian in the year of Our Lord 1519, after the 4th day before the Nones of June, when the sun was at 21° of Gemini. The midtime of the eclipse was 11³/₅ equatorial hours after midday; and the moon was eclipsed for approximately eight-twelfths of its diameter, from the south, at an ascending section.

Accordingly, from the beginning of the years of Alexander (to the first eclipse) there are 149 Egyptian years 206 days 14¹/₃ hours at Alexandria, but at Cracow 13¹/₃ hours according to apparent time, but 13¹/₂ upon correction. At that time the position of anomaly by our calculation, which agreed approximately with Ptolemy's, was at 163°33' of regular movement; and there was a subtractive additosubtraction of 1°23', by which the true position of the moon was exceeded by the regular. But from the established beginning of the years of Alexander to the second eclipse [116ª] there are 1832 Egyptian years 295 days 11 hours 45 minutes by apparent time, but by equal time 11 hours 55 minutes: whence the regular movement of the moon was 182°18'. The position of anomaly was 159°55', but as corrected it was 161°13'; and the additive additosubtraction, by which the regular movement was exceeded by the apparent, was 1°44'.

Accordingly it is clear that in both eclipses the distance of the moon from the Earth was equal, and the sun was approximately at the apogee in both cases, but there was a difference of one-twelfth in the eclipses. But since the diameter of the moon usually occupies approximately ¹/₂°, as we will show afterwards, its twelfth part will be 2¹/₂', which corresponds to approximately ¹/₂° in the oblique circle of the moon at the ecliptic sections. And so the moon was ¹/₂° farther away from the ascending section at the second eclipse than from the descending section at the first eclipse. Hence it is clear that the true movement in latitude of the moon was 179¹/₂° after the complete revolutions. But the lunar anomaly between the first and second eclipse adds 21'—which is the difference between the additosubtractions—to the regular (movement). Accordingly we shall have a regular lunar movement in latitude of 179°51' after the full circles. Now the time between the two eclipses was 1683 years 88 days 22 hours 35 minutes by apparent time, which agreed with the equal (time). During that time there were 40,577 complete equal revolutions and 179°51', which agree with the numbers which we have already set down.

14. ON THE POSITIONS OF LUNAR ANOMALY IN LATITUDE

However, in order to determine the positions of the moon's movement in relation to the established beginnings of calendar years, we have here also assumed two lunar eclipses, not at the same section and not at diametrically opposite parts, as in the foregoing, but at equal distances north or south, and fulfilling all the other requirements,

[116^b] as we said, in accordance with Ptolemy's rule, and in this way we shall solve our problem without any error.

Accordingly, the first eclipse, which we have already used in investigating other movements of the moon, is the one which we said was observed by Claud Ptolemy in the 19th year of Hadrian when two days of the month Chiach had passed, one equatorial hour before midnight at Alexandria, but at Cracow two hours before midnight, which the third day followed. The moon was eclipsed at the midpoint of the eclipse to the extent of ten-twelfths of the diameter, *i.e.*, ten-twelfths from the north, while the sun was at 25°10' of Libra, and the position of lunar anomaly was 64°38' and its subtractive additosubtraction was 4°20' around the descending section.

We made careful observations of the other eclipse at Rome, in the year of Our Lord 1500, after the Nones of November, 2 hours after midnight, and it was the 8th daybreak before the Ides of November. But at Cracow which is 5° to the east, it was 2²/₅ hours after midnight, while the sun was at 23°16' of Scorpio; and once more there was a ten-twelfths eclipse from the north.

Therefore, since the death of Alexander there have passed 1824 Egyptian years 84 days 14 hours 20 minutes by apparent time, but by equal time 14 hours 16 minutes. Accordingly the mean movement of the moon was 174°14', and the lunar anomaly was 294°44', but as corrected it was 291°35'; and there was an additive additosubtraction of 4°27'.

Accordingly it is clear that at both these eclipses the distances of the moon from the highest apsis was approximately equal, and at both times the sun was at its mean apsis, and the magnitude of the shadows was equal. All that makes clear that the latitude of the moon was southern and equal; and hence that the moon was at an equal distance from the sections but was ascending at the second eclipse and descending at the first. Accordingly between both eclipses there are 1366 Egyptian years 358 days 4 hours 20 minutes by apparent time, but by equal time 4 hours 24 minutes, wherein the movement in latitude was 159°55'.

Now let *ABCD* be the oblique circle of the moon; and let *AB* be its diameter and common section with the ecliptic. Let *C* be the northern limit, and *D* the southern; [117^a] *A* the ecliptic section descending, and *B* the ecliptic section ascending. Now let there be taken *AF* and *BE* two equal arcs in the south, according as the first eclipse was at point *F* and the second at *E*. And again let *FK* be the subtractive additosubtraction at the first eclipse, and *EL* the additive additosubtraction at the second.

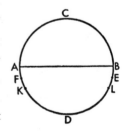

Accordingly, since
$$\text{arc } KL = 159°55'$$
and
$$\text{arc } FK = 4°20'$$
and
$$\text{arc } EL = 4°27';$$
$$\text{arc } FKLE = \text{arcs } FK + KL + LE = 168°42'.$$
And
$$180° - 168°42' = 11°18'.$$
Now
$$\text{arc } AF = \text{arc } BE = \frac{1}{2}(11°18') = 5°39',$$
which is the true distance of the moon from section AB, and on that account
$$\text{arc } AFK = 9°59'.$$
Hence it is clear that K the mean position in latitude is 99°59' away from the northern limit.

From the death of Alexander to this position and time of Ptolemy's observation there are 457 Egyptian years 91 days 10 hours by apparent time, but by equal time 9 hours 54 minutes, during which the mean movement in latitude is 50°59'. And when 50°59' is subtracted from 99°59', there remain 49° for noon on the first day of Thoth the first month by the Egyptian calendar at the beginning of the years of Alexander but on the Cracow meridian. Hence for each of the other beginnings there are given in accordance with the differences of time the positions of the course of the moon in latitude as taken in relation to the northern limit, from which we measure the movement.

Now from the first Olympiad to the death of Alexander there are 451 Egyptian years 247 days—from which in accordance with equality of time 7 minutes of an hour are subtracted—and during that time the progress in latitude was 136°57'. Again from the first Olympiad to Caesar there are 780 Egyptian years 12 hours, but 10 minutes of an hour are added to the equal time; and during that time the movement is 206°53'. Then come 45 years 12 days up to Christ. Accordingly if 136°57' are subtracted from 49° plus the 360° of the circle, there remain 272°3' for noon of the first day of the month Hekatombaion of the first Olympiad.

Now if 206°53' are added to 272°3', the sum will be 118°56' for midnight before the Kalends of January [117$^\text{b}$] at the beginning of the Julian years.

Finally, with the addition of 10°49' the sum becomes 129°45' as the position (at the beginning of the years) of Our Lord similarly at midnight before the Kalends of January.

15. Construction of the Instrument for
Observing Parallaxes

But chance and the hindrance of the lunar parallaxes did not grant to us, as it had to Ptolemy, the occasion of discovering experimentally that the greatest latitude of the moon—in accordance with the angle of section of its orbital circle and the ecliptic—is 5°, whereof the circle is 360°. For he was watching at Alexandria—which has 30°58' as the elevation of the north pole—until the moon should come most near to the vertex of the horizon, namely when it was at the beginning of Cancer and at the northern limit, which can be foreknown by means of calculations. Therefore at that time by means of an instrument which he called the parallacticon and which was constructed for measuring the parallaxes of the moon, he found that the least distance was only $2^1/_8°$ from the vertex, and if any parallax had occurred at this distance it would necessarily have been very slight in such a small spatial interval. Accordingly by the subtraction of $2^1/_8°$ from 30°58'[1] the remainder is 28°50$^1/_2$', which exceeds the greatest obliquity of the ecliptic—which at that time was 23°51'20"—by approximately 5°; and this latitude for the moon is found to agree in every respect with the other particulars.

But the instrument for observing parallaxes consists of three straight-edges. Two of them are of equal length and are at least eight or nine feet long; the third is somewhat longer. The latter and one of the former two are joined to both extremities of the remaining straight-edge, by carefully making holes and fitting cylinders or pivots into them in such a way that while the straight-edges are movable in a plane surface they do not wobble at all at the joints. Now in the longer straight-edge a straight line should be drawn from the centre of its place of joining through its total length, and the line is made equal to the distance between the places of joining (on the other straight-edge) measured as accurately as possible. This line is divided into 1,000 equal parts—or into more, if that can be done; and the division of the remainder should be carried on [118ª] in the same unit parts, until it reaches 1414 parts, which subtend the side of a square that may be inscribed in a circle whose radius has 1,000 parts. It will be all right to cut off as superfluous the remainder of the straight-edge over and above this. In the other ruler too, there should be drawn from the centre of the joining-place a line equal to those 1,000 parts or to the distance between the centres of the two joining-places; the ruler should have eyepieces fastened to it on one side, as in a dioptra, which sight may have passage through. The eyepieces should be so adjusted that the sight-passages do not at all swerve away from the line already drawn the length of the straight-edge, but keep at an equal distance; and provided also that the line as extended from its terminus to the longer ruler can touch the divided line. And in this way by means of the rulers an isosceles triangle is made, the base of which will be along the parts of the divided line.

[1] The elevation of the north pole above the horizon is equal to the declination of the vertex of the horizon from the equator.

Then a pole, which has been divided crosswise in the best manner and well smoothed, should be erected on a firm base. The ruler which has the two joining-places should be affixed to this pole by means of pivots, around which the instrument may swing, like a swinging door, but in such a way that the straight line through the centres of the joining-places will always correspond to the plumbline of the ruler and point towards the vertex of the horizon, as if its axis. Accordingly when a person who wishes to find the distance of some star from the vertex of the horizon has the star itself in full view along a straight line through the eyepieces, then by the application underneath of the ruler with the divided line, he should learn how many unit parts—whereof the diameter of the circle has 20,000—subtend the angle between the line of vision and the axis of the horizon; and by means of the table he will get the sought arc of the great circle passing through the star and the vertex of the horizon.

16. HOW THE PARALLAXES OF THE MOON ARE DETERMINED

By means of this instrument, as we said, Ptolemy found the greatest latitude of the moon to be 5°. Next he turned to observing the parallax and said he discovered that at Alexandria it was 1°7', while the sun was at 5°28' of Libra and the mean movement of the moon away from the sun was 78°13', the regular anomaly was 262°20'; the movement in latitude was 354°40'; the additive additosubtraction was 7°26'; [118^b] and accordingly the position of the moon was at 3°9' of Capricorn. The corrected movement in latitude was 2°6'; the northern latitude of the moon was 4°59'; its declination from the equator was 23°49'; and the latitude of Alexandria was 30°58'. The moon, he says, as seen through the instrument, was approximately in the meridian circle at 50°45' from the vertex of the horizon, *i.e.*, 1°7' more than the computation demanded. Hence by the rule of the ancients concerning the eccentric circle and the epicycle, he shows that the distance of the moon from the centre of the Earth was then 39^{P}45' whereof the radius of the Earth is 1^P; and what next follows from the ratio of the circles, namely that the greatest distance of the moon from the Earth—which they say occurs at a new and at a full moon in the apogee of the epicycle—is 64^{P}10', but the least distance—at the quadratures and at the half moon in the perigee of the epicycle— is only 33^{P}33'. Hence he even evaluated the parallaxes, which occur at about 90° from the vertex (of the horizon): the least at 53'34", and the greatest at 1°43'—as it is possible to see in a broad outline what he built up concerning them. But now it is perfectly obvious to those wishing to consider the question that these things are far otherwise, as we have found out experimentally very often.

However, we shall review two observations, by which it is again made clear that our hypotheses as to the moon have more certitude than his, because they are found to agree better with the appearances and to leave nothing in doubt. In the 1522nd year since the birth of Christ, on the 5th day before the Kalends of October, after the passage of $5^2/_3$ equal hours since midday, at about sunset at Frauenburg we found by means of the parallactic instrument that the centre of the moon, which was in the meridian circle, was 82°50' distant from the vertex of the horizon. Accordingly from the beginning of the years of Our Lord to this hour there were 1522 Egyptian years 284 days $17^2/_3$ hours by apparent time but by equal time 17 hours 24 minutes. Wherefore the apparent position of the sun was by calculation at 13°29' of Libra and the regular movement of the moon away from the sun [119ª] was 87°6'; the regular anomaly was 357°39'; but the true (the corrected) anomaly was 358°40', and it added 7'; and thus the true position of the moon was at 12°32' of Aries. The mean movement in latitude was 197°1' from the northern limit, the true was 197°8'; the southern latitude of the moon was 4°47'; the moon had a declination of 27°41' from the equator; the latitude of the place of our observation was 54°19', and the addition of 54°19' to the lunar declination makes the true distance of the moon from the pole of the horizon to be 82°. Accordingly the 50' not accounted for belong to the parallax, which by Ptolemy's teaching should be 1°17'.

Once more we made another observation at the same place in the 1524th year of Our Lord on the 7th day before the Ides of August 6 hours after midday; and we saw through the same instrument the moon at 82° from the vertex of the horizon. Accordingly from the beginning of the years of Our Lord to this hour there were 1524 Egyptian years 234 days 18 hours (by apparent time) and also 18 hours by exact time. The position of the sun was by calculation at 24°14' of Leo; the mean movement of the moon away from the sun was 97°6'; the regular anomaly was 242°10'; the corrected anomaly was 239°43', adding approximately 7° to the mean movement; wherefore the true position of the moon was at 9°39' of Sagittarius; the mean movement of latitude was 193°19'; the true, 200°17'; the southern latitude of the moon was 4°41'; the southern declination was 26°36', and the addition of 26°36' to 54°19' of the latitude of the place of observation makes the distance of the moon from the pole of the horizon to be 80°55'. But there appeared to be 82°. Accordingly the difference of 1°5' came from the lunar parallax, which according to Ptolemy should have been 1°38' and also according to the theory of the ancients, as the harmonic ratio, which follows from their hypotheses, forces you to admit.

17. Distance of the Moon from the Earth and Demonstration of Their Ratio in Parts Whereof the Radius of the Earth Is the Unit

[119ᵇ] From this it will now be made apparent how great the distance of the moon from the earth is. And without this distance a sure ratio cannot be given for the parallaxes, for they are mutually related. And it will be established in this way. Let AB be a

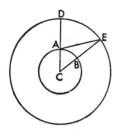

great circle of the Earth, and let C be its centre. Around C let another circle be described in comparison with which the Earth has considerable magnitude, and let this circle be DE. Let D be the pole of the horizon, and let the centre of the moon be at E, so that DE its distance from the vertex is known. Accordingly, since at the first observation,

$$\text{angle } DAE = 82°50'$$

and by calculation

$$\text{angle } ACE = 82°$$

and hence

$$\text{angle } DAE - \text{angle } ACE = 50',$$

which belonged to the parallax; we have triangle ACE with its angles given and therefore with its sides given. For since

$$\text{angle } CAE \text{ is given,}$$
$$\text{side } CE = 99,219$$

where diameter of circle circumscribing triangle $AEC = 100,000$ and

$$AC = 1,454;$$

and

$$CE \doteqdot 68\text{P},$$

where AC, radius of Earth, $= 1\text{P}$.

And this was the distance of the moon from the centre of the Earth at the first observation.

But at the second observation

$$\text{angle } DAE = 82°,$$

as the apparent movement; and, by calculation,

$$\text{angle } ACE = 80°55;$$

and, by subtraction,

$$\text{angle } AEC = 1°5'.$$

Accordingly,

$$\text{side } EC = 99,027$$

and

$$\text{side } AC = 1894$$

where diameter of circle circumscribing triangle = 100,000.

And so

$$CB = 56\text{P}42',$$

where the radius of Earth = 1P.

And that was the distance of the moon.

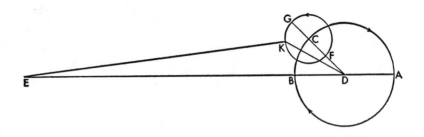

But now let *ABC* be the greater epicycle of the moon; and let its centre be *D*. Let *E* be taken as the centre of the Earth, and from *E* let the straight line *EBDA* be drawn, so that *A* is the apogee and *B* the perigee. Now let

$$\text{arc } ABC = 242°10',$$

in accordance with the computed regularity of the lunar anomaly.
And with *C* as centre let epicycle *FGK* be described, whereon

$$\text{arc } FGK = 194°10',$$

twice the distance of the moon from the sun. And let *DK* be joined.
Thus,

$$\text{angle } GDK = - \text{ add. } [120^{\text{a}}]\ 2°27';$$

and, by subtraction,

$$\text{corr. anomaly} = 59°43',$$

since

$$\text{arc } CDB = \text{arc } ABC - 180° = 62°10',$$

and

$$\text{angle } BEK = 7°,$$

Therefore, in triangle *KDE* the angles are given in the degrees whereof 2 rt. angles = 180°; and the ratio of the sides is also given:

$$DE = 91,856$$

and

$$EK = 86,354,$$

where diameter of circle circumscribing triangle *KDE* = 100,000.
But

$$KE = 94,010$$

$$\text{where } DE = 100,000.$$

Now it was shown above that

$$DF = 8,600$$

and

$$DFG = 13,340.$$

Accordingly it follows from the given ratio that when, as was shown,

$$EK = 56^P42',$$

where radius of the Earth = 1P;

then

$$DE = 60^P18',$$
$$DF = 5^P11',$$
$$DFG = 8^P2';$$

and hence, as extended in a straight line.

$$EDG = 68^1/_3{}^P;$$

and that is the greatest altitude of the half moon. Furthermore,

$$ED - DG = 52°17',$$

which is its least distance. And thus, at its greatest,

$$EDF = 65^1/_2{}^P,$$

which is the altitude occurring at the bright, full moon; and at its least,

$$EDF - DF = 55^P8'.$$

And we should not be moved by the fact that others—and especially those to whom the parallaxes of the moon could not become known except partially, on account of the location of their places—estimate the greatest distance of the new moon and the full moon to be 64P10'. But the greater nearness of the moon to the horizon—for it is clear that the parallaxes are filled out in relation to the horizon—has allowed us to perceive them more perfectly, and we have not found the parallaxes to differ by more than 1' on account of the difference caused by the nearness of the moon to the horizon.

18. ON THE DIAMETER OF THE MOON AND ON THE DIAMETER OF THE TERRESTRIAL SHADOW IN THE PLACE OF PASSAGE OF THE MOON

[120^b] Moreover, the apparent diameters of the moon and the shadow vary with the distance of the moon from the Earth. Wherefore it is pertinent to speak of them. And although the diameters of the sun and the moon are rightly determined through the dioptra of Hipparchus, nevertheless in the case of the moon astronomers judge that this is done with more certainty through some particular eclipses of the moon, in which the moon is at an equal distance from its highest or lowest apsis, especially if at that time the sun too is in the same relative situation, so that the circle of shadow which the moon passes through is found equal—unless the eclipses themselves are unequal in extent. For it is clear that the comparison of the difference in extent of the eclipses with the latitude of the moon shows how much of the circle around the centre of the Earth the diameter of the moon subtends. When that has been perceived, the semidiameter of the shadow is also known.

All this will be made clearer by an example. In this way at the midpoint of the first eclipse $3/_{12}$ of the diameter of the moon was eclipsed; and the moon had a latitude of 47'54"; but at the other eclipse $10/_{12}$ of the diameter was eclipsed, and the latitude was 29'37". The difference between the extent of the eclipses is $7/_{12}$ of the diameter; the difference in latitude is 18'17"; and the $12/_{12}$ are proportional to the 31'20" which the diameter of the moon subtends. Accordingly it is clear that the centre of the moon at the midpoint of the first eclipse was about a quarter of the moon's diameter—or 7'50" of latitude—beyond the shadow if these 7'50" are subtracted from the 47'54" of the total latitude, 40'4" remain as the semidiameter of the shadow; just as at the other eclipse the shadow occupied—in proportion to $1/_3$ of the lunar diameter—10'27" more than the latitude of the centre of the moon. The addition of 29'37" to 10'27" similarly makes the semidiameter of the shadow to be 40'4". And so in accordance with Ptolemy's conclusion, when the sun and moon are in conjunction or opposition at their greatest distance from the Earth, the diameter of the moon is 31'20"—[121ᵃ] as he admits he found the sun's diameter to be through the dioptra of Hipparchus—but the diameter of the shadow is 1°21'20"; and he believed that the diameters were in the ratio of 13 to 5, *i.e.,* the ratio of double plus three-fifths.

19. HOW THE DISTANCES OF THE SUN AND MOON FROM THE EARTH, THEIR DIAMETERS AND THAT OF THE SHADOW AT THE PLACE OF CROSSING OF THE MOON, AND THE AXIS OF THE SHADOW ARE DEMONSTRATED SIMULTANEOUSLY

But even the sun has some parallax; and since it is very slight, it is not perceived so easily, except that the following things are related reciprocally: namely the distance of the sun and moon from the Earth, their diameters and that of the shadow at the crossing of the moon, and the axis of the shadow; and for that reason they are mutually productive of one another in analytical demonstrations. First we shall review Ptolemy's conclusions on these things, and how he demonstrated them, and we shall draw out from them what seems the most true. He assumed $31^1/_3'$ as the apparent diameter of the sun, which he employed without any qualification. He assumed as equal to that the diameter of the full and new moon when at its apogee, which he says was at a distance of 64ᴾ10', whereof the radius of the Earth is 1ᴾ.

From that he demonstrated the rest in this way: Let *ABC* be the circle of the solar globe around centre *D*; and let *EFG* be the circle of the terrestrial globe around its centre *K* at its greatest distance from the sun. Let *AG* and *CE* be straight lines touching both circles, and let them as extended meet at the apex of the shadow, as at point *S*. And let *DKS* be a line through the centres of the sun and the Earth. Moreover, let *AK* and *KC* be drawn, and let *AC* and *GE* be joined, which should hardly differ at all from the diameters on account of the great distance between them. Now on *DKS* let equal segments *LK* and *KM* be taken in proportion to the distance of the moon in the apogee when new and when full: in his opinion 64P10', where *EK* is 1P. Let *QMR* be the diameter of the shadow at this crossing of the moon; and let *NLO* be the diameter of the moon at right angles to *DK*, and let it be extended as *LOP*.

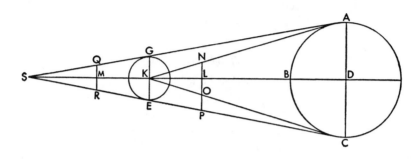

The first problem is to find

$$DK : KE.$$

Accordingly, since

$$\text{angle } NKD = 31^{1}/_{3}',$$
$$\text{where 4 rt. angles} = 360°$$

and [121b]

$$\text{angle } LKO = {}^{1}/_{2} \text{ angle } NKO,$$
$$\text{angle } LKO = 15^{2}/_{3}'.$$

And

$$\text{angle } L = 90°.$$

Accordingly, in triangle *LKO*, which has its angles given,

$$KL : LO \text{ is given,}$$

and

$$LO = 17\text{P}33',$$

and

$$LK = 64\text{P}10',$$
$$\text{where } KE = 1\text{P}$$

And because

$$LO : MR = 5 : 13,$$
$$MR = 45'38''.$$

But since *LOP* and *MR* are parallel to *KE* at equal intervals, on that account

$$LOP + MR = 2 \ KE.$$

And

$$OP = LOP - (MR + LO) - 56'49''.$$

Now by Euclid, VI, 2,

$$EC : PC = KC : OC = KD : LD = KE : OP = 60' : 56'49'.$$

Hence

$$LD = 56'49',$$
$$\text{where } DLK = 1\text{P}.$$

And accordingly, by subtraction,

$$KL = 3'11'.$$

But according as

$$KL = 64\text{P}10',$$
$$\text{where } FK = 1\text{P},$$
$$KD = 1210\text{P}.$$

Now it is also clear that

$$MR = 45'38''.$$

Hence

$$KE : MR \text{ is given}$$

and

$$KMS : MS \text{ is given}.$$

And of the whole KMS

$$KM = 14'22''.$$

And, *separando*,

$$KMS = 268\text{P}$$
$$\text{where } KM = 64\text{P}10'.$$

So in truth Ptolemy.

But others after Ptolemy, since they found that these things did not agree sufficiently with the appearances, published other things concerning all this. None the less they admit that the greatest distance of the full moon and the new moon from the Earth is 64P10'; and that the apparent diameter of the sun at its apogee is $31^1/_3$'. They even grant that the diameter of the shadow at the place of crossing of the moon is as 13 to 5, even as Ptolemy himself. Nevertheless they deny that the apparent diameter of the moon at that time is greater than $29^1/_2$'; and for that reason they put the diameter of the shadow at approximately $1°16^3/_4$'. They hold that it follows from this that at its apogee the distance of the sun from the Earth is 1146P and that the axis of the shadow is 254P, whereof the radius of the Earth is 1P. [122ª] And astronomers attribute these things to the Harranite philosopher (al-Battani) as the discoverer, although they cannot be joined together at all reasonably. We considered that these things must be adjusted and corrected as follows, since we put the apparent diameter of the sun in its apogee at 31'40"—for it should be somewhat greater now than before Ptolemy—and that of the full or the new moon in its highest apsis at 30' and the diameter of the shadow in its crossing at $80^3/_5'$. For astronomers should have a slightly greater ratio than that of 5 to 13, that is to say 150 to 403. And the whole sun is not covered by the moon, unless the moon is at a lesser distance from the Earth than 62P, whereof the radius of the Earth is 1P. For when these things are put down in this way they seem to be connected with one another and the rest in a sure fashion and to be consonant with the apparent eclipses of the sun and moon. And in accordance with the foregoing demonstration:

$$LO = 17'85''$$

where KE radius of Earth = 1P.

And for that reason

$$MR = 46'1'',$$

and

$$OP = 56'51''.$$

And

$$DLK = 1179 \text{ P},$$

the distance from the Earth of the sun at its apogee; and

$$KMS = 265,$$

which is the axis of the shadow.

20. ON THE MAGNITUDE OF THESE THREE CELESTIAL BODIES: THE SUN, MOON, AND EARTH, AND THEIR COMPARISON WITH ONE ANOTHER

Hence it is also manifest that

$$LK : KD = 1 : 18$$

and

$$LO : DC = 1 : 18.$$

Now

$$1 : 18 = 17'8'' : 5P27'$$

where KE = 1P.

And

$$SK : KE = 265\text{P} : 1\text{P} = SKD : DC = 1444\text{P} : 5\text{P}27'.$$

For they are all proportional; and that will be the ratio of the diameters of the sun and Earth. But, as globes are in the ratio of the cubes[1] of their diameters, accordingly

$$(5\text{P}27')^3 - 1617/_8\text{P};$$

and the sun is $1617/_8$ greater than the terrestrial globe.

Again, since

$$\text{moon's radius} = 17'9''$$

[1] *literally*, in the triplicate ratio.

where KE = 1P,

[122b] Earth's diameter: moon's diameter = 7 : 2, *i.e.*, in the triple sesquialter ratio. When the cube[1] of that ratio is taken, it shows that the Earth is $42^7/_8$ greater than the moon. And hence the sun will be $6,999^{62}/_{63}$ greater than the moon.

21. ON THE APPARENT DIAMETER OF THE SUN AND ITS PARALLAXES

But since the same magnitude when farther away appears smaller than when nearer; for that reason it happens that the sun, moon, and the shadow of the Earth vary with their unequal distances from the Earth no less than do their parallaxes. By means of the aforesaid, all these things are easily determinable for any elongation whatsoever. That is first made manifest in the case of the sun. For since we have shown that the Earth at its farthest is 10,323 parts distant from the sun, whereof the radius of the orbital circle of annual revolution = 10,000; and at its nearest the Earth has a distance of 9,678 parts of the remainder of the diameter: accordingly the highest apsis is 1179P whereof the radius of the Earth is 1P, the lowest apsis will be 1105P, and so the mean apsis will be 1142P. Accordingly in the right triangle[2]

$$1,000,000 \div 1179 = 848^3 = {}^1/_2 \text{ ch. } 2 \ (2'55''),$$

which is the small angle of greatest parallax, and that is found around the horizon. Similarly, as the least distance is 1105P,

$$1,000,000 - 1105 = 905^4 = {}^1/_2 \text{ ch. } 2 \ (3'7'');$$

and 3'7" measures the angle of greatest parallax of the lowest apsis. Now it was shown that the diameter of the sun is 5P27', whereof the diameter of the Earth is 1P, and that it appears at the highest apsis as 31'48". For

$$1179\text{P} : 5\text{P}27' = 2,000,000 : 9,245;$$

$$\text{where diameter of circle} = 2,000,000,$$

and

$${}^1/_2 \text{ ch. } 2(31'48'') = 9245.$$

It follows that at the least distance of 1105P there is an apparent diameter of 33'54". Therefore the difference between them is 2'6"; but there is a difference of only 12" [123a] between the parallaxes. Ptolemy considered that both of these differences should be ignored on account of their smallness; for 1' or 2' is not easily perceptible to the senses, much less than are a few seconds perceptible. Wherefore if we keep the greatest parallax of the sun at 3' everywhere, we shall be seen to have made no error.

---

[1] *literally,* the triplicate.
[2] *i.e.,* the right triangle formed by the line joining the centres of the sun and the Earth, the tangent from the centre of the sun to the Earth's surface, and the radius of the Earth to that point of tangency.
[3] *i.e.,* when the highest apsis = 1179P, 1P = 848 whereof radius of circle = 1,000,000.
[4] Similarily, where the lowest apsis = 1105P, 1P = 905 whereof radius of circle = 1,000,000.

Now we shall determine the mean apparent diameters of the sun through its mean distances; or, as do others, through the apparent hourly movement of the sun, which they believe to be to its diameter as 5 to 66 or as 1 to 14$^1/_5$. For its hourly movement is approximately proportional to its distance.

22. ON THE UNEQUAL APPARENT DIAMETER OF THE MOON AND ITS PARALLAXES

A greater diversity in the apparent diameter and parallaxes appears in the case of the moon as being the nearest planet. For since its greatest distance from the Earth is 65$^1/_2$P at new moon and full moon, its least distance—will by the above demonstrations be 55P8'; and the greatest (altitudinal) elongation of the half moon will be 68P21', and the least 52P17'. Accordingly we shall have the parallaxes of the setting or rising moon at these four termini, when we have divided the radius of the circle by the distances of the moon from the Earth: the parallax of the farthest half moon will be 50'18" and that of the farthest new or full moon will be 52'24"; the parallax of the nearest full or new moon will be 62'21" and that of the nearest half moon 65'45".

Furthermore by this the apparent diameters of the moon are established. For it was shown that the diameter of the Earth is to the diameter of the moon as 7 to 2, and the radius of the Earth will be to the diameter of the moon as 7 to 4. Moreover, the parallaxes are in that ratio to the apparent diameters of the moon, since the straight lines, which comprehend the angles of the greater parallaxes, do not differ at all from the apparent diameters at the same crossing of the moon; and the angles, (or arcs of parallax) are approximately proportional to the chords subtending them; and their difference is not perceptible to sense. By this summary it is clear that at the first limit of the parallaxes which have been already set forth the apparent [123^b] diameter of the moon will be 28$^3/_4$'; at the second, approximately 30'; at the third, 35'38"; and at the last limit, 37'34". By the hypothesis of Ptolemy and others the diameter would have been approximately 1°, and so it ought to have been, as the half moon at that time was shedding as much light on the Earth as the full moon would.

23. WHAT THE RATIO OF DIFFERENCE BETWEEN THE SHADOWS OF THE EARTH IS

We have already made clear that
shadow's diameter; moon's diameter = 403 : 150.

For that reason at a full or a new moon, when the sun is at its apogee, the shadow's is found to be 80'36" at its least and 95'44" at its greatest; and the greatest difference is 15'8". Moreover, the shadow of the Earth varies, even in the same place of crossing of the moon, on account of the unequal distance of the Earth from the sun, as follows:

For, as in the foregoing diagram, let *DKS* the straight line through the centres of the sun and the Earth be drawn again, and also *CES* the line of tangency. As was shown, when

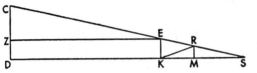

distance $DK = 1179^P$,

where $KE = 1^P$

and

$KM = 62^P$

then the semidiameter of the shadow

$$MR = 46'1''$$

where $KE = 1^P$;

and (*KR* being joined)

$$\text{angle } MKR = 42'32'',$$

which is the angle of sight, and the axis of the shadow

$$KMS = 265^P.$$

Now when the Earth is nearest to the sun, so that

$$DK = 1105^P,$$

we shall evaluate the shadow of the Earth at the same crossing of the moon, as follows: For let *EZ* be drawn parallel to *DK.* Then

$$CZ : ZE = EK : KS.$$

But

$$CZ = 4^P27'$$

and

$$ZE = 1105^P.$$

For

$$ZE = DK$$

and

$$DZ = KE,$$

as *KZ* is a parallelogram. Accordingly

$$KS = 248^P19'$$

where $KE = 1^P.$

Now

$$KM = 62;$$

and accordingly, by subtraction,

$$MS = 186^P19'.$$

But since

$$SM : MR = SK : KE,$$

therefore

$$MR = 45'1'',$$
where [124ª] $KE = 1^P$.

And hence

$$\text{angle } MKR = 41'35'',$$

which is the angle of sight. Whence it happens that on account of the approach and withdrawal of the sun and the Earth, the greatest difference in the diameters of the shadow at the same place of crossing of the moon is 1', whereof $EK = 1^P$, in proportion to an angle of sight of 57'', whereof 4 rt. angles = 360°.

Furthermore, in the first case

$$\text{shadow's diameter : moon's diameter} > 13 : 5;$$

but here

$$\text{shadow's diameter : moon's diameter} < 13 : 5,$$

as 13 : 5 is a sort of mean ratio. Wherefore we shall make but slight error if we employ it as everywhere the same, thus saving labour and following the judgment of the ancients.

24. On the Table of the Particular Parallaxes in the Circle Passing Through the Poles of the Horizon

Moreover, it will not be difficult now to determine all the single parallaxes of the sun and the moon. For let there be drawn again AB the terrestrial circle through 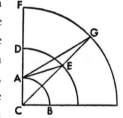 the vertex of the horizon, with C as its centre. And in the same plane let DE be the orbital circle of the moon, FG that of the sun, CDF the line through the vertex of the horizon; and let line CEG be drawn, in which the true positions of the sun and the moon are understood to be, and let the lines of sight AG and AE be joined to those points. Therefore the parallaxes of the sun are measured by angle AGC, those of the moon by angle AEC. Moreover, there is a parallax between the sun and moon which is measured by angle GAE, which is determined according to the difference between angles AGC and AEC. Now let us take angle ACG, with which we wish to compare those angles; and, for example, let

$$\text{angle } ACG = 30°.$$

Now it is clear from what we have shown concerning plane triangles that when

$$\text{line } CG = 1142^P,$$
$$\text{where } AC = 1^P,$$
$$\text{angle } AGC = 1\tfrac{1}{2}',$$

which is the difference between the true and the seeming altitude of the sun. But when

$$\text{angle } ACG = 60°,$$
$$\text{angle } AGC = 2'36''.$$

Everything will be similarly clear as regards the remaining angles.

But in the case of the moon at its four limits: If at the greatest lunar distance from the Earth, wherein, as we said,

$$CE = 68^P21',$$
[124b] where $CA = 1^P$,
angle $DCE = 30°$,
where 4 rt. angles = 360°,

we shall have triangle ACE in which the two sides AC and CE together with angle ACE have been given. From that we find that

$$\text{parallax } AEC = 25'28''.$$

And when

$$CE = 65^1/_2{}^P$$
angle $AEC = 26'36''.$

Similarly in the third case when

$$CE = 55^P8',$$
parallax $AEC = 31'42''.$

Finally, at the least distance when

$$CE = 52^P17',$$
angle $AEC = 33'27''.$

Again, when

$$\text{arc } DE = 60°,$$

the parallaxes in the same order will be as follows:

First parallax = 43'55'',
second parallax = 45'51'',
third parallax = 54$^1/_2$',

and

fourth parallax = 57$^1/_2$'.

We shall inscribe all these things after the order of the subjoined table, which for the sake of convenience we shall extend like all the other tables into a series of thirty rows but proceeding by 6°'s by which twice the arcs from the vertex of the horizon— of which the greatest is 90°—are given to be understood. But we have divided the table into nine columns. For in the first and second will be found the common numbers of the circle. We shall put the parallaxes of the sun in the third, and in the next the lunar parallaxes, and in the fifth column the differences, by which the least parallaxes, which occur at the half moon and at the apogee, are deficient as measured by the parallaxes occurring at the apogee of the full moon or the new moon. The sixth column will contain the parallaxes which the full or bright moon produces at its perigee; and in the next column are the minutes of difference, by which the parallaxes which occur at half

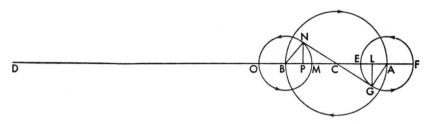

moon when the moon is nearest to us exceed those occurring at half moon in the apogee. Then, the two spaces which are left are reserved for the proportional minutes, by which the parallaxes between these four limits can be computed. We shall set forth these parallaxes, and first in connection with the apogee and the parallaxes which are between the first two limits—as follows.

Let circle [125ª] *AB* be the first epicycle of the moon, and let *C* be its centre. With *D* taken as the centre of the Earth, let the straight line *DBCA* be drawn; and with apogee *A* as centre let the second epicycle *EFG* be described. Now let

arc *EG* = 60°,

and let *AG* and *CG* be joined. Accordingly, since in the foregoing it was shown that

straight line *CE* = 5P11',

where radius of Earth = 1P,

straight line *DC* = 60P 18',

and

straight line *EF* = 2P51',

then in triangle *ACG*

side *GA* = 1P25'

and

side *AC* = 6P36';

and

angle *CAG* is given,

which is the angle comprehended by *GA* and *AC*. Accordingly, by what has been shown concerning plane triangles,

side *CG* = 6P7'.

Accordingly, as extended in a straight line,

DCG = *DCL* = 66P25'.

But

$$DCE = 65^1/_2\text{P}.$$

Therefore, by subtraction,

$$EL \doteqdot 55^1/_2{}',$$

and that is the excess. Moreover, by this given ratio, when

$$DCE = 60\text{P};$$
$$EF = 2\text{P}37',$$

and

$$EL = 46'.$$

Therefore, according as,

$$EF = 60',$$
$$\text{excess } EL \doteqdot 18'.$$

We shall mark these down in the eighth column of the table as corresponding to 60° (in the first column).

We shall show something similar in the case of perigee *B*. Let the second epicycle *MNO* be drawn again around centre *B*, and let

$$\text{angle } MBN = 60°.$$

For, as before, triangle *BCN* will have its sides and angles given, and similarly

$$\text{excess } MP = 55^1/_2{}'$$
$$\text{where Earth's radius} = 1\text{P}.$$

But that is because

$$DBM = 55\text{P}8'.$$

If

$$DBM = 60\text{P};$$
$$MBO = 3\text{P}7',$$

and

$$\text{excess } MP = 55'.$$

Now

$$3\text{P}7' : 55' = 60' : 18';$$

and so on, the same as before. Nevertheless there is a difference of a few seconds. We shall do this for the rest; and thus we shall fill out the eighth column of the table. But if we were to employ instead of them, those (proportional minutes) which were set

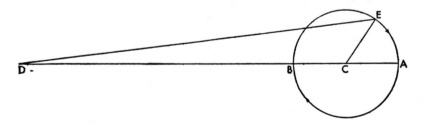

out in the table of additosubtractions, we shall make no error. For they are approximately the same—and it is a question of very small numbers. [125^b] There remain the proportional minutes which occur at the mean termini, namely between the second and the third termini.

Now let circle AB be the first epicycle at the new or full moon. Let C be its centre; and let D be taken as the centre of the Earth. And let the straight line $DBCA$ be extended. Now from apogee A let some arc be taken: for instance, let

$$\text{arc } AE = 60°;$$

and let DE and CE be joined. For we shall have the triangle DCE, in which two sides are given:

$$CD = 60\text{P}19'$$

and

$$CE = 5\text{P}11'.$$

Now angle DCE is an interior angle, and

$$\text{angle } DCE = 180° - \text{angle } ACE.$$

Accordingly, by what we have shown concerning triangles,

$$DE = 63\text{P}4'.$$

But

$$DBA = 65\tfrac{1}{2}\text{P},$$

and

$$DBA - ED = 2\text{P}26'.$$

Now

$$AB = 10\text{P}22';$$

and

$$10\text{P}22' : 2\text{P}26' = 60' : 14'.$$

And they are inscribed in the table in the ninth column opposite 60°. Following this example, we have completed the rest and filled out the table, which follows. And we have added another table of the semidiameters of the sun and the moon, and the shadow of the Earth, so that as far as possible they may be at hand.

TABLE OF THE PARALLAXES OF THE SUN AND MOON

Common Numbers	Parallaxes of the sun	Parallax of the moon at the second limit		Difference between the first and second limit of the moon: to be subtracted		Parallax of the moon at the third limit		Differences between the third and fourth lunar limit: to be added		Proportional minutes of the smaller epicycle	Proportional minutes of the greater epicycle
Deg.	Deg.	Min.	Sec.	Min.	Sec.	Min.	Sec.	Min.	Sec.		
6	354	0	10	2	46	0	7	3	18	0	12	0	0
12	348	0	19	5	33	0	14	6	36	0	23	1	0
18	342	0	29	8	19	0	21	9	53	0	34	3	1
24	336	0	38	11	4	0	28	13	10	0	45	4	2
30	330	0	47	13	49	0	35	16	26	0	56	5	3
36	324	0	56	16	32	0	42	19	40	1	6	7	5
42	318	1	5	19	5	0	48	22	47	1	16	10	7
48	312	1	13	21	39	0	55	25	47	1	26	12	9
54	306	1	22	24	9	1	1	28	49	1	35	15	12
60	300	1	31	26	36	1	8	31	42	1	45	18	14
66	294	1	39	28	57	1	14	34	31	1	54	21	17
72	288	1	46	31	14	1	19	37	14	2	3	24	20
78	282	1	53	33	25	1	24	39	50	2	11	27	23
84	276	2	0	35	31	1	29	42	19	2	19	30	26
90	270	2	7	37	31	1	34	44	40	2	26	34	29
96	264	2	13	39	24	1	39	46	54	2	33	37	32
102	258	2	20	41	10	1	44	49	0	2	40	39	35
108	252	2	26	42	50	1	48	50	59	2	46	42	38
114	246	2	31	44	24	1	52	52	49	2	53	45	41
120	240	2	36	45	51	1	56	54	30	3	0	47	44
126	234	2	40	47	8	2	0	56	2	3	6	49	47
132	228	2	44	48	15	2	2	57	23	3	11	51	49
138	222	2	49	49	15	2	3	58	36	3	14	53	52
134	216	2	52	50	10	2	4	59	39	3	17	55	54
150	210	2	54	50	55	2	4	60	31	3	20	57	56
156	204	2	56	51	29	2	5	61	12	3	22	58	57
162	198	2	58	51	56	2	5	61	47	3	23	59	58
168	192	2	59	52	13	2	6	62	9	3	23	59	59
174	186	3	0	52	22	2	6	62	19	3	24	60	60
180	180	3	0	52	24	2	6	62	21	3	24	60	60

TABLE OF THE SEMIDIAMETERS OF THE SUN, MOON, AND SHADOW

Common Numbers		Semidiameter of the Sun		Semidiameter of the Moon		Semidiameter of the Shadow		Variation of the Shadow
Deg.	Deg.	Min.	Sec.	Min.	Sec.	Min.	Sec.	Min.
6	354	15	50	15	0	40	18	0
12	358	15	50	15	1	40	21	0
18	342	15	51	15	3	40	26	1
24	336	15	52	15	6	40	34	2
30	330	15	53	15	9	40	42	3
36	324	15	55	15	14	40	56	4
42	318	15	57	15	19	41	10	6
48	312	16	0	15	25	41	26	9
54	306	16	3	15	32	41	44	11
60	300	16	6	15	39	42	2	14
66	294	16	9	15	47	42	24	16
72	288	16	12	15	56	42	40	19
78	282	16	15	16	5	43	13	22
84	276	16	19	16	13	43	34	25
90	270	16	22	16	22	43	58	27
96	264	16	26	16	30	44	20	31
102	258	16	29	16	39	44	44	33
108	252	16	32	16	47	45	6	36
114	246	16	36	16	55	45	20	39
120	240	16	39	17	4	45	52	42
126	234	16	42	17	12	46	13	45
132	228	16	45	17	19	46	32	47
138	222	16	48	17	26	46	51	49
144	216	16	50	17	32	47	7	51
150	210	16	53	17	38	47	23	53
156	204	16	54	17	41	47	31	54
162	198	16	55	17	44	47	39	55
168	192	16	56	17	46	47	44	56
174	186	16	57	17	48	47	49	56
180	180	16	57	17	49	47	52	57

25. On Computing the Parallax of the Sun and Moon

[127ª] We shall also set out briefly the mode of computing the parallaxes of the sun and moon by the table. If for the distance of the sun or twice the distance of the moon from the vertex of the horizon we take the corresponding parallaxes in the table—the solar parallaxes simply but the lunar parallaxes at the four limits—and if we take the first proportional minutes corresponding to twice the movement of the moon or twice its distance from the sun; by means of these minutes we shall determine the parts of the difference between the first and the last terminus which are proportional to sixty minutes; we shall always subtract these parts from the parallaxes following next, and we shall always add the later parts to the parallax at the next to the last limit. And we shall have two corrected parallaxes of the moon at the apogee and perigee; the lesser epicycle increases or decreases these parallaxes. Then we shall take the last proportional minutes corresponding to the lunar anomaly; and by means of them we shall determine the proportional part of the difference between the two parallaxes found nearest; we shall always add this proportional part to the first corrected parallax, the parallax at the apogee; and the result will be the parallax of the moon sought for that place and time, as in the following example.

Let the distance of the moon from the vertex (of the horizon) be 54°, the mean movement of the moon 15°, and the corrected anomaly 100°: I wish to find from them by means of the table the lunar parallax. I double the degrees of distance, and the result is 108°, to which in the table there correspond a difference of 1'48" between the first and second limit, a parallax of 42'50" at the second limit, a parallax of 50'69" at the third limit, and a difference of 2'46" between the third and the fourth limit—which I shall mark down separately. Doubling the movement of the moon makes 30°; I find five of the first proportional minutes corresponding to it, and with them I determine 9" to be the part of the first difference which is proportional to sixty minutes: I subtract these 9" from the 42'50" of the parallax, and the remainder is 42'41". Similarly the proportional part of the second difference—which was 2'46"—is 14"; and I add it to the 50'59" of the parallax at the third limit; the sum is 51'13". The difference between these parallaxes is 8'32". After this I take the last proportional minutes corresponding to the corrected anomaly, and there are 39'. By means of them I take 4'50" as the proportional part of the difference of 8'32"; [127ᵇ] I add this 4'50" to the first corrected parallax, and the sum is 47'31", which will be the sought parallax of the moon in the circle of altitude.

But since any other parallaxes of the moon differ very little from the parallaxes at full moon and new moon, it would seem to be sufficient if we kept within the mean limits everywhere, for we have great need of them for the sake of predicting eclipses. The rest do not require such great examination, which will be held to offer perhaps less in the way of utility than in the satisfaction of curiosity.

26. HOW THE PARALLAXES OF LONGITUDE AND LATITUDE ARE DISTINGUISHED

Now the parallax is divided simply into the parallax of longitude and that of latitude, or the parallax between the sun and moon is distinguished according to the arcs and angles of the intersection of the ecliptic and the circle through the poles of the horizon; since it is clear that when this circle falls at right angles upon the ecliptic, it makes no parallax in longitude, but the parallax is transferred wholly to latitude, as the circle is wholly a circle of latitude and altitude. But where conversely the ecliptic falls at right angles upon the horizon and becomes wholly the same as the circle of altitude; then, if the moon has no latitude, it does not admit anything except a parallax in longitude, but if it has a digression in latitude, it does not escape some parallax in latitude.

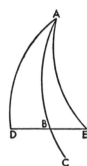

In this way let circle *ABC* be the ecliptic, and let it be at right angles to the horizon, and let *A* be the pole of the horizon. Accordingly circle *ABC* will be the same as the circle of altitude of a moon without latitude. Let *B* be the position of the moon, and *BC* its total parallax in longitude.

But when it also has latitude, let *DBE* be the circle described through the poles of the ecliptic and with *DB* or *BE* as the latitude of the moon, it is clear that side *AD* or *AE* will not be equal to *AB*; and the angle at *D* or *E* will not be right, since *DA* and *EA* are not circles through the poles of *DBE*; and the parallax will participate in latitude, and it will do so all the more the nearer the moon is to the vertex. For let triangle *ADE* keep the same base, but let sides *AD* and *AE* be shorter and comprehend acuter angles at the base; the greater the distance of the moon from the vertex is, the more like right angles will the angles be.

Now let *ABC* be the ecliptic, and *DBE* the oblique circle of altitude of a moon not having latitude, as being at an ecliptic section. [128ᵃ] Let *B* be the ecliptic section, and *BE* the parallax in the circle of altitude. Let there be drawn *EF* the arc of a circle through the poles of *ABC*. Accordingly since in triangle *BEF* angle *EBF* is given, as was shown above, and

angle *F* = 90°,

and side *BE* is also given: by what has been shown concerning spherical triangles, the remaining sides are given: *BF* the parallax in longitude and *FE* the parallax in latitude, which agree with parallax *BE*. But since *BE*, *EF*, and *FB* on account of their shortness differ but slightly and imperceptibly from straight lines, we shall not make an error if we use the right triangle as rectilinear; and on that account the ratio will become easy.

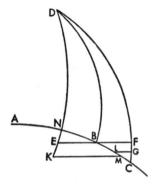

For let circle *ABC* be drawn as the ecliptic, and let *DB* the oblique circle through the poles of the horizon fall upon it. Let *B* be the position in longitude of the moon and *FB* the northern latitude or *BE* the southern. Let the vertex of the horizon be *D*, and from *D* let fall on the moon the circles of altitude *DEK* or *DFC*, whereon are the parallaxes *EK* and *FG*. For the true positions of the moon in longitude and latitude will be at points *E* or *F*, but the seeming positions will be at *K* or *G*. And from *K* and *G* let arcs *KM* and *LG* be drawn at right angles to the ecliptic *ABC*. Accordingly, since the longitude and latitude of the moon have been established together with the latitude of the region; in triangle *DEB* two sides *DB* and *BE* will be known and also *ABD* the angle of section and

angle *DBE* = angle *ABD* + 90°.

Accordingly the remaining side *DE* will be given together with angle *DEB*.

Similarly in triangle *DBF* since two sides *DB* and *BF* are given together with angle *DBF*, which with angle *ABD* makes up a right angle, *DF* will also be given together with angle *DFB*. Accordingly parallaxes *EK* and *FG* on arcs *DE* and *DF* are given by the table, and so is *DE* or *DF*, the true distance of the moon from the vertex, and similarly the seeming distance *DEK* or *DFG*. But in triangle *EBN*, which has the intersection of *DE* with the ecliptic at point *N*, angle *NEB* is given, and base *BE* is given, and angle *NBE* is right: the remaining angle *BNE* will become known too together with the remaining sides *BN* and *NE*. And similarly in the whole triangle *NKM*, as angles *M* and *N* and the whole side *KEN* are given, the base *KM* will be established. *KM* is the seeming southern latitude of the moon, and its excess over *EB* is the parallax of latitude: and the remaining side *NBM* is given: by the subtraction of *NB* from *NBM* the remainder *BM* will be the parallax in longitude.

Moreover, thus in the northern triangle *BFC*, since side *BF* is given together with angle *BFC* [128ᵇ] and angle *B* is right, the remaining sides *BLC* and *FGC* together with the remaining angle *C* are given; and after the subtraction of *FG* from *FGC* there remains *GC*, which is a side given in triangle *GLC* together with the angle *LCG* and angle *CLG*, which is right. Accordingly the remaining sides are given: *GL* and *LC*, and hence *BL* which is the remainder of *BC* and is the parallax in longitude; and *GL* is the seeming latitude, and its parallax is the excess of *BF* the true latitude over *GL*.

As you see however, this computation, which deals with very small magnitudes, contains more labour than fruitfulness. For it will be sufficient if we use angle *ABD* instead of *DCB* and angle *DBF* instead of *DEB*, and as before, simply the mean *DB* instead of arcs *DE* and *DF*—neglecting the latitude of the moon. For no error will appear because of that, especially in regions of the northern part of the Earth; but in very southern regions, where *B* touches the vertex of the horizon at the greatest latitude of 5° and when the moon is at its perigee, there is a difference of approximately 6'. But in the ecliptic conjunctions with the sun, where the latitude of the moon cannot exceed $1/_2$°, there can be a difference of merely $1^3/_4$'. Accordingly it is clear from this that the parallax is always added to the true position of the moon in the eastern quadrant of the ecliptic and is always subtracted in the other quadrant, so that we may have the seeming longitude of the moon; and may have the seeming latitude through the parallax in latitude, since if the true latitude and the parallax are in the same direction, they are added together; if in different directions, the lesser is subtracted from the greater, and the remainder is the seeming latitude of that same part, to which the greater falls away.

27. CONFIRMATION OF WHAT HAS BEEN EXPOUNDED CONCERNING THE PARALLAXES OF THE MOON

Accordingly we can confirm by many other observations (such as the following one) that the parallaxes of the moon as set forth above are in conformity with the appearances. We made this observation at Bologna, after sunset on the seventh day before the Ides of March, in the year of Christ 1497. For we observed how long [129ª] the moon would occult the bright star of the Hyades (which the Romans call Paliticium), and with this in mind, we saw the star brought into contact with the shadowy part of the lunar body and already lying hidden between the horns of the moon at the end of the fifth hour of the night, though the star was nearer the southern horn by three quarters as it were of the width or diameter of the moon. And since according to the tables the star was at 2°52' of Gemini with a southerly latitude of $5^1/_6$°, it was clear that to eyesight the centre of the moon was half a diameter to the west of the star, and accordingly its seen position was 2°36' in longitude and approximately 5°6' in latitude.

Accordingly from the beginning of the years of Christ there have been 1497 Egyptian years 76 days 23 hours at Bologna; but at Cracow, which is approximately 8° farther east, 23 hours 36 minutes, to which equal time adds 4 minutes. For the sun was at $28^1/_2$° of Pisces and therefore the regular movement of the moon away from the sun was 74°, the regular anomaly was 111°10', the true position of the moon was at 3°24' of Gemini, the southerly latitude 4°35', for the true movement of latitude was 203°41'. Moreover, at that time at Bologna, 26° of Scorpio was rising, with an angle of $57^1/_2$°;

and the moon was 83° from the vertex of the horizon, and the angle of section between the circle of altitude and the ecliptic was approximately 29°, the parallax of the moon was 1°51' in longitude and 30' in latitude. Those things agree perfectly with the observations; and all the less will anyone doubt that our hypotheses and what results from them are correct.

28. On the Mean Oppositions and Conjunctions of the Sun and Moon

The method of investigating the conjunctions and oppositions of the sun and the moon is clear from what has been said so far concerning their movement. For in relation to that approaching time at which we think this or that conjunction or opposition will take place, we shall seek the regular movement of the moon; and if we find that the regular movement has already completed a circle, we understand a full conjunction at the semicircle. [129^b] But since that rarely presents itself, we shall have to observe the distance between the sun and the moon; and when we have divided it by the daily movement of the moon, we shall know by how much time the one of them is in advance of the other, or how far off in the future the conjunction or the opposition is. Therefore we shall seek out the movements and positions for this time, and with them we shall set up ratios for the true new moons and full moons; and we shall distinguish the ecliptic conjunctions from the others, as we shall indicate below. When we have got these things set up, it will be possible to go on into any number of months and continue through some number of years by means of the table of twelve months, which contains the time and the regular movements of the anomaly of the sun and the moon and the regular movement of the moon in latitude—joining single movements to the single movements already found. But, we shall put down the anomaly of the sun as true, so as to have it as corrected immediately. For its difference is not perceptible to sense in one or more years on account of its slowness at its beginning, *i.e.*, at its highest apsis.

TABLE OF THE CONJUNCTION AND OPPOSITION OF SUN AND MOON

Months	Divisions of Time				Movements of Lunar Anomaly				Movement in Latitude of the Moon			
	Days	Min. of Day	Sec.	Thirds	60°	Deg.	Min.	Sec.	60°	Deg.	Min.	Sec.
1	29	31	50	8	0	25	49	0	0	30	40	13
2	59	3	40	16	0	51	38	0	1	1	20	27
3	88	35	30	24	1	17	27	0	1	32	0	41
4	118	7	20	32	1	43	16	0	2	2	40	55
5	147	39	10	40	2	9	5	0	2	33	21	9
6	177	11	0	48	2	34	34	0	3	4	1	23
7	206	42	50	57	3	0	43	0	3	34	41	36
8	236	14	41	25	3	26	32	0	4	5	21	50
9	265	46	31	13	3	52	21	0	4	36	2	4
10	295	18	21	21	4	18	10	0	5	6	42	18
11	324	50	11	29	4	4	59	0	5	37	22	32
12	354	22	1	37	5	9	48	0	0	8	2	46

THE HALF MONTH BETWEEN THE FULL AND NEW MOON

1/2	14	45	55	4	3	12	54	30	3	15	20	6

MOVEMENT OF SOLAR ANOMALY

Months	60°	°	′	″	Months	60°	°	′	″
1	0	29	6	18	7	3	23	44	6
2	0	58	12	36	8	3	52	50	24
3	1	27	18	54	9	4	21	56	42
4	1	56	25	12	10	4	51	3	0
5	2	25	31	30	11	5	20	9	19
6	2	54	57	48	12	5	49	15	37

THE HALF MONTH

	1/2	0	14	33	9

29. ON THE CLOSE EXAMINATION OF THE TRUE CONJUNCTIONS AND OPPOSITIONS OF THE SUN AND MOON

[130b] Since we possess, as was said, the time of mean conjunction or opposition of these heavenly bodies together with their movements, then the true distance between them, whereby they precede or follow one another, will be necessary in order to find their true (conjunctions and oppositions). For if the (true) moon is prior to the sun in (mean) conjunction or opposition, it is clear that the true one will be in the future; but if the sun, then it is already past the true one which we are seeking. This is made clear by the additosubtractions in the case of both of them, since if there were no additosubtractions, or if they were equal and of the same quality, *viz.*, both additive or both subtractive, it is clear that at the same moment the true conjunctions or oppositions and the mean ones coincide. But if they are unequal, the difference indicates what their distance is and that the star to which the additive or subtractive difference belongs precedes or follows. But when they are in different parts (of their circles) that star all the more precedes whose additosubtraction is subtractive; and the adding together of the additosubtractions shows what the distance between them is. In connection with this we shall decide how many whole hours can be is traversed by the moon—taking two hours for every degree of distance.

In this way, if there were about 6° of distance, we should take 12 hours as corresponding to them. Therefore we shall seek the true movement of the moon away from the sun for the interval of time thus set up; and we shall do that easily, when we know that the mean movement of the moon is 1°1' per 2 hours, but that the true hourly movement of anomaly around the full moon or the new moon is approximately 50'. In 6 hours that makes the regular movement to be 3°3', and the true movement of anomaly 5°; and in the table of lunar additosubtractions we shall note the difference between the additosubtractions and add it to the mean movement—if the anomaly is in the lower part of the circle—and subtract it if the anomaly is in the upper. For the sum or the remainder is the true movement of the moon for the hours taken. Therefore that movement, if equal to the distance first existing, is sufficient. Otherwise the distance multiplied by the number of estimated hours should be divided by this movement; or else we shall divide the true simple distance by the hourly movement taken. [131a] For the quotient will be the true difference in time in hours and minutes between the mean and the true conjunction or opposition. We shall add this difference to the mean time of conjunction or opposition, if the moon is west of the sun, or to the position of the sun diametrically opposite: or we shall subtract, if the moon is eastward; and we shall have the time of true conjunction or opposition, although we must

confess that the anomaly of the sun too adds or subtracts something, but it is rightly neglected, as in the whole tract and at greatest elongation—which extends beyond
7°—the anomaly cannot fill 1'; and the method of evaluating the lunar movements
is more certain.

For those who rely only upon the hourly movement of the moon, which they call
the hourly excelling movement, make mistakes sometimes and are forced rather often
to repeat their calculations. For the moon is changeable even from hour to hour and
does not stay like itself. Accordingly, for the time of true conjunction or opposition, we
shall work out the true movement in latitude, so as to learn the latitude of the moon
and work out the true position of the sun in relation to the spring equinox, *i.e.*, in the
signs, whereby the true position of the moon is known to lie the same or opposite to
it. And since time is here understood as mean and equal with respect to the Cracow
meridian, we shall reduce it to apparent time by the method described above. But if we
should wish to set this up for any other place than Cracow, we shall note its longitude
and take four minutes of an hour for each degree of longitude and four seconds of an
hour for each minute of longitude; and we shall add them to the Cracow time, if the
other place is to the east, and subtract them, if it is to the west. And the sum or the
remainder will be the time of conjunction or opposition of the sun and moon.

30. How the Ecliptic Conjunctions and
Oppositions of the Sun and Moon are
Distinguished from the Others

In the case of the moon it is easily discernible whether or not they are ecliptic;
since, if the latitude of the moon is less than half the diameters of the moon and the
shadow, it will undergo an eclipse, but if greater, it will not. But there is more than
enough bother in the case of the sun, as the parallax of each of them, by which for the
most part the visible conjunction differs from the true, is mixed up in it. Accordingly
when we have examined [131ᵇ] what the parallax in longitude between the sun and
moon is at the time of true conjunction, similarly we shall look for the apparent
(angular) elongation of the moon from the sun at the interval of an hour before the
true conjunction in the eastern quarter of the ecliptic or after the true conjunction in
the western quarter, in order to understand how far the moon seems to move away
from the sun in one hour. Therefore when we have divided the parallax by this hourly
movement, we shall have the difference in time between the true and the seen
conjunction, When that is subtracted from the time of the true conjunction in the
eastern part of the ecliptic or added in the western—for in the eastern part the seen
conjunction precedes the true, and in the western it follows it—the result will be the time of
seen conjunction which we were looking for. Therefore we shall reckon the seen latitude of the

moon in relation to the sun for this time, or the distance between the centres of the sun and the moon at the seen conjunction, after deducting the parallax of the sun. If this latitude is greater than half the diameters of the sun and moon, the sun will not undergo an eclipse; but if smaller, it will. From this it is clear that if the moon at the time of true conjunction does not have any parallax in longitude, the seen and the true conjunction will be the same, and the conjunction will take place at 90° of the ecliptic as measured from the east or the west.

31. How Great an Eclipse of the Sun or Moon Will Be

Therefore, after we have learned that the sun or moon will undergo an eclipse, we shall easily come to know how great the eclipse will be—in the case of the sun by means of the seen latitude between the sun and moon at the time of seen conjunction. For if we subtract the latitude from half the diameters of the sun and the moon, the remainder is the eclipse of the sun as measured along its diameter; and when we have multiplied that by twelve and divided the product by the diameter of the sun, we shall have the number of twelfths of the eclipse of the sun. But if there is no latitude between the sun and the moon, there will be a total eclipse of the sun or as much of it as the moon can cover.

Approximately the same method (is used) in the case of a lunar eclipse, except that instead of the seen latitude we employ the simple latitude. When the latitude is subtracted from half the diameters of the moon and shadow, the remainder is the part of [132ᵃ] the moon eclipsed, provided the latitude of the moon is not less than half the diameters of the moon and shadow, as taken along the diameter of the moon. For then there will be a total eclipse. And furthermore the lesser latitude even adds some delay in the darkness; and the delay will be greatest when there is no latitude—as I think is perfectly clear to those who consider it. Accordingly, in the case of a particular eclipse of the moon, when we have multiplied the eclipsed part by twelve and divided the product by the diameter of the moon, we shall have the number of twelfths of the eclipse—just as in the case of the sun.

32. How to Know Beforehand How Long an Eclipse Will Last

It remains to see how long an eclipse will last. It should be noted that we use the arcs which occur in the case of the sun, moon, and shadow as straight lines; for they are so small that they do not seem to be different from straight lines.

Accordingly let us take point *A* as the centre of the sun or of the shadow, and line *BC* as the passage of the orb of the moon. And let *B* be the centre of the moon touching the sun or shadow at the beginning of incidence and C at the end of its transit. Let AB and BC be joined, and let fall AD perpendicular to BC.

It is clear that when the centre of the moon is at D, it will be the middle of the eclipse. For AD is the shortest of the lines falling from A, and

$$BD = DC,$$

since

$$AB = AC,$$

and AB or AC is equal to half the sum of the diameters of the sun and the moon in a solar eclipse and to that of the diameters of the moon and shadow in a lunar eclipse; and AD is the true latitude of the moon or the seen latitude at the middle of the eclipse. Accordingly, when we have subtracted the square on AD from the square on AB, the remainder is the square on BD. Therefore BD will be given in length. When we have divided it by the true hourly movement of the moon during the eclipse of the moon, or by the visible movement in the case of a solar eclipse, we shall have the time of half the duration. But the moon very often delays in the middle of the darkness—that happens when half the sum of the diameters of the moon and the shadow exceeds the latitude of the moon by more than the moon's diameter, as we said. Accordingly when we have placed E the centre of the moon at the starting-point of the total [132b] obscuration, when the moon touches the concave circumference of the shadow, and F at the other point of contact, when the moon first emerges, and have joined AE and AF, it will be made clear in the same way as before that ED and DF are the halves of the delay in darkness, because AD is the known latitude of the moon, and AE or AF is that whereby the half of the diameter of the shadow is greater than half the diameter of the moon. Therefore DE or DF will be established; and once more when we have divided it by the true hourly movement of the moon, we shall have the time of half the delay, which we were looking for.

Nevertheless we must notice here that since the moon moves in its own orbital circle, it does not, by the mediation of the circles passing through the poles of the ecliptic, cut arcs of longitude on the ecliptic wholly equal to the arcs in its own orbital circle. But the difference is very slight, so that at the total distance of 12° from the ecliptic section, which is approximately the farthest limit of the eclipses of the sun and moon, the arcs of the circles do not differ from one another by 2', which makes $1/15$ hour; on that account we often use one instead of the other as if they were the same. So too we use the same latitude of the moon at the limits of the eclipses as at the middle of the eclipse, although the latitude of the moon is always increasing or decreasing, and on that account the intervals of incidence and withdrawal are not wholly equal, but the difference is so slight that it seems a waste of time to examine them more closely.

In this way the times, durations, and magnitudes of eclipses have been unfolded with respect to the diameters.

But since it is the opinion of many persons that the parts eclipsed should be distinguished not with respect to the diameters but with respect to the surfaces, for it is not lines but surfaces which are eclipsed:

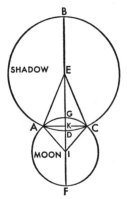

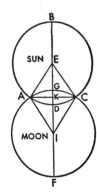

accordingly let *ABCD* be the circle of the sun or of the shadow, and let *E* be its centre. Let *AFCG* be the lunar circle, and let *I* be its centre. Let the circles cut one another in points *A* and *C*, let the straight line *BEIF* be drawn through the centres of both, and let *AE*, *EC*, *IA*, *IC* be joined and line *AKC* at right angles to *BF*. By means of this we wish to examine how great *ADCG* the surface obscured is, or how many twelfths of the whole surface of the orb of the moon or sun belong to the part eclipsed. Accordingly since the semidiameters *AE* and *AI* of each circle are given by the above, and also *EI* the distance between their centres or the lunar latitude, we shall have [133a] triangle *AEI* with its sides given; and for that reason with its angles given by the demonstrations above; and angle *AEI* is similar and equal to angle *EIC*. Accordingly

arcs *ADC* and *AGC* will be given,

where the circumference = 360°.

Furthermore, in the measurement of the circle Archimedes of Syracuse records that

circumference : diameter < $3^1/_7$: 1

but

circumference : diameter > $3^{10}/_{71}$: 1.

Ptolemy assumed as a mean between these

3p8'30" : 1p.

By means of this ratio

arcs *AGC* and *ADG* will be given,

in terms of the same parts as the semidiameters *AE* and *AI*.
And

quad. *EA*, *AD* = sector *AEC*,

and

quad. *IA*, *AG* = sector *AIC*.

But in the isosceles triangles *AEC* and *AIC* the common base *AKC* and the perpendiculars *EK* and *KI* are given. And accordingly the quadrilateral *AK*, *KE* is given, which is the area of triangle *AEC*—and similarly the quadrilateral *AK*, *KI* is the area of triangle *ACI*. Accordingly sect. *AFCK* – trgl. *AIC* = seg. circ. *AFC*

and

sect. *ABCK* – trgl. *AEC* = seg. circ. *ABC*

and hence,

figure *ADCG* is given,

which was sought.

And moreover, the total area of the circle—which is comprehended by *BE* and *BAD* in a solar eclipse or by *FI* and *FAG* in a lunar eclipse—was given. Accordingly it will be manifest how many twelfths of the total circle of the sun or moon was eclipsed in *ADCG*. Let all this—which has been treated in more detail by others—be enough now concerning the moon: for we are in a hurry to get to the revolutions of the remaining five planets, which will be spoken of in the books following.

BOOK FIVE

[133^b] Up to now we have been explaining to the best of our ability the revolutions of the Earth around the sun and of the moon around the Earth. Now we are turning to the movements of the five wandering stars: the mobility of the Earth binds together the order and magnitude of their orbital circles in a wonderful harmony and sure commensurability, as we said in our brief survey in the first book, when we showed that the orbital circles do not have their centres around the Earth but rather around the sun. Accordingly it remains for us to demonstrate all these things singly and with greater clarity; and let us fulfil our promises adequately, in so far as we can, particularly by measuring the appearances by the experiments which we have got from the ancients or from our own times, in order that the ratio of the movements may be held with greater certainty. Now in Plato's *Timaeus* each of these five stars is named in accordance with its visible aspect: Saturn, Phaenon—as if to say "shining" or "appearing," for Saturn is hidden less than the others, and emerges more quickly after undergoing occultation by the sun; Jupiter, Phaeton from his radiance; Mars, Pyrois from his fiery glow; Venus sometimes φωσφόρος and sometimes ἕοπερος, *i.e.*, Lucifer and Vesperugo, according as she shines at morning or evening; and finally Mercury, Stilbon from his sparkling and twinkling light. Moreover the planets have greater irregularities in longitude and in latitude than the moon.

1. On Their Revolutions and Mean Movements

Two longitudinal movements which are quite different appear in the planets. One of them is on account of the movement of the Earth, as we said; and the other is proper to each planet. We may rightly call the first the movement of parallax, since it is the one which makes the planets appear to have stoppings, progressions, and retrogradations— [134^a] not that the planet which always progresses by its own movement, is pulled in different directions, but that it appears to do so by reason of the parallax caused by the movement of the Earth taken in relation to the differing magnitudes of their orbital circles.

Accordingly it is clear that the true position of Saturn, Jupiter, and Mars become visible to us only at the time when they are in opposition to the sun; and that occurs approximately in the middle of their retrogradations. For at that time they fall on a straight line with the mean position of the sun, and lay aside their parallax.

Furthermore there is a different ratio in the case of Venus and Mercury: for they are hidden at the time they are in conjunction with the sun, and they show only the digressions which they make on either side away from the sun: hence they are never found without parallax.

Therefore the revolution of parallax—I mean the movement of the Earth with respect to the planet—is private to each planet; and the planet and the Earth are mutually explanatory of it. For we say that the movement of parallax is nothing except that wherein the regular movement of the Earth exceeds their movement, as in the case of Saturn, Jupiter, and Mars, or is exceeded by it, as in the case of Venus and Mercury. But since such periods of parallax are found unequal by a manifest difference, the ancients recognized that the movements of these planets too were irregular and had apsides; of circles to which their irregularity returned, and the ancients supposed that these apsides had perpetual seats in the sphere of the fixed stars. By that argument the road is opened for learning their mean movements and equal periods. For when the ancients had recorded in memory the position of some planet with respect to its exact distance from the sun and a fixed star, and after an interval of time found that it had arrived at the same position with an equal distance from the sun; the planet was seen to have completed its whole movement of irregularity and to have returned through all to its former relationship with the Earth. And so by means of the time which intervened they calculated the number of whole and equal revolutions and from them the particular movements of the planet. Ptolemy surveyed the circuits through a number of years, according as, he acknowledged, he got them from Hipparchus. Now he means solar years to be understood as the years measured from an equinox or solstice. But it has already been made clear that such years are not quite equal; on that account we shall use years measured from the fixed stars, and by means of them the movements of these five planets have been reconstituted more correctly by us, according as in our time [134^b] we found there was some deficiency in them or excess, as follows.

For the Earth has 57 revolutions in respect to Saturn—we call this the movement of parallax—in 59 of our solar years 1 day 6 minutes of a day 48 seconds approximately: during this time the planet has by its own movement completed two circuits plus 1°6'6".

Jupiter is outrun by the Earth 65 times in 71 solar years minus 5 days 45 minutes 27 seconds: during this time the planet by its own movement has 6 revolutions minus 5°41'2$^1/_2$".

Mars has 37 revolutions of parallax in 79 solar years 2 days 27 minutes 3 seconds: during this time the planet by its own movement completes 42 periods plus 2°24'56".

Venus outruns the movement of the Earth 5 times in 8 solar years minus 2 days 26 minutes 46 seconds. And during this time it has 13 revolutions minus 2°24'40" around the sun.

Finally, Mercury completes 145 periods of parallax, by which it outruns the movement of the Earth, in 46 solar years plus 34 minutes of a day 23 seconds. And it has 191 revolutions around the sun in that time plus 34 minutes of a day 23 seconds approximately.

Accordingly the single circuits of parallax are as follows: for the single planets;

Saturn:	378 days	5 min.	32 sec.	11 thirds
Jupiter:	398 days	23 min.	25 sec.	56 thirds
Mars:	779 days	56 min.	19 sec.	7 thirds
Venus:	583 days	45 min.	17 sec.	24 thirds
Mercury:	115 days	52 min.	42 sec.	12 thirds

When we have reduced these circuits to the degrees of a circle and multiplied by the ratio of 365 to the number of days and minutes, we shall have as the annual movements (of parallax):

Saturn:	347°32'2"54'''12''''
Jupiter:	329°25'8"15'''6''''
Mars:	168°28'29"13'''12''''
Venus:	225°1'48"54'''30''''
Mercury:	3 (360°) + 53°56'46"54'''40''''

[135ᵃ] The three-hundred-sixty-fifth part of these is the daily movement:

Saturn:	57'7"44'''
Jupiter:	54'9"3'''49''''
Mars:	27'41"40'''8''''
Venus:	36'49"28'''35''''
Mercury:	3°6'24"7'''43''''

according as they are set out in the following tables, like the mean movement of the sun and moon.

But we thought it unnecessary to set down their proper movements in this way. For the proper movements are determined by the subtraction of the movements of parallax from the mean movement of the sun, as the proper movement of the planet and the mean movement of parallax compose the mean movement of the sun. For the proper annual movements in relation to the sphere of the fixed stars are as follows for the upper planets:

Saturn:	12°12'46"12'''52''''
Jupiter:	30°19'40"51'''58''''
Mars:	191°16'19"53'''52''''

But in the case of Venus and Mercury, since their proper movements are not apparent to us,[1] the movement of the sun itself is used by us instead; and it furnishes a way of investigating and demonstrating their apparent movements, in the following tables.

---

[1] The proper movements of Venus and Mercury are not apparent to us in that their positions are never viewed without parallax.

SATURN'S MOVEMENT OF PARALLAX IN YEARS AND PERIODS OF SIXTY YEARS

Egyptian Years	Movement					Egyptian Years	Movement				
	60°	°	'	"	'''		60°	°	'	"	'''
1	5	47	32	3	9	31	5	33	33	37	59
2	5	35	4	6	19	32	5	11	5	41	9
3	5	22	36	9	29	33	5	8	37	44	19
4	5	10	8	12	38	34	4	56	9	47	28
5	4	57	40	15	48	35	4	43	41	50	38
6	4	45	12	18	58	36	4	31	13	53	48
7	4	32	44	22	7	37	4	18	45	56	57
8	4	20	16	25	17	38	4	6	18	0	7
9	4	7	48	28	27	39	3	53	50	3	17
10	3	55	20	31	36	40	3	41	22	6	26
11	3	42	52	34	46	41	3	18	54	9	36
12	3	30	24	37	56	42	3	16	26	12	46
13	3	17	56	41	5	43	3	3	58	15	55
14	3	5	28	44	15	44	2	51	30	19	5
15	2	53	0	47	25	45	2	39	2	22	15
16	2	40	32	50	34	46	2	26	34	25	24
17	2	28	4	53	44	47	2	14	6	28	34
18	2	15	36	56	54	48	2	1	38	31	44
19	2	3	9	0	3	49	1	49	10	34	53
20	1	50	41	3	13	50	1	36	42	38	3
21	1	38	13	6	23	51	1	24	14	41	13
22	1	25	45	9	32	52	1	11	46	44	22
23	1	13	17	12	42	53	0	59	18	47	32
24	1	0	49	15	52	54	0	46	50	50	42
25	0	48	21	19	1	55	0	34	22	43	51
26	0	35	53	22	11	56	0	21	54	57	1
27	0	23	25	25	21	57	0	9	27	0	11
28	0	10	57	28	30	58	5	56	59	3	20
29	5	58	29	31	40	59	5	44	31	6	30
30	5	46	1	34	50	60	5	32	3	9	40

SATURN'S MOVEMENT OF PARALLAX IN PERIODS OF SIXTY DAYS

Days	Movement					Days	Movement				
	60°	°	'	"	'''		60°	°	'	"	'''
1	0	0	57	7	44	31	0	29	30	59	46
2	0	1	54	15	28	32	0	30	28	7	30
3	0	2	51	23	12	33	0	31	25	15	14
4	0	3	48	30	56	34	0	32	22	22	58
5	0	4	45	38	40	35	0	33	19	30	42
6	0	5	42	46	24	36	0	34	16	38	26
7	0	6	39	54	8	37	0	35	13	46	1
8	0	7	37	1	52	38	0	36	10	53	55
9	0	8	34	9	36	39	0	37	8	1	39
10	0	9	31	17	20	40	0	38	5	9	23
11	0	10	28	25	4	41	0	39	2	17	7
12	0	11	25	32	49	42	0	39	59	24	51
13	0	12	22	40	33	43	0	40	56	32	35
14	0	13	19	48	17	44	0	41	53	40	19
15	0	14	16	56	1	45	0	42	50	48	3
16	0	15	14	3	45	46	0	43	47	55	47
17	0	16	11	11	29	47	0	44	45	3	31
18	0	17	8	19	13	48	0	45	42	11	16
19	0	18	5	26	57	49	0	46	39	19	0
20	0	19	2	34	41	50	0	47	36	26	44
21	0	19	59	42	25	51	0	48	33	34	28
22	0	20	56	50	9	52	0	49	30	42	12
23	0	21	53	57	53	53	0	50	27	49	56
24	0	22	51	5	38	54	0	51	24	57	40
25	0	23	48	13	22	55	0	52	22	5	24
26	0	24	45	21	6	56	0	53	19	13	8
27	0	25	42	28	50	57	0	54	16	20	52
28	0	26	39	36	34	58	0	55	13	28	36
29	0	27	36	44	18	59	0	56	10	36	20
30	0	28	33	52	2	60	0	57	7	44	5

Jupiter's Movement of Parallax in Years and Periods of Sixty Years

Egyptian Years	Movement 60°	°	'	"	'''	Egyptian Years	Movement 60°	°	'	"	'''
1	5	29	25	8	15	31	2	11	59	15	48
2	4	58	50	16	30	32	1	41	24	24	3
3	4	28	15	24	45	33	1	10	49	32	18
4	3	57	40	33	0	34	0	40	14	40	33
5	3	27	5	41	15	35	0	9	39	48	48
6	2	56	30	49	30	36	5	39	4	57	8
7	2	25	55	57	45	37	5	8	30	5	18
8	1	55	21	6	0	38	4	37	55	13	33
9	1	24	46	14	15	39	4	7	20	21	48
10	0	54	11	22	31	40	3	36	45	30	4
11	0	23	36	30	46	41	3	6	10	38	19
12	5	53	1	39	1	42	2	35	35	46	34
13	5	22	25	47	16	43	2	5	0	54	49
14	4	51	51	55	31	44	1	34	26	3	4
15	4	21	17	3	46	45	1	3	51	11	19
16	3	50	42	12	1	46	0	33	16	19	34
17	3	20	7	20	16	47	0	2	41	27	49
18	2	49	32	28	31	48	5	32	6	36	4
19	2	18	57	35	46	49	5	1	31	44	19
20	1	48	22	45	2	50	4	30	56	52	34
21	1	17	47	58	17	51	4	0	22	0	50
22	0	47	13	1	32	52	3	29	47	9	5
23	0	16	38	9	47	53	3	59	12	17	20
24	5	45	3	18	2	54	2	28	37	25	33
25	5	15	28	26	17	55	2	58	2	33	50
26	4	44	53	34	32	56	1	27	27	42	5
27	4	14	18	42	47	57	1	56	52	50	20
28	3	43	43	51	2	58	0	26	17	58	35
29	3	13	8	59	17	59	0	55	43	6	50
30	2	42	34	7	33	60	5	25	8	15	6

Jupiter's Movement of Parallax in Periods of Sixty Days

Days	Movement 60°	°	'	"	'''	Days	Movement 60°	°	'	"	'''
1	0	0	54	9	3	31	0	27	58	40	58
2	0	1	49	18	7	32	0	28	52	50	2
3	0	2	42	27	11	33	0	29	46	59	5
4	0	3	36	36	15	34	0	30	41	8	9
5	0	4	30	45	19	35	0	31	35	17	13
6	0	5	24	54	22	36	0	32	29	26	17
7	0	6	19	3	26	37	0	33	23	35	21
8	0	7	13	12	30	38	0	34	17	44	25
9	0	8	7	21	34	39	0	35	11	53	29
10	0	9	1	30	38	40	0	36	6	2	32
11	0	9	55	39	41	41	0	37	0	11	36
12	0	10	49	48	45	42	0	37	54	20	40
13	0	11	43	57	49	43	0	38	48	29	44
14	0	12	38	6	53	44	0	39	42	38	47
15	0	13	32	15	57	45	0	40	36	47	51
16	0	14	26	25	1	46	0	41	30	56	55
17	0	15	20	34	4	47	0	42	25	5	59
18	0	16	14	43	8	48	0	43	19	15	3
19	0	17	8	52	12	49	0	44	13	24	6
20	0	18	3	1	16	60	0	45	7	33	10
21	0	18	57	10	20	51	0	46	1	42	14
22	0	19	51	19	23	52	0	46	55	51	18
23	0	20	45	28	27	53	0	47	50	0	22
24	0	21	39	37	31	64	0	48	44	9	26
25	0	22	33	46	35	55	0	49	38	18	29
26	0	23	27	55	39	56	0	50	32	27	33
27	0	24	22	4	43	57	0	51	26	36	37
28	0	25	16	13	46	58	0	52	20	45	41
29	0	26	10	22	50	59	0	53	14	54	45
30	0	27	4	31	54	60	0	54	9	3	49

Mars' Movement of Parallax in Years and Periods of Sixty Years

Egyptian Years	60°	°	'	"	'''	Egyptian Years	60°	°	'	"	'''
1	2	48	28	30	36	31	3	2	43	48	38
2	5	36	57	1	12	32	5	51	12	19	14
3	2	25	25	31	48	33	2	39	40	49	50
4	5	13	54	2	24	34	5	28	9	20	26
5	2	2	22	33	0	35	2	16	37	51	2
6	4	50	51	3	36	36	5	5	6	21	38
7	1	39	19	34	12	37	1	53	34	52	14
8	4	27	48	4	48	38	4	42	3	22	50
9	1	16	16	35	24	39	1	30	31	53	26
10	4	4	45	6	0	40	4	19	0	24	2
11	0	53	13	36	36	41	1	7	28	54	38
12	3	41	42	7	12	42	3	55	57	25	14
13	0	30	10	37	46	43	0	44	25	55	50
14	3	18	39	8	24	44	3	32	54	26	26
15	0	7	7	39	1	45	0	21	22	57	3
16	2	55	36	9	37	46	3	9	51	27	39
17	5	44	4	40	13	47	5	58	19	58	15
18	2	32	33	10	49	48	2	46	48	28	51
19	5	21	1	41	25	49	5	35	16	59	27
20	2	9	30	12	1	50	2	23	45	30	3
21	4	57	58	42	37	51	5	12	14	0	39
22	1	46	27	13	13	52	2	0	42	31	15
23	4	34	55	43	49	53	4	49	11	1	51
24	1	23	24	14	25	54	1	37	39	32	27
25	4	11	52	45	1	55	4	26	8	3	3
26	1	0	21	15	37	56	1	14	36	33	39
27	3	48	49	46	13	57	4	3	5	4	15
28	0	37	18	16	49	58	0	51	33	34	51
29	3	25	46	47	25	59	3	40	2	5	27
30	0	14	15	18	2	60	0	28	30	36	4

Mars' Movement of Parallax in Periods of Sixty Days

Days	60°	°	'	"	'''	Days	60°	°	'	"	'''
1	0	0	27	41	40	31	0	14	18	31	51
2	0	0	55	23	20	32	0	14	46	13	31
3	0	1	23	5	1	33	0	15	14	55	12
4	0	1	50	46	41	34	0	15	41	36	52
5	0	2	18	28	21	35	0	16	9	18	32
6	0	2	46	10	2	36	0	16	37	0	13
7	0	3	13	51	42	37	0	17	4	41	53
8	0	3	41	33	22	38	0	17	32	23	33
9	0	4	9	15	3	39	0	18	0	5	14
10	0	4	36	56	43	40	0	18	27	46	54
11	0	5	4	38	24	41	0	18	55	28	35
12	0	5	32	20	4	42	0	19	23	10	15
13	0	6	0	1	44	43	0	19	50	51	55
14	0	6	27	43	25	44	0	20	18	33	36
15	0	6	55	25	5	45	0	20	46	15	16
16	0	7	23	6	45	46	0	21	13	56	56
17	0	7	50	48	26	47	0	21	41	38	37
18	0	8	18	30	6	48	0	22	9	20	17
19	0	8	46	11	47	49	0	22	37	1	57
20	0	9	13	53	27	50	0	23	4	43	38
21	0	9	41	35	7	51	0	23	32	25	18
22	0	10	9	16	48	52	0	24	0	6	59
23	0	10	36	58	28	53	0	24	27	48	39
24	0	11	4	40	8	54	0	24	55	30	19
25	0	11	32	21	48	55	0	25	23	12	0
26	0	12	0	3	29	56	0	25	50	53	40
27	0	12	27	45	9	57	0	26	18	35	20
28	0	12	55	26	50	58	0	26	46	17	1
29	0	13	23	8	30	59	0	27	13	58	41
30	0	13	50	50	11	60	0	27	41	40	22

Venus' Movement of Parallax in Years and Periods of Sixty Years

Egyptian Years	60°	°	'	"	'''
1	3	45	1	45	3
2	1	30	3	30	7
3	5	15	5	15	11
4	3	0	7	0	14
5	0	45	8	45	18
6	4	30	10	30	22
7	2	15	12	15	25
8	0	0	14	0	29
9	3	45	15	45	33
10	1	30	17	30	36
11	5	15	19	15	40
12	3	0	21	0	44
13	0	45	22	45	47
14	4	30	24	30	51
15	2	15	26	15	55
16	0	0	28	0	58
17	3	45	29	46	2
18	1	30	31	31	6
19	5	15	33	16	9
20	3	0	35	1	13
21	0	45	36	46	17
22	4	30	38	31	20
23	2	15	40	16	24
24	0	0	42	1	28
25	3	45	43	46	31
26	1	30	45	31	35
27	5	15	47	16	39
28	3	0	49	1	42
29	0	45	50	46	46
30	4	20	52	31	50
31	2	15	54	16	53
32	0	0	56	1	57
33	3	45	57	47	1
34	1	30	59	32	4
35	5	16	1	17	8
36	3	1	3	2	12
37	0	46	4	47	15
38	4	31	6	32	19
39	2	16	8	17	23
40	0	1	10	2	26
41	3	46	11	47	30
42	1	31	13	32	34
43	5	16	15	17	37
44	3	1	17	2	41
45	0	46	18	47	45
46	4	31	20	32	48
47	2	16	22	17	52
48	0	1	24	2	56
49	3	46	25	47	59
50	1	31	27	33	3
51	5	16	29	18	7
52	3	1	31	3	10
53	0	46	32	48	14
54	4	31	34	33	18
55	2	16	36	18	21
56	0	1	38	3	25
57	3	46	39	48	29
58	1	31	41	33	32
59	5	16	43	18	36
60	3	1	45	3	40

Venus' Movement of Parallax in Periods of Sixty Days

Days	60°	°	'	"	'''
1	0	0	36	59	28
2	0	1	13	58	57
3	0	1	50	58	25
4	0	2	27	57	54
5	0	3	4	57	22
6	0	3	41	56	51
7	0	4	18	56	20
8	0	4	55	55	48
9	0	5	32	55	17
10	0	6	9	54	45
11	0	6	46	54	14
12	0	7	23	53	43
13	0	8	0	53	11
14	0	8	37	52	40
15	0	9	14	52	8
16	0	9	51	51	37
17	0	10	28	51	5
18	0	11	5	50	34
19	0	11	42	50	2
20	0	12	19	49	31
21	0	12	56	48	59
22	0	13	33	48	28
23	0	14	10	47	57
24	0	14	47	47	26
25	0	15	24	46	54
26	0	16	1	46	23
27	0	16	38	45	51
28	0	17	15	45	20
29	0	17	52	44	48
30	0	18	29	44	17
31	0	19	6	43	46
32	0	19	43	43	14
33	0	20	20	42	43
34	0	20	57	42	11
35	0	21	34	41	40
36	0	22	11	41	9
37	0	22	48	40	37
38	0	23	25	40	6
39	0	24	2	39	34
40	0	24	39	39	3
41	0	25	16	38	31
42	0	25	53	38	0
43	0	26	30	37	29
44	0	27	7	36	57
45	0	27	44	36	26
46	0	28	21	35	54
47	0	28	58	35	23
48	0	29	35	34	52
49	0	30	12	34	20
50	0	30	49	33	49
51	0	31	26	33	17
52	0	32	3	32	46
53	0	32	40	32	14
54	0	33	17	31	43
55	0	33	54	31	12
56	0	34	31	30	40
57	0	35	8	30	9
58	0	35	45	29	37
59	0	36	22	29	6
60	0	36	59	28	35

MERCURY'S MOVEMENT OF PARALLAX IN YEARS AND PERIODS OF SIXTY YEARS

Egyptian Years	Movement					Egyptian Years	Movement				
	60°	°	′	″	‴		60°	°	′	″	‴
1	0	53	57	23	6	31	3	52	38	56	21
2	1	47	54	46	13	32	4	46	36	19	28
3	2	41	52	9	19	33	5	40	33	42	34
4	3	35	49	32	26	34	0	34	31	5	41
5	4	29	46	55	32	35	1	28	28	28	47
6	5	23	44	18	39	36	2	22	25	51	54
7	0	17	41	41	45	37	3	16	23	15	0
8	1	11	39	4	52	38	4	10	20	38	7
9	2	5	36	27	58	39	5	4	18	1	13
10	2	59	33	51	5	40	5	58	15	24	20
11	3	53	31	14	11	41	0	52	12	47	26
12	4	47	28	37	18	42	1	46	10	10	33
13	5	41	26	0	24	43	2	40	7	33	39
14	0	35	23	23	31	44	3	34	4	56	46
15	1	29	20	46	37	45	4	28	2	19	52
16	2	23	18	9	44	46	5	21	59	42	59
17	3	17	15	32	50	47	0	15	57	6	5
18	4	11	12	55	57	48	1	9	54	29	12
19	5	5	10	19	3	49	2	3	51	52	18
20	5	59	7	42	10	50	2	57	49	15	25
21	0	53	5	5	16	51	3	51	46	38	31
22	1	47	2	28	23	52	4	45	44	1	38
23	2	40	59	51	29	53	5	39	41	24	44
24	3	34	57	14	36	54	0	33	38	47	51
25	4	28	54	37	42	55	1	27	36	10	57
26	5	22	52	0	49	56	2	21	33	34	4
27	0	16	49	23	55	57	3	15	30	57	10
28	1	10	46	47	2	58	4	9	28	20	17
29	2	4	44	10	8	59	5	3	25	43	23
30	2	58	41	33	15	60	5	57	23	6	30

MERCURY'S MOVEMENT OF PARALLAX IN PERIODS OF SIXTY DAYS

Days	Movement					Days	Movement				
	60°	°	′	″	‴		60°	°	′	″	‴
1	0	3	6	24	13	31	1	36	18	31	3
2	0	6	12	48	27	32	1	39	24	55	17
3	0	9	19	12	41	33	1	42	31	19	31
4	0	12	25	36	54	34	1	45	37	43	44
5	0	15	32	1	8	35	1	48	44	7	58
6	0	18	38	25	22	36	1	51	50	32	12
7	0	21	44	49	35	37	1	54	56	56	25
8	0	24	51	13	49	38	1	58	3	20	39
9	0	27	57	38	3	39	2	1	9	44	53
10	0	31	4	2	16	40	2	4	16	9	6
11	0	34	10	26	30	41	2	7	22	33	20
12	0	37	16	50	44	42	2	10	28	57	34
13	0	40	23	14	57	43	2	13	35	21	47
14	0	43	29	39	11	44	2	16	41	46	1
15	0	46	36	3	25	45	2	19	48	10	15
16	0	49	42	27	38	46	2	22	54	34	28
17	0	52	48	51	52	47	2	26	0	58	42
18	0	55	55	16	6	48	2	29	7	22	56
19	0	59	1	40	19	49	2	32	13	47	9
20	1	2	8	4	33	50	2	35	20	11	23
21	1	5	14	28	47	51	2	38	26	35	37
22	1	8	20	53	0	52	2	41	32	59	50
23	1	11	27	17	14	53	2	44	39	24	4
24	1	14	33	41	28	54	2	47	45	48	18
25	1	17	40	5	41	55	2	50	52	12	31
26	1	20	46	29	55	56	2	53	58	36	45
27	1	23	52	54	9	57	2	57	5	0	59
28	1	26	59	18	22	58	3	0	11	25	12
29	1	30	5	42	36	59	3	3	17	49	26
30	1	33	12	6	50	60	3	6	24	13	40

2. DEMONSTRATION OF THE REGULAR AND APPARENT MOVEMENTS OF THESE PLANETS ACCORDING TO THE THEORY OF THE ANCIENTS

[140^b] Accordingly this is the way the mean movements are. Now let us turn to the apparent irregularity. The ancient mathematicians who kept the earth immobile imagined in the case of Saturn, Jupiter, Mars, and Venus eccentric circles bearing epicycles, and another further eccentric circle, with respect to which the epicycle and the planet in the epicycle should move regularly.

In this way let *AB* be an eccentric circle, and let its centre be at *C*. Let *ABC* be its diameter, whereon *D* is the centre of the Earth, so that *A* is the apogee and *B* the perigee. Let *DC* be bisected at *E*, and with *E* as centre let *FG* another eccentric circle to the first be described.

Let *H* be anywhere on this eccentric circle, and with *H* as centre let the epicycle *IK* be described. Through its centre let there be drawn the straight line *IHKC* and similarly *LHME*. Now let it be understood that on account of the latitudes of the planet the eccentric circles are inclined to the plane of the ecliptic and similarly the epicycle to the plane of the eccentric circle; but here they are represented as if in one plane for the sake of ease of demonstration. Accordingly the ancients say that this whole plane together with points *E* and *C* moves around *D*—the centre of the ecliptic—in the movement of the sphere of the fixed stars: by this they mean that these points have unchanging positions in the sphere of the fixed stars. And they say that the epicycle moves eastward in circle *FHG* but in accordance with line *IHC*, and in relation to this line the planet revolves regularly in epicycle *IK*. But it is clear that the regularity of the epicycle should occur in relation to *E* the centre of its deferent,[1] and the revolution of the planet in relation to line *LME*. Accordingly they concede that in this case the regularity of the circular movement can occur with respect to a foreign and not the proper centre; similarly and more so in the case of Mercury. But I think I have already made a sufficient refutation of that in the case of the moon. These and similar things furnished us with an occasion for working out the mobility of the Earth and some other ways by which regularity and the principles of this art might be preserved, and the ratio of apparent irregularity rendered more constant.

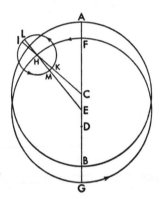

3. GENERAL DEMONSTRATION OF APPARENT IRREGULARITY ON ACCOUNT OF THE MOVEMENT OF THE EARTH

[141^a] Accordingly there are two reasons why the regular movement of a planet should appear irregular: on account of the movement of the Earth and on account of

[1]The deferent of an epicycle is the circle on the circumference of which the centre of the epicycle moves.

its proper movement. We shall make both of them clear generally and separately by ocular demonstration, whereby they can be better distinguished from one another; and we shall begin with the movement which mixes itself with all of them on account of the movement of the Earth: and first in the case of Venus and Mercury, which are comprehended by the (orbital) circle of the Earth.

Therefore let *AB* be the circle eccentric to the sun, which the centre of the Earth describes during its annual circuit in the way we explained above; and let *C* be its centre. But now let us put down that the planet has no other irregularity except this one;

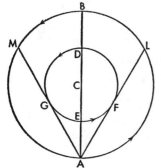

and that will be the case if we make *DE*, the orbital circle of Venus or Mercury, homocentric with *AB*; and *DE* should be inclined to *AB* on account of its latitude. But for the sake of ease of demonstration they can be thought of as if in the same plane. Let the Earth be assumed at point *A*; and from *A* let there be drawn the lines of sight *AFL* and *AGM* touching the circle of the planet at points *F* and *G*; and let *ACB* the diameter common to both circles be drawn.

Now let the movement of both the Earth and the planet be in the same direction, *i.e.*, eastward, but with greater velocity in the case of the planet than in that of the Earth. Therefore *C* and line *ACB* will appear to the eye borne along at *A* to move in accordance with the mean movement of the sun; but the planet in circle *DFG* as in an epicycle will traverse arc *FDG* eastward in greater time than it will the remaining arc *GEF* westward; and in the upper arc it will add the total angle *FAG* to the mean movement of the sun, and in the lower arc will subtract the same. Accordingly where the subtractive movement of the planet, especially around *E* the perigee, is greater than the additive (movement) of *C*, it will seem to *A* to retrograde in proportion to the excelling (movement)—as happens in these planets, when line *CE* has a greater ratio to line *AE* than the movement at *A* has to the movement of the planet, according to the demonstrations of Apollonius of Perga, as will be said later. But where the additive movement is equal to the subtractive, [141^b] the planet will seem to come to a stop on account of the mutual equilibrium; all this agrees with the appearances.

Accordingly if there were no other irregularity in the movement of the planet, as Apollonius opined, this would be sufficient. But the greatest angular elongations from the mean movement of the sun, which these planets have in the morning and evening and which are understood by angles *FAE* and *GAE*, are not everywhere equal, neither the one to the other, nor are the sums of the two equal; for the apparent reason that the route of these planets is not along circles homocentric with the terrestrial circle but along certain others, by which they effect the second irregularity.

The same thing is also demonstrated in the case of the three upper planets, Saturn, Jupiter, and Mars, which circle around the Earth. For let the former circle of the Earth be drawn again, and let *DE* be as an exterior homocentric circle in the same plane: let the position of the planet be taken anywhere, at point *D*; and from *D* let there be drawn *DACBE* the common diameter and *DF* and *DG* straight lines touching the orbital circle of the Earth at points *F* and *G*. It is manifest that from point *A* only will the true position of the planet in *DE*

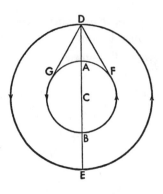

the line of mean movement of the sun be apparent, when the planet is opposite the sun and is nearest to the Earth. For when the Earth is in the opposite position at *B*, the opposition (of the planet and the sun), although in the same straight line, will not be at all apparent on account of the closeness of the sun to *C*. But as the movement of the Earth is speedier, so that it outruns the movement of the planet, it will seem along *FBG* the arc of apogee to add the total angle *GDF* to the movement of the planet and along the remaining arc *GAF* to subtract the same, according as arc *GAF* is smaller. But where the subtractive movement of the Earth excels the additive movement of the planet, especially in the neighbourhood of *A*, the planet will seem to be left behind by the Earth, to move westward and to come to a stop at the place where there is least difference between the movements which are contrary according to sight.

And so it is once more manifest that all these apparent movements—which the ancients were looking into by means of the epicycles of the individual planets—occur on account of the movement of the Earth. But since in spite of the opinion of Apollonius and the ancients the movement of the planet is not found regular, as the irregular revolution of the Earth with respect to the planet produces that; accordingly the planets are not carried in a homocentric circle but in some other which we shall demonstrate straightway.

4. WHY THE PROPER MOVEMENTS OF THE PLANETS APPEAR IRREGULAR

[142ᵃ] But since their proper movements in longitude follow approximately the same mode except for Mercury, which is seen to differ from them, we shall treat of those four planets together, but another place has been given over to Mercury. Accordingly as the ancients placed one movement in two eccentric circles, as was shown, we have decreed two regular movements out of which the apparent irregularity is compounded either by a circle eccentric to an eccentric circle, or by the epicycle of an epicycle or by a combination of an eccentric circle carrying an epicycle. For they can all effect the same irregularity, as we demonstrated above in the case of the sun and the moon.

Accordingly let *AB* be an eccentric circle around centre *C*. Let *ACB* be the diameter drawn through the highest and lowest apsis of the planet and containing the mean position of the sun. On *ACB* let *D* be the centre of the orbital circle of the Earth; and with the highest apsis *A* as centre and the third part of *CD* as radius, let epicycle *EF* be described. Let *F* be its perigee, and let the planet be placed there. Now let the movement of the

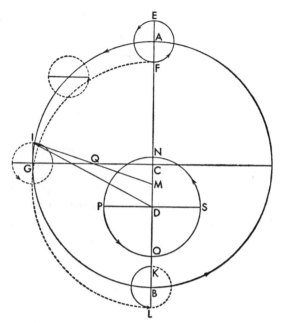

epicycle along eccentric circle *AB* take place eastward; and let the movement of the planet in the upper arc (of the epicycle) take place similarly eastward [142ᵇ] but in the remaining arc westward; and let the revolutions of the epicycle and the planet be proportionately equal to one another.

On that account when the epicycle is at the highest apsis of the eccentric circle and the planet on the contrary is at the perigee of the epicycle, the relation between their movements is reversed[1] with respect to one another, since both the planet and the epicycle have traversed their semicircle. But in both mean quadrants each will have its mean apsis, and then only will the diameter[2] of the epicycle be parallel to line *AB*; and at the midpoints (between the mean quadrants and the perigee or apogee) the diameter will be perpendicular to *AB*: the rest of the time always moving towards *AB* or moving away. All that is easily understood as following from the movements.

Hence it will also be demonstrated that by this composite movement the planet does not describe a perfect circle in accordance with the theory of the ancient mathematicians but a curve differing imperceptibly from one.

For let the same epicycle *KL* be drawn again, and let *B* be its centre. Let *AG* the quadrant of a circle be assumed, and let *HI* be an epicycle around *G*. Let *CD* be cut into three equal parts, and let

---

[1]That is to say, during the hemicycle of movement wherein the epicycle is passing from the lowest to the highest apsis of the eccentric circle and the planet is passing from the apogee to the perigee of the epicycle, the movement on the epicycle adds to the movement on the eccentric circle; but during the hemicycle wherein the epicycle is passing from the highest to the lowest apsis, the movement on the epicycle subtracts from the movement on the eccentric circle.

[2]In this passage Copernicus is speaking as if the planet were borne around the epicycle by the revolving diameter, although he usually speaks as if the diameter of the epicycle pointed perpetually at the centre of the homocentric circle.

$$CM = \tfrac{1}{3}CD = GI.$$

And let GC and IM, which cut one another at Q, be joined.

Accordingly since, by hypothesis

$$\text{arc } AG = \text{arc } HI$$

and

$$\text{angle } ACG = 90°;$$

then

$$\text{angle } HGI = 90°.$$

And

$$\text{angle } IQG = \text{angle } MQC,$$

because they are vertical angles. Therefore triangles GIQ and QCM are equiangular; and they have correspondingly equal sides, since by hypothesis

$$\text{base } GI = \text{base } CM.$$

And

$$QI > QC = QI > QG;$$

therefore

$$IQM > GQC,$$

but

$$FM = ML = AC = CG.$$

Therefore the circle which is described around centre M through points F and L and is hence equal to circle AB will cut line IM. The same demonstration will hold in the opposite quadrant. Accordingly by the regular movements of the epicycle in the eccentric circle the planet in the epicycle will not describe a perfect circle but a quasi-circle—as was to be demonstrated.[1]

Now around centre D let NO the annual orbital circle of the Earth be described; let IDR be extended; and let PDS be drawn parallel to CG. Accordingly IDR will be the straight line of the true movement of the planet; GC, the straight line of the mean and regular movement. And R will be the true apogee of the Earth with respect to the planet; and S, the mean apogee. Accordingly angle RDS or IDP is the difference between the regular and the apparent movement of both, namely between angle ACG and angle CDI.

But in place of eccentric circle AB we may take an equal homocentric circle around D as the deferent of the epicycle, whose radius is equal to DC and which is the deferent of the other epicycle, whose semi-diameter is half MD.[2] New let the first epicycle be moved [143^a] eastward, but the second in the opposite direction; and

---

[1] As has been pointed out, if in the foregoing diagram we consider a point X so situated on semi-diameter CA that CX is equal to GI (and consequently DM is equal to MX), then since the planet on reaching point I has expended one quarter of its periodic time and has traversed one quarter of a full revolution about point X, evidently point X is analogous to the centre of a Ptolemaic equant, point M (the centre of the quasi-circle) to the centre of the Ptolemaic deferent, and point D (the centre of the sun for Copernicus) to the centre of the Earth.

[2] As in the accompanying diagram:

lastly let the planet on it (*i.e.*, on the second epicycle) be deflected by the twofold movement. The same things will happen as before and no differently from in the moon, or by some other of the aforesaid modes.

But here we have chosen the eccentric circle bearing the epicycle, because by remaining always between the sun and *C* centre *D* is meantime found to have changed, as was shown in the case of solar appearances. But as the remaining appearances do not accord proportionately with this change there must be some other irregularity in those planetary movements: this irregularity, although very slight, is perceptible in the case of Mars and Venus, as will be seen in the right place.

Accordingly we shall soon demonstrate from observations that these hypotheses are sufficient for the appearances; and we shall do that first in the case of Saturn, Jupiter, and Mars: in them the position of the apogee and the distance *CD* are very difficult to find and of the greatest importance, since the rest is easily demonstrable by means of the apogee and the distance *CD*. Now in this case we shall use the method we used concerning the moon, namely a comparison of three ancient solar oppositions with the same number of modern ones, which the Greeks call their "acronychial gleams" and we the "deeps of the night," namely when the planet opposite the sun falls upon the straight line of the mean movement of the sun, where it throws off all that irregularity which the movement of the Earth brings to it. Such positions are determined by observations with an astrolabe and by computation of the oppositions of the sun, until it is clear that the planet has arrived at a point opposite the Sun.

5. DEMONSTRATIONS OF THE MOVEMENT OF SATURN

Accordingly we shall begin with Saturn by taking three oppositions once observed by Ptolemy. The first of them occurred in the 11th year of Hadrian on the 7th day of the month Pachom at the first hour of night; in the year of Our Lord 127 on the 7th day before the Kalends of April, 17 equal hours after midnight in relation to the Cracow meridian, which we find an hour distant from Alexandria. Now the position of the planet in relation to the sphere of the fixed stars, to which as to the starting-point of the regular movement we are referring all these things, was found to be at approximately 174°40', since [143^b] the sun by its simple movement was then opposite at 354°40' from the horn of Aries, the starting-point assumed.

The second opposition was in the 17th year of Hadrian on the 18th day of the month Epiphi by the Egyptian calendar; but by the Roman, in the year of Our Lord 133 on the 3rd day before the Nones of June, 11 equatorial hours after midnight: he found the planet at 243°3', while by its mean movement the sun was at 63°3', 15 hours after midnight.

He recorded the third as occurring in the 20th year of Hadrian on the 24th day of the month Mesori by the Egyptian calendar; which was in the year of Our Lord 136 on the 8th day before the Ides of July, 11 hours after midnight (similarly according to the Cracow meridian) at 277°37', while by its mean movement the sun was at 97°37'.

Accordingly in the first interval there are 6 years 70 days 55 minutes (of a day), during which the planet is moved 62°23' in relation to sight, and the mean movement of the Earth with respect to the planet, *i.e.*, the movement of parallax, is 352°44'. Accordingly the 7°16' in which the circle is deficient belong to the mean movement of the planet, so that it is 75°39'.

In the second interval there are 3 Egyptian years 35 days 50 minutes; the apparent movement of the planet is 34°34', (the movement) of parallax 356°43'; and the remaining 3°17' of a circle are added to the apparent movement of the planet, so that the mean movement is 37°51'.

After this survey let *ABC* the eccentric circle of the planet be described. Let *D* be its centre, and *FDG* its diameter, whereon *E* is the centre of the great orbital circle of the Earth. Now let *A* be the centre of the epicycle at the first opposition to the sun; *B*, at the second; and *C*, at the third; and around them let the same epicycle be described with a radius equal to one-third of *DE*. Let the centres *A*, *B*, and *C* be joined to *D* and *E* by straight lines, which will cut the

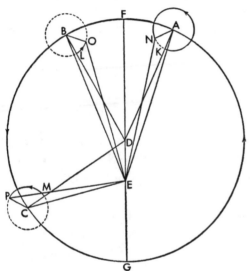

circumference of the epicycle at points *K*, *L*, and *M*. And let there be taken arc *KN* similar to *AF*, arc *LO* similar to *BF*, and *MP* similar to *FBC*; and let *EN*, *EO*, and *EP* be joined. Therefore by computation

arc *AB* = 75°39'

and

arc *BC* = 37°51';

and of the angles of apparent movement,

angle *NEO* = 68°23'

and

angle *OEP* = 34°34'.

Our problem is to examine the positions of highest and lowest apsis, *i.e.*, of *F* and *G*, together with the distance *DE* between the centres, without which there is no way of discerning the regular and the apparent [144ᵃ] movement.

But here too we run into as great a difficulty as in this part of Ptolemy, since, if the given angle *NEO* comprehended the given arc *AB*, and (angle) *OEP* (arc) *BC*, the entrance to demonstrating what we are looking for would be already opened. But the known arc *AB* subtends the unknown angle *AEB*, and similarly the unknown angle

BEC is subtended by the known arc *BC*; for it was necessary for both of them to be known. But *AEN*, *BEO*, and *CEP*, the differences between the angles, cannot be perceived, unless arcs *AF*, *FB*, and *FBC* are first set up as similar to those on the epicycle; accordingly these things are mutually dependent so as to be simultaneously known or unknown. Therefore those who were destitute of the means of demonstration relied upon detours and the *a posteriori* method, as the straightforward and *a priori* approach was not open. So Ptolemy in this investigation expended his energies in a prolix argument and a great multitude of calculations, which I judge boring and supererogatory to review, especially as in our calculations, which follow, we shall copy the same method approximately.

Finally in going over his calculations again he found that

$$\text{arc } AF = 57°1',$$
$$\text{arc } BF = 18°37',$$

and

$$\text{arc } FBC = 56\frac{1}{2}°.$$

But

$$\text{ecc.} = 6^P50',$$
$$\text{where } DF = 60^P.$$

But

$$\text{ecc.} = 1139,$$
$$\text{where } DF = 10,000.$$

Now

$$\frac{3}{4}(1139) \doteqdot 854,$$

and

$$\frac{1}{4}(1139) \doteqdot 1285.$$

Hence

$$DE = 854$$

and

$$\text{rad. ep.} = 285.$$

Making these assumptions and borrowings for our hypothesis, [144^b] we shall show that these things agree with the appearances observed.

Now at the first solar opposition, in triangle *ADE*,

$$\text{side } AD = 10,000,$$

and

$$\text{side } DE = 854;$$

and

$$\text{angle } ADE = 180° - \text{angle } ADF.$$

Hence, by means of what we have shown concerning plane triangles, side AE = 10,489,

and

angle DEA = 53°6',

and

angle DAE = 3°55',
where 4 rt. angles = 360°.

But

angle KAN = angle ADF = 57°1'.

Therefore by addition

angle NAE = 60°56'.

Accordingly in triangle NAE two sides are given:
side AE = 10,489,
side NA = 285
where AD = 10,000,

and

angle NAE is given.

Hence

angle AEN = 1°22';

and, by subtraction,

angle NED = 51°44',
where 4 rt. angles = 360°.

Similarly at the second solar opposition. For in triangle BDE
side DE = 854,
where BD = 10,000;

and

angle BDE = 180° − BDF = 161°22'.

So triangle BDE too has its sides and angles given:
side BE = 10,812,
where BD = 10,000,

and

angle DBE = 1°27',

and

angle BED = 17°11'.

But

angle OBL = angle BDF = 18°36'.

Therefore, by addition

angle EBO = 20°3'.

Accordingly in triangle EBO two sides are given together with angle EBO:

$$BE = 10,812$$

and

$$BO = 285.$$

By what we have shown concerning plane triangles

angle $BEO = 32'$.

Hence

angle $BED = 16°39'$.

Moreover, in the third solar opposition, in triangle CDE, as before,

side CD is given

and

side DE is given;

and

angle $CDE = 180° - 56°29'$.

By the fourth rule for plane triangles

base $CE = 10,512$,

where $CD = 10,000$;

and

angle $DCE = 3°53'$

and, by subtraction,

angle $CED = 52°36'$.

Therefore, by addition,

angle $ECP = 60°22'$,

where 4 rt. angles = 360°.

So also in triangle ECP two sides are given together with angle ECP; furthermore,

angle $CEP = 1°22'$,

whence, by subtraction,

angle $PED = 51°14'$.

Hence, of the total angles of apparent movement,

angle $OEN = 68°23'$,

and

angle $OEP = 34°35'$,

which agree with the observations. And the position of the highest apsis of the eccentric circle

$$F = 226°20'$$

from the head of Aries. And as the then existing precession of [145ᵃ] the spring equinox was 6°40',

$$226°20' + 6°40' = 23° \text{ of Scorpio,}$$

in accordance with Ptolemy's conclusion. For the apparent position of the planet at this third solar opposition, as was reported above, was 227°37'. And as the angle of apparent movement,

angle PED = 51°14'.

Hence

$$227°37' - 51°14' = 226°23',$$

which is the position of the highest apsis of the eccentric circle.

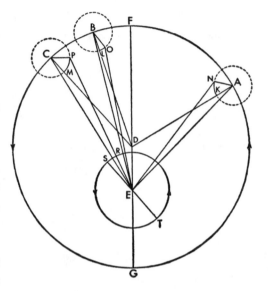

Now let there be described RST the annual orbital circle of the Earth, which will cut line PE at point R; and let the diameter SET be drawn parallel to the line of mean movement of the planet. Accordingly, as

angle SED = angle CDF,

angle SER will be the difference and the additosubtraction between the apparent and mean movement, *i.e.*, between angles CDF and PED, and

angle SER = 5°16'.

And there is the same difference between the mean and the true movements of parallax. Now

arc RT = 180° – arc SER = 174°44',

which is the regular movement of parallax from starting-point T, *i.e.*, from the mean conjunction of the sun and the planet, to this third solar opposition or true opposition of the Earth and the planet.

Accordingly at the time of this observation, namely in the 20th year of the reign of Hadrian, but in the 136th year of Our Lord on the 8th day before the Ides of July, 11 hours after midnight, we have the movement of anomaly of Saturn from the highest apsis of its eccentric circle as $56^1/_2$°, and the mean movement of parallax as 174°44', as was timely to demonstrate on account of what follows.

6. ON THREE OTHER SOLAR OPPOSITIONS OF SATURN RECENTLY OBSERVED

[145b] Now since the computation of the movement of Saturn handed down by Ptolemy has no small discrepancy with our times, and since it cannot be understood right away in what quarter the error lies, we are forced to make new observations, out of which we have again taken three solar oppositions. The first opposition was in the

year of Our Lord 1514, on the 3rd day before the Nones of May $1^1/_5$ hours before midnight, at which time Saturn was discovered at 205°24'.

The second was in the year of Our Lord 1520 on the third day before the Ides of July at midday, and the planet was at 273°25'.

The third was in the year of Our Lord 1527 on the 6th day before the Ides of October $6^2/_5$ hours after midnight; and Saturn appeared at 7' from the horn of Aries.

Accordingly between the first and second solar oppositions there are 6 Egyptian years 70 days 33 minutes (of a day), during which time the apparent movement of Saturn is 68°1'.

From the second to the third there are 7 Egyptian years 89 days 46 minutes, and the apparent movement of the planet is 86°42'; and the mean movement during the first interval is 75°39'; and during the second, 88°29'. Accordingly in investigating the

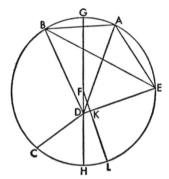

highest apsis and the eccentricity, we must at first abide by the rule of Ptolemy, just as if the planet moved in a simple eccentric circle; and although that is not sufficient, nevertheless we shall be led fairly near and shall arrive at the truth more easily.

Accordingly, let ABC be the circle in which the planet is moved regularly: and let the first opposition be at A, the second at B, and the third at C. Let the centre of the orbital circle of the Earth be taken within it as D. Let AD, BD, and CD be joined, and let any one of them be extended in a straight line to the opposite part of the circumference—say CDE—and let AE and BE be joined.

Accordingly, since

$$\text{angle } BDC = 86°42',$$
$$\text{where 2 rt. angles} = 180°;$$
$$\text{angle } BDE = [146^a]\ 93°18';$$

but

$$\text{angle } BDE = 186°36',$$
$$\text{where 2 rt. angles} = 360°.$$

And, as intercepting arc BC,

$$\text{angle } BED = 88°29',$$

and

$$\text{angle } DBE = 84°55'.$$

Accordingly, as the angles of triangle BDE are given, the sides are given by the table:

$$BE = 19,953$$

and

$$DE = 13,501$$

where diameter of circle circumscribing triangle = 20,000.

Similarly in triangle *ADE*, since

$$\text{angle } ADC = 154°43',$$

where 2 rt. angles = 180°;

$$\text{angle } ADE = 180° - \text{angle } ADC = 25°17';$$

but

$$\text{angle } ADE = 50°34',$$

where 2 rt. angles = 360°.

And, as intercepting arc *ABC*,

$$\text{angle } AED = 164°8',$$

and

$$\text{angle } DAE = 145°18' :$$

hence the sides are established:

$$DE = 19,090$$

and

$$AE = 8,542,$$

where diameter of circle circumscribing triangle *ADE* = 20,000.

But

$$AE = 6,043$$

where *DE* = 13,501 and

$$BE = 19,953.$$

Hence too, in triangle *ABE*, these two sides *BE* and *EA* have been given; and, as intercepting arc *AB*,

$$\text{angle } AEB = 75°39'.$$

Accordingly by what we have shown concerning plane triangles,

$$AB = 15,647$$

where *BE* = 19,968.

But according as

$$\text{ch. } AB = 12,266,$$

where diameter of eccentric circle = 20,000;

$$EB = 15,664$$

and

$$DE = 10,599.$$

Accordingly, in proportion to chord *BE*,

$$\text{arc } BAE = 103°7'.$$

Hence, by addition,

$$\text{arc } EABC = 191°36';$$

and
$$\text{arc } CE = 360° - \text{arc } EABC = 168°24';$$
and hence
$$\text{ch. } CDE = 19,898.$$
And
$$CD = CDE - DE = 9,299.$$

And now it is manifest that, if *CDE* were the diameter of the eccentric circle, the positions of highest and lowest apsis would fall upon it, and the distance between the centres would be evident; but because segment *EABC* is greater, the centre will be in it. Let *F* be the centre, and let the diameter *GFDG* be extended through *F* and *D*, and let *FKL* be drawn at right angles to *CDE*.

Now it is manifest that
$$\text{rect. } CD, DE = \text{rect. } GD, DH.$$
But
$$\text{rect. } GD, DH + \text{sq. } FD = \text{sq. } (^1/_2 GDH) = \text{sq. } FDH.$$
Accordingly
$$\text{sq. } FDH - \text{rect. } CD, DE = \text{sq. } FD.$$
Therefore
$$FD = 1,200$$
$$\text{where radius } GF = 10,000;$$
but
$$FD = 7\text{P}12'$$
$$\text{where radius} = 60\text{P},$$
[146^b] which differs little from Ptolemy.

But since
$$CDK = ^1/_2 CDE = 9,949$$
and
$$CD = 9,299,$$
therefore
$$DK = CDK - CD = 650,$$
$$\text{where } GF = 10,000$$
$$\text{and } FD = 1,200.$$
But
$$DK = 5,411$$
$$\text{where } FD = 10,000.$$
And since
$$DK = ^1/_2 \text{ ch. } 2 \text{ } DFK,$$
$$\text{angle } DFK = 32°45',$$

where 4 rt. angles = 360°;

and as standing at the centre of the circle it intercepts a similar chord and arc *HL* on the circumference.

But

$$\text{arc } CHL = \frac{1}{2}CLE = 84°13';$$

therefore

$$\text{arc } CH = CHL - HL = 51°28',$$

which is the distance from the third opposition to the perigee.

Now

$$180° - 51°28' = CBG = 128°32'$$

from the highest apsis to the third opposition. And since

$$\text{arc } CB = 88°29',$$

$$\text{arc } BG = CBG - CB = 40°3',$$

from the highest apsis to the second solar opposition. Then, as

$$\text{arc } BGA = 75°39',$$

$$\text{arc } GA = BGA - BG = 35°36'$$

from the first opposition to the apogee *G*.

Now let *ABC* be a circle with diameter *FDEG*, centre *D*, apogee *F*, perigee *G*. Let

arc *AF* = 35°36',

arc *FB* = 40°3',

and

arc *FBC* = 128°32'.

Now let *DE* be taken as three quarters of what has already been shown to be the distance between the centres, *i.e.*, let

DE = 900;

and

quarter distance = 300

where radius = 10,000.

And with that quarter distance as radius, let the epicycle be described around centres *A*, *B*, and *C*—and let

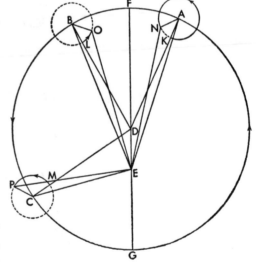

the figure be completed according to the hypothesis set before us. But if with this layout we wish to elicit the observed positions of Saturn [147ª] by the method handed down above and soon to be repeated, we shall find some discrepancy.

And—to speak briefly, so as not to burden the reader with many words or seem to have laboured more in indicating by-ways than in pointing out the high road—these things will, by means of what we have shown concerning triangles, necessarily lead us to the conclusion that

$$\text{angle } NEO = 67°35'$$

and

$$\text{angle } OEP = 87°12'.$$

But angle OEP is $1/2°$ greater than the apparent angle, and the angle NEO is 26' smaller. And we find that they square with one another only if we move the apogee forward a little, and set up

$$\text{arc } AF = 38°50',$$
$$\text{arc } FB = 36°49',$$
$$\text{arc } FBC = 125°18',$$
$$DE = 854,$$

which is the distance between the centres, and

$$\text{rad. ep.} = 285,$$
$$\text{where } FD = 10,000;$$

and that agrees approximately with Ptolemy, as set out above. For it is clear from this that these magnitudes agree with the three apparent solar oppositions observed.

Since at the first opposition, in triangle ADE,

$$\text{side } DE = 854,$$
$$\text{where } AD = 10,000,$$

and

$$\text{angle } ADE = 141°10',$$
$$\text{where angle } ADE + \text{angle } ADF = 2 \text{ rt. angles;}$$

hence it is shown that

$$\text{side } AE = 10,679,$$
$$\text{where radius } FD = 10,000,$$
$$\text{angle } DAE = 2°52',$$

and

$$\text{angle } DEA = 35°58'.$$

Similarly in triangle AEN, since

$$\text{angle } KAN = \text{angle } ADF,$$
$$\text{angle } EAN = 41°42',$$

and

$$\text{side } AN = 285,$$
$$\text{where } AE = 10,679.$$

Hence

$$\text{angle } AEN = 1°3'.$$

But

angle $DEA = 35°58'$;

accordingly, by subtraction,

angle $DEN = 34°55'$.

In the second solar opposition triangle DEB has two sides given:

$$DE = 854,$$

where $DB = 10,000$

and

angle $BDE = 153°11'$.

Accordingly

$$BE = 10,697,$$
angle $DBE = 2°45'$,

and

angle $BED = 34°4'$.

But

angle LBO = angle BDF;

therefore, as at the centre,

angle $EBO = 39°34'$.

Now this angle is comprehended by the given sides

$$BO = 285$$

and

$$BE = 10,697;$$

hence

angle $BEO = 59'$.

And

angle OED = angle BED − angle BEO 33°5'.

But in the first solar opposition it has already been shown that

angle $DEN = 34°55'$.

Therefore by addition

angle OEN = 68°

by which the distance of the first solar opposition from the second becomes apparent; and it harmonizes with the observations.

The same thing will be shown at the third opposition.

In triangle CDE

angle $CDE = 54°42'$,
side $CD = 10,000$,

and

side $DE = 854$;

[147^b] hence

282

$$\text{side } EC = 9{,}532,$$
$$\text{angle } CED = 121°5',$$

and

$$\text{angle } DCE = 4°13';$$

therefore by addition

$$\text{angle } PCE = 129°31'.$$

So again in triangle *EPC*

$$\text{side } CE = 9{,}532$$

and

$$\text{side } PC = 285,$$

and

$$\text{angle } PCE = 129°31' :$$

hence

$$\text{angle } PEC = 1°18'.$$

And

$$\text{angle } PED = \text{angle } CED - \text{angle } PEO = 119°47'$$

from the highest apsis of the eccentric circle to the position of the planet at the third opposition.

Now it was shown that there were 33°5' to the second solar opposition: accordingly between the second and third solar oppositions of Saturn there remain 86°42', which agree with the observations. Now the position of Saturn was found by observation at that time to be at 7' from the assumed starting-point of the first star of Aries, and it was shown that there were 60°13' from it to the lowest apsis of the eccentric circle: accordingly the lowest apsis is approximately $60^1/_3°$, and the position of the highest apsis is diametrically opposite at $240^1/_3°$.

Now let *RST* the great orbital circle of the Earth be set around its centre *E*, and let its diameter *SET* be parallel to *CD* the line of mean movement; and let

$$\text{angle } FDC = \text{angle } DES.$$

Therefore the Earth and our point of sight will be on line
PE, namely at point *R*. Now

$$\text{angle } PES = 5°31',$$

and angle *PES* or arc *RS* is the difference between *FDC*
the angle of regular movement and *DEP* the angle of
apparent movement.
Now

$$\text{arc } RT = 180° - 5°31' = 174°29'$$

which is the distance of the planet from the apogee of the
orbital circle, *i.e.*, from *T*, as if from the mean position
of the sun.

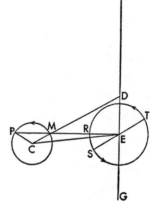

And so we have demonstrated that in the year of Our Lord 1527 on the sixth day before the Ides of October at $6^2/_5$ hours after midnight, the movement of anomaly of Saturn from the highest apsis of the eccentric circle was 125°18', the movement of parallax was 174°29', and the position of the highest apsis was at 240°21' from the first star of Aries in the sphere of the fixed stars.

7. ON THE EXAMINATION OF THE MOVEMENT OF SATURN

[148ª] Now it was shown that Saturn at the time of the last of the three observations of Ptolemy was by its movement of parallax at 174°44', and the position of the highest apsis of the eccentric circle was at 226°23', from the head of the constellation of Aries. Accordingly it is clear that in the midtime between the two observations Saturn has completed 1344 revolutions minus $^1/_4$° of regular parallaxes.

Now from the 20th year of Hadrian on the 24th day of the Egyptian month Mesori one hour before midday to the year of Our Lord 1527 on the 6th day before the Ides of October at 6 hours (after midnight, the time) of this observation, there are 1392 Egyptian years 75 days 48 minutes (of a day).

Hence if we wish to get the movement itself from the table, we shall similarly find 359°45', the movement beyond the 1343 revolutions of parallax. Accordingly what was set down concerning the mean movements of Saturn is correct. Moreover during that time the simple movement of the sun is 82°30'. If 359°45' are subtracted from 82°30', the remainder is the 82°45' of the mean movement of Saturn, which are already being added up in its 47th revolution, in harmony with the computation. Meanwhile too the position of the highest apsis of the eccentric circle has been moved forward to 13°58' in the sphere of the fixed stars. Ptolemy believed it to be fixed in the same way, but now it appears to move approximately 1° per 100 years.

8. ON DETERMINING THE POSITIONS OF SATURN

Now from the beginning of the years of Our Lord to the 20th of Hadrian on the 24th day of the month Mesori at 1 hour before midday, the time of Ptolemy's observation, there are 135 Egyptian years 222 days 27 minutes (of a day), during which time Saturn's movement of parallax was 328°55'. The subtraction of 328°55' from 174°44' leaves 205°49' [148ᵇ] as the locus of distance of the mean position of the sun from the mean (position) of Saturn, and as its movement of parallax at midnight before the Kalends of January.

From the first Olympiad to this locus 775 Egyptian years $12^1/_2$ days comprehend a movement of 70°55' besides the whole revolutions. The subtraction of 70°55' from 205°49' leaves 134°54' for the beginning of the Olympiads at noon on the 1st day of the month Hekatombaion.

Then after 451 years 247 days there are 13°7' besides the whole revolutions: the addition of 13°7' to 134°54' puts the locus (of the years) of Alexander the Great at 148°1' on noon of the 1st day of the month Thoth by the Egyptian calendar; and there are 278 years 118¹/₂ days to (years of) Caesar; the movement is 247°20', and it sets up the locus at 35°21' on midnight before the Kalends of January.

9. On the Parallaxes of Saturn, Which Arise from the Annual Orbital Circle of the Earth, and How Great the Distance of Saturn Is (from the Earth)

In this way it has been demonstrated that the regular movements of Saturn in longitude are at one with the apparent. For the other apparent movements which occur in the case of Saturn are, as we said, parallaxes arising from the annual orbital circle of the Earth, since, as the magnitude of the Earth in relation to the distance of the moon causes parallaxes, so too its orbital circle, in which it revolves annually, should in the case of the five wandering stars cause (parallaxes) which are far more evident in proportion to the magnitude of the orbital circle. Now such parallaxes cannot be determined, unless the altitude of the planet—which, however, it is possible to apprehend through any one observation of a parallax—becomes known first.

We have such (an observation) in the case of Saturn in the year of Our Lord 1514 on the sixth day before the Kalends of May 5 equatorial hours after the preceding midnight. For Saturn was seen to be in a straight line with the stars in the forehead of Scorpio, namely with the second and third stars, which have the same longitude and are at 209° of the sphere of the fixed stars. Accordingly the position of Saturn is made evident through them. Now there are 1514 Egyptian years 61 days 13 minutes (of a day) from the beginning of the years of Our Lord to this time; and according to [149a] calculation the mean position of the sun was at 315°41', the anomaly of parallax of Saturn was at 116°31', and for that reason the mean position of Saturn was 199°10' and that of the highest apsis of the eccentric circle was at approximately 240¹/₃°.

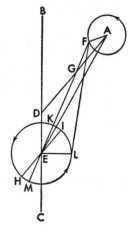

Now in accordance with our problem, let ABC be the eccentric circle: let D be its centre, and on the diameter BDC let B be the apogee, C the perigee, and E the centre of the orbital circle of the Earth. Let AD and AE be joined, and with A as centre and ¹/₃ DE as radius let the epicycle be drawn. On the epicycle let F be the position of the planet; and let

angle DAF = angle ADB.

And through E the centre of the orbital circle of the Earth let HI be drawn, as if in the same plane with circle ABC, and as a diameter, parallel to *AD*, so as to have it understood that with respect to the planet the apogee of the orbital circle is at *H* and the perigee at *I*.

Now on the orbital circle let

$$\text{arc } HL = 116°31'$$

in accordance with the computation of the anomaly of parallax; let *FL* and *EL* be joined, and let *FKEM* produced cut both arcs of the orbital circle.

Accordingly since by hypothesis

$$\text{angle } ADB = \text{angle } DAF = 41°10',$$

and

$$\text{angle } ADE = 180° - ADB = 138°50';$$

and

$$DE = 854$$
$$\text{where } AD = 10,000:$$

whence in triangle *ADE*

$$\text{side } AE = 10,667,$$
$$\text{angle } DEA = 38°9',$$

and

$$\text{angle } EAD = 3°1':$$

therefore by addition

$$\text{angle } EAF = 44°12'.$$

So again in triangle *FAE*

$$\text{side } FA = 285$$
$$\text{where } AE = 10,667,$$
$$\text{side } FKE = 10,465,$$

and

$$\text{angle } AEF = 1°5':$$

accordingly it is manifest that

$$\text{angle } AEF + \text{angle } DAE = 4°6',$$

which is the total difference or additosubtraction between the mean and the true position of the planet. Wherefore if the position of the Earth had been at *K* or *M*, the position of Saturn would have been apparent as if from centre *E* and would have been seen to be at 203°16' from the constellation of Aries. But with the Earth at *L*, Saturn is seen to be at 209°. The difference [149b] of 5°44' goes to the parallax in accord with angle *KFL*. But by calculation of the regular movement

$$\text{arc } HL = 116°31',$$

and

$$\text{arc } ML = \text{arc } HL - \text{add. } HM = 112°25'.$$

And by subtraction[1]

---

[1]Arc *MLIK* = 180°.

$$\text{arc } LIK = 67°35' :$$

hence

$$\text{angle } KEL = 67°35'.$$

Wherefore in triangle *FEL* the angles are given, and the ratio of the sides is given too: Hence

$$EL = 1,090$$
$$\text{where } EF = 10,465,$$
$$\text{and } AD = BD = 10,000;$$

but

$$EL = 6\text{P}32',$$
$$\text{where } BD = 60\text{P},$$
$$\text{by usage of the ancients;}$$

and there is very little difference between that and what Ptolemy gave.

Accordingly

$$BDE = 10,854,$$

and, as the remainder of the diameter

$$CE = 9,146.$$

But since the epicycle when at *B* always subtracts 285 from the altitude of the planet, but adds the same amount, *i.e.*, its radius, when at *C*; on that account the greatest distance of Saturn from centre *E* will be 10,569, and the least 9,431, where *BD* = 10,000. By this ratio the altitude of the apogee of Saturn is 9P42', where the radius of the orbital circle of the Earth = 1P; and the altitude of the perigee is 8P39': hence it is quite evident by the mode set forth above in the case of the small parallaxes of the moon that the parallaxes of Saturn can be greater. And when Saturn is at the apogee,

$$\text{greatest parallax} = 5°45';$$

and when at the perigee,

$$\text{greatest parallax} = 6°39';$$

and they differ from one another by 44'—measuring the angles by the lines coming from the planet and tangent to the orbital circle of the Earth. In this way the particular differences in the movement of Saturn have been found, and we shall afterwards set them out simultaneously and in conjunction with those of the five planets.

10. DEMONSTRATIONS OF THE MOVEMENT OF JUPITER

Having solved the problems concerning Saturn, we shall use the same method and order of demonstration in the case of the movement of Jupiter too, and first we shall repeat three positions reported and demonstrated by Ptolemy, and by the foreshown transformation of circles we shall reconstitute them as the same or as very little different.

The first of the solar oppositions was in the 17th year of Hadrian on the 1st day of the month Epiphi by the Egyptian calendar 1 hour before the following midnight [150ª] at 23°11' of Scorpio, as he says, but after deducting the precession of the equinoxes, at 226°33'.

He recorded the second as occurring on the 21st year of Hadrian on the 13th day of the month Phaophi by the Egyptian calendar 2 hours before the following midnight, at 7°54' of Pisces; but with respect to the sphere of the fixed stars it was 331°16'.

The third was during the 1st year of Antoninus in the month Athyr during the night following the 20th day of the month 5 hours after midnight, at 7°45' in the sphere of the fixed stars.

Accordingly from the first opposition to the second there were 3 Egyptian years 106 days 23 hours, and the apparent movement of the planet was 104°43'. From the second to the third opposition there was 1 year 37 days 7 hours, and the apparent movement of the planet was 36°29'. During the first interval of time the mean movement was 99°55'; during the second it was 33°26'.

Now he found that the arc of the eccentric circle from the highest apsis to the first opposition was 77°15'; and next, 2°50' from the second opposition to the lowest apsis; and from that to the third opposition, 30°36'. Now the eccentricity of the whole circle was $5^1/_2$P whereof the radius is 60P; but it is 917, whereof the radius would be 10,000; and all that corresponds approximately to the observations.

Now let *ABC* be the circle; and from the first opposition to the second let

      arc *AB* = 99°55';

and let

      arc *BC* = 33°26'.

Through the centre *D* let diameter *FDG* be drawn, so that from the highest apsis *F*

      FA = 77°15',

      FAB = 177°10',

and

      GC = 30°36'.

Now let *E* be taken as the centre of the orbital circle of the Earth. Let the distance between the centres be equal to three-quarters 917, *i.e.*, let

      DE = 687;

let

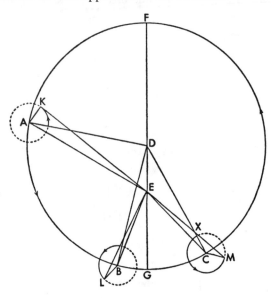

rad. ep. = 229,

which is one-quarter distance, and let the epicycle be described at points *A*, *B*, and *C*.
Let *AD*, *BD*, *CD*, *AE*, *BE*, and *CE* be joined; and in the epicycles let *AK*, *BL*, and *BM*
be joined in such a way that

angle *DAK* = angle *ADF*,

angle *DBL* = angle *FDB*,

and

angle *DCM* = angle *FDC*.

Finally let *K*, *L*, and *M* be joined to *E* by straight lines.

Accordingly, since in triangle *ADE*

angle *ADE* = 102°45',

because angle *ADF* is given; and

side *DE* = 687,

where *AD* = 10,000;

side *AE* = 10,174,

angle *EAD* = 3°48',

and

angle *DEA* = 73°27';

and by addition

angle *EAK* = 81°3'.

Accordingly in [150ᵇ] triangle *AEK* two sides have been given:

EA = 10,174

and

AK = 229,

and

angle *EAK* = 81°3';

it will be clear that

angle *AEK* = 1°17'.

Hence, by subtraction,

angle *KEO* = 72°10'.

Something similar will be shown in triangle *BED*. For the sides *BD* and *DE* always
remain equal to the corresponding sides in the first triangle; but

angle *BDE* = 2°50'.

For that reason

base *BE* = 9,314,

where *DB* = 10,000;

and

angle *DBE* = 12'.

So once more, in triangle *ELB* two sides are given; and

angle *EBL* = 177°22';

moreover

angle *LEB* = 4'.

But

angle *FEL* = angle *FDB* – 16' = 176°54'.

And as

angle *KED* = 72°10';

angle *KEL* = angle *FEL* – angle *KED* = 104°44',

which is the angle of apparent movement between the first and the second termini observed; and there is approximate agreement.

Similarly at the third opposition, in triangle *CDE* two sides *CD* and *DE* have been given, and

angle *CDE* = 30°36';

base *EC* = 9,410

and

angle *DCE* = 2°8'.

Whence in triangle *ECM*

angle *ECM* = 147°49';

hence

angle *CEM* = 39';

and because the exterior angle is equal to the sum of the interior and opposite angles

angle *DXE* = angle *ECX* + angle *CEX* = 2°47'

and

angle *FDC* – angle *DEM* = 2°47'.

Hence

angle *GEM* = 180° – angle *DEM* = 33°23';

and, by addition,

angle *LEM* = [151ª] 36°29',

which is the distance from the second opposition to the third; and that agrees with the observations. But since this third solar opposition was found to be at 7°45' (in the sphere of the fixed stars) and 33°23' to the east of the lowest apsis; the remainder of the semicircle gives us the position of the highest apsis as 154°22' in the sphere of the fixed stars.

Now around *E* let there be drawn *RST* the annual orbital circle of the Earth with diameter *SET* parallel to line *DC*. Now it has been made clear that

angle *GDC* = angle *GER* = 30°36';

and

angle *DXE* = angle *RES* = arc *RS* = 2°47',
the distance of the planet from the mean perigee of the
orbital circle. Hence by addition

arc *TSR* = 182°47',

which is the distance from the highest apsis of the orbital circle.

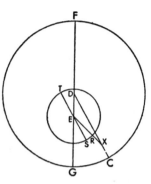

And by this we have confirmation of the fact that at
the time of the third opposition of Jupiter during the first
year of Antoninus on the 20th day of the month Athyr
by the Egyptian calendar 5 hours after the following midnight the planet Jupiter by its anomaly of parallax was at
182°47'. Its regular position in longitude was at 4°58', and the position of the highest
apsis of the eccentric circle was at 154°22'. All these things are in perfect agreement
with our hypothesis of the mobility of the Earth and absolute regularity (of movement).

11. ON THREE OTHER OPPOSITIONS OF JUPITER RECENTLY OBSERVED

Having recorded three positions of the planet Jupiter and evaluated them in this
way, we shall set up three others in their place, which we observed with greatest care at
the solar oppositions of Jupiter.

The first was in the year of Our Lord 1520 on the day before the Kalends of May
11 hours after the preceding midnight, at 220°18' of the sphere of the fixed stars.

The second was in the year of Our Lord 1526 on the fourth day before the
Kalends of December 3 hours after midnight, at 48°34'.

But the third opposition was in the year of Our Lord 1529 on the Kalends of
February 18 hours after midnight, at 113°44'.

From the first [151ᵇ] to the second there are 6 years 212 days 40 minutes (of a
day), during which time the apparent movement of Jupiter was 208°6'. From the second to the third opposition there are 2 Egyptian years 66 days 39 minutes (of a day),
and the apparent movement of the planet is 65°10'. But the regular movement of the
planet during the first interval is 199°40', and during
the second 66°10'.

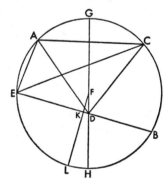

With this as a paradigm let eccentric circle *ABC* be
described, in which the planet is assumed to move simply and regularly. And let the three positions observed
be designated in the order of the letters *A*, *B*, and *C* in
such a way that

arc *AB* = 199°40'

and

arc BC = 66°10',

on that account

arc AC = 360° − (AB + BC) = 94°10'.

Moreover let D be taken as the centre of the annual orbit of the Earth. Let AD, BD and CD be joined; and let any of them, say DB, be extended in a straight line BDE to both arcs of the circle; and let AC, AE and CE be joined.

Accordingly, since

angle BDC = 65°10',

where 4 rt. angles at centre = 360°;

and that is the angle of apparent movement, and since

angle CDE = 180° − 65°10' = 114°50',

but

angle CDE = 229°40',

where 2 rt. angles at circumference 360°;

and since, as standing on arc BC of circumference,

angle CED = 66°10',

and accordingly

angle DCE = 64°10';

therefore, as triangle CDE has its angles given, it has its sides given too:

CE = 18,150

and

ED = 10,918

where diameter of circle circumscribing triangle = 20,000.

Similarly, in triangle ADE, since

angle ADB = 151°54',

which is the remainder of the circle after the subtraction of the given distance between the first opposition and the second; accordingly

angle ADE = 180° − 151°54' = 28°6',

as at the centre, but as on the circumference

angle ADE = 56°12';

and, as on arc BCA of the circumference

angle AED = 160°20';

and

angle EAD = 143°28'.

Hence

side AE = 9,420

and

side ED = 18,992

where diameter of circle circumscribing triangle ADE = 20,000.

But

$$AE = 5,415$$
where ED = 10,918
and CE = 18,150

Again therefore we shall have triangle EAC, of which the two sides EA and EC are given; and, as standing on arc AC of the circumference

angle AEC = 94°10'.

[152ª] Hence it will be shown that, as standing on arc AE,

angle ACE = 30°40',

angle ACE + arc AC = 124°50',

and

$$CE = \text{ch. } EAC = 17,727$$
where diameter of eccentric circle = 20,000.

And by the ratio given before,

$$DE = 10,665,$$

and

arc $BCAE$ = 191°.

It follows that

arc EB = 360° − 191° = 169°

and

$$BDE = \text{ch. } EB = 19,908$$

and by subtraction

$$BD = 9,243.$$

Accordingly, since $BCAE$ is the greater segment, it will contain F the centre of the circle. Now let the diameter $GFDH$ be drawn. It is manifest that

rect. ED, DB = rect. $GD, DH,$

which is therefore also given. But

rect. GD, DH + sq. FD = sq. $FDH.$

Now

sq. FDH − rect. GD, DH = sq. $FD.$

Therefore

$$FD = 1,193,$$
where FG = 10,000,

but

$$FD = 7^\text{P}9',$$
where FG = 60P.

Now let *BE* be bisected at *K*, and let *FKL* be extended; accordingly *FKL* will be at rt. angles to *BE*. And since

$$BDK = \frac{1}{2} BE = 9,954$$

and

$$DB = 9,243,$$

then, by subtraction,

$$DK = 711.$$

Accordingly in triangle *DFK*, which has its sides given,

$$\text{angle } DFK = 36°35',$$

and similarly

$$\text{arc } HL = 36°35'.$$

But

$$\text{arc } LHB = 84\frac{1}{2}°;$$

and, by subtraction,

$$\text{arc } BH = 47°55',$$

which is the distance of the second position from the perigee. And

$$\text{arc } BCG = 180° - 47°55' = 132°5',$$

which is the distance of the apogee from the second position. And

$$\text{arc } BCG - \text{arc } BC = 132°5' - 66°10' = 65°55',$$

which is the distance from the third position to the apogee *G*. Now

$$99°10' - 65°55' = 28°15',$$

which is the distance from the apogee to the first position of the epicycle. That harmonizes too little with the appearances, as the planet does not run through the proposed eccentric circle: hence this method of demonstration which is based upon an uncertain principle cannot give us any certainty. One sign of this among others is that Ptolemy in the case of Saturn recorded a too great distance between the centres and in the case of Jupiter a too small distance; but the same thing seemed a great enough distance to us, so that evidently upon the assumption of different arcs of circles for the same planet [152b] that which is sought does not come about in the same way. Not otherwise was it possible to compound the apparent and the regular movements at the three proposed termini and then at all the termini, unless we kept the total egression of eccentricity of the centres which was recorded by Ptolemy as 5P30', whereof the radius of the eccentric circle is 60P, but which is 917 parts, whereof the radius is 10,000. And let the arc from the highest apsis to the first opposition be 45°2'; from the lowest apsis to the second opposition 64°42'; and from the third opposition to the highest apsis 49°8'.

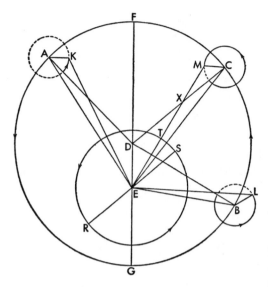

For let the above figure of the eccentric circle carrying an epicycle be repeated, inasmuch as it fits this example. So by our hypothesis

$$DE = 687,$$

which is three-quarters of the total distance between the centres. And

radius of epicycle = 229,

where $FD = 10,000$,

which is the remaining quarter of the distance. Accordingly, since

angle $ADF = 45°2'$,

triangle ADE will have the two sides AD and DE given, together with angle ADE; hence it is shown that

side $AE = 10,496$,

where $AD = 10,000$;

and

angle $DAE = 2°39'$.

And since

angle DAK = angle ADF,

by addition

angle $EAK = 47°41'$.

Also in triangle AEK the two sides AK and AE are given. Hence

angle $AEK = 57'$.

Now

angle KED = angle ADF − (angle AEK + angle DAE) = 41°26',

as the angle of apparent movement at the first solar opposition.

[153ª] In triangle BDE a similar thing will be shown. Since the two sides BD and DE are given, and

angle $BDE = 64°42'$;

side $BE = 9,725$

where $BD = 10,000$

and

angle $BDE = 3°40'$.

Furthermore, in triangle BEL the two sides BE and BL are also given, and

angle $EBL = 118°58'$:

angle $BEL = 1°10'$;

and hence
$$\text{angle } DEL = 110°28'.$$
But it has already been made clear that
$$\text{angle } AED = 41°26',$$
therefore, by addition,
$$\text{angle } KEL = 151°54'.$$
Hence
$$360° - 151°54' = 208°6',$$
the angle of apparent movement between the first and the second solar oppositions; and that agrees with the observations.

Finally, at the third opposition, in triangle *CDE*, sides *DC* and *DE* are given in the same way; and
$$\text{angle } CDE = 130°52'.$$
On account of angle *FDC* being given,
$$\text{side } CE = 10,463,$$
$$\text{where } CD = 10,000,$$
and
$$\text{angle } DCE = 2°51'.$$
Therefore, by addition,
$$\text{angle } ECM = 51°59'.$$
Now in triangle *ECM* the two sides *CM* and *CE* are given together with angle *MCE*:
$$\text{angle } MEC = 1°,$$
and
$$\text{angle } MEC + \text{angle } DCE = \text{angle } FDC - \text{angle } DEM,$$
and angles *FDC* and *DEM* are the angles of regular and apparent movement. And hence, at the third solar opposition,
$$\text{angle } DEM = 45°17'.$$
But it has already been shown that
$$\text{angle } DEL = 90°28';$$
accordingly
$$\text{angle } LEM = 65°10',$$
which is the distance between the second and the third solar oppositions observed; and that agrees with the observations. But since the third position of Jupiter was viewed at 113°44' of the sphere of the fixed stars, it shows that the position of the highest Jovial apsis is at approximately 159°.

But if around centre *E* we now describe *RST* the orbital circle of the Earth, of which the diameter *RES* is parallel to *DC*, then it will be manifest that at the third opposition of Jupiter angle FDX = angle DES = 49°8',

and that the apogee of the regular movement in parallax is at R.

But now that the Earth has passed through 180° plus arc ST, it is in conjunction with Jupiter at its solar opposition; and

arc ST = 3°51',

according as angle SET has been shown to be of the same magnitude.

And so it is clear from this that in the year of Our Lord, 1529, on the Kalends of February 19 hours after midnight, [153^b] the regular movement of anomaly of parallax of Jupiter was at 183°51', but by its proper movement Jupiter was at 109°52'; and the apogee of the eccentric circle is approximately 159° from the horn of the constellation of the Ram, as was to be investigated.

12. CONFIRMATION OF THE REGULAR MOVEMENT OF JUPITER

But it has already been seen above that at the last of the three solar oppositions observed by Ptolemy the planet Jupiter by its proper movement was at 4°58' with an anomaly of parallax of 182°47'. Hence it is clear that during the time between the two observations the movement of parallax of Jupiter was 1°5' besides the full revolutions and its proper movement was approximately 104°54'. The time, however, which flowed between the 1st year of Antoninus on the 20th day of the month Athyr by the Egyptian calendar at 5 hours after the following midnight and the year of Our Lord 1529 on the Kalends of February 18 hours after the preceding midnight was 1392 Egyptian years 99 days 37 minutes of a day, to which time there similarly corresponds according to the above calculation 1°5' besides the whole revolutions, by which the regular revolutions of the Earth has anticipated Jupiter 1267 times; and so the number is seen to harmonize with the observations and is held as certain and exact.

And it is also manifest that during this time the highest and lowest apsides of the eccentric circle moved $4^1/_2$° to the east. The equal distribution (of the movement) yields approximately 1° per 300 years.

13. POSITIONS TO BE ASSIGNED TO THE MOVEMENT OF JUPITER

But the time from the last of the three observations, in the 1st year of Antoninus on the 20th day of the month Athyr at 4 hours after the following midnight, going back to the beginning of the years of Our Lord, amounts to 136 Egyptian years 314 days 10 minutes (of a day), during which time the mean movement of [154^a] parallax was 84°31'. The subtraction of 84°31' from 182°47' leaves 98°16' for the movement up to midnight on the Kalends of January at the beginning of the years of Our Lord.

Backward to the first Olympiad there were 775 Egyptian years $12^1/_2$ days, during which time a movement of 70°58' was reckoned besides the whole revolutions. The subtraction of 70°58' from 98°16' leaves 27°18' as the position for the Olympiad.

Coming down from there for 451 years 247 days, there are 110°52', which together with the movement for the first Olympiad amount to 138°10' for the position of the years of Alexander at noon of the 1st day of the month Thoth by the Egyptian calendar. And so for any others.

14. ON INVESTIGATING THE PARALLAXES OF JUPITER AND ITS ALTITUDE IN RELATION TO THE ORBITAL CIRCLE OF TERRESTRIAL REVOLUTION

In order to investigate the remaining apparent movements of parallax in the case of Jupiter, we carefully observed its position in the year of Our Lord 1520 on the 12th day before the Kalends of March 6 hours before noon, and we perceived through the instrument that Jupiter was 4°31' to the west of the first bright star in the forehead of Scorpio; and since the position of the fixed star was at 209°40', it is clear that the position of Jupiter was at 205°9' in the sphere of the fixed stars.

Accordingly from the beginning of the years of Christ to the time of this observation there were 1520 equal years 62 days 15 minutes (of a day), during which time the mean movement of the sun is calculated to have been 309°16', and the anomaly of parallax 111°15', whereby the mean position of the planet Jupiter is put at 198°1'. And since at this our time the position of the highest apsis of the eccentric circle was found to be at 159°, the anomaly of the eccentric circle of Jupiter was 39°1'.

Following this example, let eccentric circle *ABC* be described with centre *D* and diameter *ADC*. Let *A* be the apogee, and *C* the perigee; and for that reason let *E* the centre of the annual orbital circle of the Earth be on *DC*. Now let

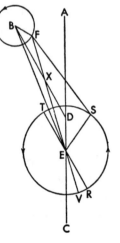

arc *AB* = 39°1';

and with *B* as centre let the epicycle be described with *BF* as radius equal to one third of distance *DE*. Let

angle *DBF* = angle [154ᵇ] *ADB*;

and let the straight lines *BD*, *BE*, and *FE* be joined. Accordingly since in triangle *BDE* two sides are given:

DE = 687,

where *BD* = 10,000;

and since these two sides comprehend the given angle *BDE*, and

angle *BDE* = 140°59' :

it will be shown that

base $BE = 10,543$

and

angle DBE = angle ADB – angle BED = 2°21',

which is the difference between angle BED and angle ADB. Therefore, by addition,

angle EBF = 41°22'.

Accordingly in triangle EBF angle EBF is given together with the two sides comprehending it:

$$EB = 10,543$$

and

$$BF = \,^1/_3\, DE = 229,$$
where $BD = 10,000$.

It follows from this that

side $FE = 10,373$

and

angle BEF = 50'.

Now as lines BD and FE out one another in point X,

angle DXE = angle BDA – angle FED

and angles FED and BDA are the angles of mean and true movement; and

angle DXE = angle DBE + angle BEF = 3°11'.

Now

angle FED = 39°1' – 3°11' = 35°50'

from the highest apsis of the eccentric circle to the planet.
But the position of the highest apsis was at 159°; and

159° + 35°50' = 194°50',

which was the true position of Jupiter with respect to centre E, but the apparent position was at 205°9'. Accordingly, the difference of 10°19' is due to the parallax.

Now let RST the orbital circle of the Earth be described around centre E; and let its diameter RET be parallel to DB, so that R is the apogee of parallax. Moreover, in accordance with the measure of the mean anomaly of parallax, let

arc RS = 111°15';

and let FEV be extended in a straight line to both arcs of the orbital circle of the Earth. The true apogee of the planet will be at V; angle REV is equal to the difference between regular and apparent movement; and

angle REV = angle DXE.

Hence, by addition,

arc VRS = 114°26',

and by subtraction

angle $FES = 65°34'$.

[155ª] But since

angle $EFS = 10°19'$;

angle $FSE = 104°7'$.

Hence, as in triangle EFS the angles are given, the ratio of the sides will be given too:

$$FE : ES = 9{,}698 : 1{,}791.$$

Accordingly

$$FE = 10{,}373$$

and

$$ES = 1{,}916,$$
$$\text{where } BD = 10{,}000.$$

For Ptolemy however

$$ES = 11^\text{P}30',$$
where radius of eccentric circle $= 60^\text{P}$;

and that is approximately the same ratio as

$$1{,}916 : 10{,}000;$$

and therein we do not seem to differ from Ptolemy at all.

Accordingly

$$\text{dmtr. } ADC : \text{dmtr. } RET = 5^\text{P}13' : 1^\text{P}.$$

Similarly

$$AD : ES = AD : RE = 5^\text{P}13'9'' : 1^\text{P}$$

thus

$$DE = 21'9''$$

and

$$BF = 7'10''.$$

Accordingly, when Jupiter is at apogee,

$(ADF - BF)$: radius of orbital circle of Earth $= 5^\text{P}27'29'' : 1^\text{P}$.

And when Jupiter is at perigee;

$(EC + BF)$: radius of orbital circle of Earth $= 4^\text{P}58'49'' : 1^\text{P}$;

and when in the mean positions, as is proportional. Hence it is gathered that Jupiter at apogee has a greatest parallax of $10°35'$, and at perigee $11°35'$: there is a difference of $1°$ between them. So the regular movements of Jupiter have been demonstrated to be at one with the apparent.

15. On the Planet Mars

We must now inspect the revolutions of Mars by taking three ancient solar oppositions, with which we shall connect the mobility of the Earth in antiquity. Accordingly of those oppositions which Ptolemy recorded, the first was in the 15th year of Hadrian on the 26th day of Tybi the 5th month by the Egyptian calendar 1 equatorial hour after the midnight following. And he says that it was at 21° of Gemini, but in relation to the sphere of the fixed stars was at 84°20'.

He noted the second opposition [155ᵇ] as occurring in the 19th year of Hadrian on the 6th day of Pharmuthi the 8th month by the Egyptian calendar 3 hours before the following midnight at 28°50' of Leo but at 142°10' in the sphere of the fixed stars.

The third was in the 2nd year of Antoninus on the 12th day of Epiphi the 11th month by the Egyptian calendar 2 equatorial hours before the following midnight at 2°34' of Sagittarius but at 235°54' of the sphere of the fixed stars.

Accordingly, between the first and second oppositions there are 4 Egyptian years 69 days 20 hours or 50 minutes of a day, and the apparent movement of the planet was 67°50' besides the whole revolutions. From the second opposition to the third there were 4 years 96 days and 1 hour, and the apparent movement of the star was 93°44'. Now during the first interval the mean movement was 81°44' besides the complete revolutions; during the second interval it was 95°28'. Then he found that the total distance between the centres was 12ᴾ, whereof the radius of the eccentric circle was 60ᴾ; but it was 2,000 whereof the radius was 10,000. And the mean movement from the first opposition to the highest apsis was 41°33'; and then it was 40°11' from the highest apsis to the second opposition; and from the third opposition to the lowest apsis it was 44°21': But by our hypothesis of regular movements there will be three-quarters of that distance, *i.e.*, 1,500 between the centres of the eccentric circle and the orbital circle of the Earth, and the remaining quarter of 500 will be the radius of the epicycle.

Now thus let eccentric circle *ABC* be described with centre *D*, and with *FDG* as the diameter through both apsides; and let *E* the centre of the orbital circle of annual

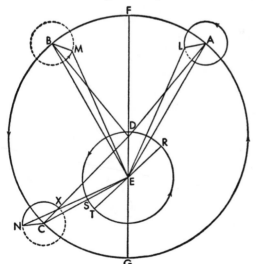

revolution be on the diameter. Let *A*, *B*, and *C* be the points of solar opposition, in that order; and let

arc *AF* = 41°34',
arc *FB* = 40°11',

and

arc *CG* = 44°21'.

At the separate points *A*, *B*, and *C* let the epicycle be described with one-third of distance *DE* as radius. And let *AD*, *BD*, *CD*, *AE*, *BE*, and *CE* be joined. On the epicycle let *AL*, *BM*, and *CN* be joined, but in such a way that

$$\text{angle } DAL = \text{angle } ADF,$$
$$\text{angle } DBM = \text{angle } BDF,$$

and

$$\text{angle } DCN = \text{angle } CDF.$$

Accordingly, since in triangle ADE

$$\text{angle } ADE = 138°26',$$

because angle FDA was given and also the two sides, *viz.*,

$$DE = 1,500$$
$$\text{where } AD = 10,000;$$

it follows from this that

$$\text{side } AE = 11,172,$$

and

$$\text{angle } DAE = 5°7'.$$

[156ª] Hence, by addition,

$$\text{angle } EAL = 46°41'.$$

So also in triangle EAL angle EAL is given, together with the two sides:

$$AE = 11,172$$

and

$$AL = 500,$$
$$\text{where } AD = 10,000.$$

Moreover,

$$\text{angle } AEL = 1°56';$$

and

$$\text{angle } AEL + \text{angle } DAE = 7°3',$$

which is the total difference between angles ADF and LED; and hence

$$\text{angle } DEL = 34^1/_2°.$$

Similarly at the second opposition: in triangle BDE

$$\text{angle } BDE = 139°49'$$

and

$$\text{side } DE = 1,500$$
$$\text{where } BD = 10,000.$$

Hence

$$\text{side } BE = 11,188,$$
$$\text{angle } BED = 35°13',$$

and

$$\text{angle } DBE = 4°58'.$$

Therefore

$$\text{angle } EBM = 45°13'$$

302

and is comprehended by the given sides *BE* and *BM*, from which it follows that
$$\text{angle } BEM = 1°53',$$
and by subtraction
$$\text{angle } DEM = 33°20'.$$
Accordingly
$$\text{angle } LEM = 47°50',$$
whereby the movement of the planet from the first solar opposition to the second is apparent; and the number is consonant with experience.

Again at the third solar opposition: triangle *CDE* has the two sides *CD* and *DE* given, which comprehend angle *CDE*. And
$$\text{angle } CDE = 44°21';$$
hence
$$\text{base } CE = 8,988,$$
$$\text{where } CD = 10,000$$
$$\text{and } DE = 1,500;$$
and
$$\text{angle } CED = 135°39',$$
and
$$\text{angle } DCE = 6°42'.$$
This again in triangle *CEN*
$$\text{angle } ECN = 142°21'$$
and is comprehended by the known sides *EC* and *CN* : hence too
$$\text{angle } CEN = 1°52'.$$
[156^b] Therefore by subtraction
$$\text{angle } NED = 127°5'$$
at the third solar opposition. But it has already been shown that
$$\text{angle } DEM = 33°20'.$$
Hence by subtraction
$$\text{angle } MEN = 93°45',$$
and is the angle of apparent movement between the second and the third solar oppositions, wherein the calculation agrees sufficiently with the observations. But at this last observed opposition of Mars the planet was seen at 235°54', being 127°5' distant from the apogee of the eccentric circle, as was shown: therefore the position of the apogee of the eccentric circle of Mars was at 108°50' in the sphere of the fixed stars.

Now let *RST* the annual orbital circle of the Earth be described around centre *E* with diameter *RET* parallel to *DC*, so that *R* is the apogee of parallax and *T* the perigee. Accordingly the planet was seen on *EX* at 235°54', in longitude, and it was shown that
$$\text{angle } DXE = 8°34',$$

the difference between the regular and the apparent movement; and on that account

$$\text{mean movement} = 244\frac{1}{2}°;$$

but, at the centre,

$$\text{angle } SET = \text{angle } DXE = 8°34'.$$

Accordingly

$$\text{arc } RS = \text{arc } RT - \text{arc } ST = 180° - 8°34' = 171°26',$$

the mean movement of parallax of the planet. Furthermore among other things we have demonstrated by this hypothesis of the mobility of the Earth that in the 2nd year of Antoninus on the 12th day of the month Epiphi by the Egyptian calendar 10 equal hours after midday the planet Mars by its mean movement in longitude was at $244\frac{1}{2}°$, and the anomaly of parallax was at $171°26'$.

16. ON THREE OTHER SOLAR OPPOSITIONS OF MARS WHICH HAVE BEEN OBSERVED RECENTLY

We have compared these three of Ptolemy's observations of Mars with three other observations, which we did not take carelessly. The first was in the year of Our Lord 1512 on the Nones of June, 1 hour after midnight, and the position of Mars was found to be at $235°33'$, according as the sun was opposite [157ª] at $55°33'$ from the first star of Aries in the sphere of the fixed stars as a starting point.

The second was in the year of Our Lord 1518 on the day before the Ides of December 8 hours after midday; and the planet was apparent at $63°2'$.

The third was in the year of Our Lord 1523 on the 8th day before the Kalends of March 8 hours before Boon, at $183°20'$.

Accordingly from the first to the second opposition there were 6 Egyptian years 191 days 45 minutes (of a day); from the second to the third 4 years 72 days 23 minutes. During the first interval of time the apparent movement was $187°29'$, and the regular movement was $168°7'$. During the second interval of time the apparent movement was $80°18'$, and the regular was $83°$.

Now let the eccentric circle of Mars be repeated again, except that here

$$\text{arc } AB = 168°7'$$

and

$$\text{arc } BC = 83°.$$

Accordingly, by the same method which we employed in the case of Saturn and Jupiter—let us pass over in silence the multitude, complication, and boredom of the calculations—we finally find that the apogee of Mars is on arc *BC*. For it is manifest that the apogee cannot be in arc *AB* because there the apparent movement is $19°22'$ greater than the mean. Again the apogee cannot be in arc *CA*, because even if *BC* the arc preceding *CA* is the lesser, nevertheless arc *BC* exceeds the apparent movement by a greater difference than arc *CA* does. But it was shown above that in the eccentric circle the lesser and decreased movement takes place around the apogee. Accordingly the apogee will be held correctly to be in arc *BC*.

Let the apogee be *F*; and let *FDG* be the diameter of the circle. And let the centre of the orbital circle of the Earth be on the diameter. Accordingly, we find that

arc *FCA* = 125°29',

arc *BF* = 66°18',

and

arc *FC* = 16°36';

but

DE = 1,460,

where radius *DE* = 10,000 which is the distance between the centres; and semi-diameter of epicycle = 500 : whence the apparent and regular movements are shown to be consonant with one another and to agree with experiments.

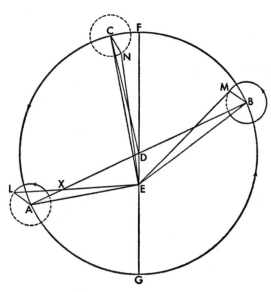

Therefore let the figure be filled out, as before. For it will be shown that, since in triangle *ADE* two sides *AD* and *DE* are known and

angle *ADE* = 54°31'

from the first opposition of Mars to the perigee;

angle *DAE* = 7°24',

and by subtraction

angle *AED* = 118°5';

and

side *AE* = 9,229.

Now by hypothesis

angle *DAL* = angle *FDA*.

Accordingly by addition

angle *EAL* = 132°53'.

So too in triangle *EAL* the two sides *EA* and *AL* comprehending the given angle at *A* are themselves given; [157b] accordingly

angle *AEL* = 2°12',

and

angle *LED* = 115°53'.

Similarly it will be shown in the case of the second opposition that, since in triangle *BDE* the two given sides *DB* and *BE* comprehend angle *BDE*, and

angle *BDE* = 113°35',

then by what we have shown concerning plane triangles

angle *DBE* = 7°11',

angle *DEB* = 59°13',

and

base *BE* = 10,668,

where *DB* = 10,000

and *BM* = 500.

And by addition

angle *EBM* = 73°36'.

So too in triangle *EBM*, since the sides comprehending the given angle are given, it will be shown that

angle *BEM* = 2°36';

and by subtraction

angle *DEM* = 56°38'.

Then

angle *MEG* = 180° − angle *DEM* = 123°22'.

But it has already been shown that

angle *LED* = 115°53';

hence

angle *LEG* = 64°7';

and

angle *LEG* + angle *GEM* = 187°29',

where 4 rt. angles = 360°.

And that agrees with the apparent distance from the first opposition to the second.

A similar thing can be seen at the third opposition. For it has been shown that

angle *DCE* = 2°6',

and

side *EC* = 11,407,

where *CD* = 10,000.

Accordingly, as

angle *ECN* = 18°42',

and as sides *CE* and *CN* of triangle *ECN* have already been given, it will be clear [158a] that

angle *CEN* = 50'.

And

angle *CEN* + angle *DCE* = 2°56',

which is the difference by which angle *DEN* of apparent movement is exceeded by angle *FDC* of regular movement. Therefore

angle *DEN* = 13°40',

which agrees approximately with the apparent movement observed between the second and the third oppositions.

Accordingly, since the planet Mars, as we told you, was apparent in this position at 133°20' from the head of the constellation of Aries; and it has been shown that

angle *FEN* ≑ 13°40';

it is manifest upon calculation backward that the position of the apogee of the eccentric circle at this last observation was at 119°40' in the sphere of the fixed stars.

At the time of Antoninus, Ptolemy found it at 108°50', and so during the time between then and now it has moved $10^{10}/_{12}°$ eastward. Moreover we have found a lesser distance between the centres, *i.e.*, 40, whereof the radius of the eccentric circle is given as 10,000—not because either Ptolemy or ourselves made a slip, but manifestly because the centre of this orbital circle of the Earth has approached the centre of the orbital circle of Mars, while the sun has remained immobile. For these things correspond approximately to one another, as will be shown below clearer than day.

Now let the annual orbital circle of the Earth be described around the centre *E*; and let its diameter *SER* be parallel to *CD* on account of the equality of revolutions. Let *R* be the regular apogee with respect to the planet, *S* the perigee, and *T* the Earth. Now let *ET* be extended; the line of sight of the planet will thus cut *CD* at point *X*. Now the line of sight along *ETX*, as was said at the last opposition, is at 133°20' of longitude.

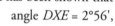

Moreover, it has been shown that

angle *DXE* = 2°56',

for angle *DXE* is the difference by which angle *XDF* of mean movement exceeds angle *XED* of apparent movement. But angle *SET* is equal to its alternate angle *DXE*, and is the additosubtraction arising from the parallax. Now

180° − 2°56' = 177°4',

which is the regular movement of the anomaly of parallax from *R* the apogee of the regular movement—and hence we have shown here that in the year of Our Lord 1523 on the 8th day before the Kalends of March, 7 equatorial hours before noon, the planet Mars by its mean movement in longitude was at 136°16'; and its regular anomaly of parallax was at 177°4', and the highest apsis of the eccentric circle was at 119°40', as was to be shown.

17. CONFIRMATION OF THE MOVEMENT OF MARS

[158^b] Now it was made clear above that in the last of Ptolemy's three observations Mars by its mean movement was at 244$^1/_2$°, and its anomaly of parallax was at 171°26'. Accordingly during the year between there was a movement of 5°38' besides the complete revolutions. Now for the 2nd year of Antoninus on the 12th day of Epiphi the 11 month by the Egyptian calendar 9 hours after midday, *i.e.*, 3 equatorial hours before the following midnight, with respect to the Cracow meridian, to the year of Our Lord 1523 on the 8th day before the Kalends of March 7 hours before noon, there were 1384 Egyptian years 251 days 19 minutes (of a day). During that time there were by the above calculation 5°38' and 648 complete revolutions of anomaly of parallax. Now the regular movement of the sun was held to be 257$^1/_2$°. The subtraction from 257$^1/_2$° of the 5°38' of the movement of parallax leaves 251°52' as the mean movement of Mars in longitude. And all that agrees approximately with what was set down just now.

18. DETERMINATION OF THE POSITION OF MARS

Now from the beginning of the years of Our Lord to the 2nd year of Antoninus on the 12th day of the month Epiphi by the Egyptian calendar at 3 hours before midnight there were 138 Egyptian years 180 days 52 minutes (of a day), and during that time the movement of parallax was 293°4'. And when 293°4' is subtracted from the 171°26' of Ptolemy's last observation—a complete revolution being borrowed—there remain 238°22' at the (beginning of) the first year of Our Lord on midnight of the Kalends of January.

From the first Olympiad to this time there were 775 Egyptian years 12$^1/_2$ days, during which the movement of parallax was 254°1'. When 254°1' has similarly been subtracted from 238°22' and a revolution borrowed, its [159^a] position at the first Olympiad remains as 344°21'.

Similarly by calculating the movements according to the other intervals of time we shall have its position at the beginning of the years of Alexander as 120°39' and at the beginning of the years of Caesar as 211°25'.

19. HOW GREAT THE ORBITAL CIRCLE OF MARS IS IN TERMS OF THE PARTS OF WHICH THE ANNUAL ORBITAL CIRCLE OF THE EARTH IS THE UNIT

Moreover we took observations of the conjunction of Mars with the first bright star of the Chelae—called the southern Claw—which occurred in 1512 on the Kalends of January, For on the morning of that day 6 equatorial hours before noon we saw Mars

$^1/_4$° distant from the fixed star but deflected towards the solstitial rising, by which it was signified that Mars was already in longitude $^1/_8$° to the east of the star but $^1/_5$° distant in northern latitude. Now it was established that the position of the star was 191°20' with a northern latitude of 40'. So it was clear that the position of Mars was at 191°28' with a northern latitude of 51'. But at this time the anomaly of parallax by calculation was 98°28'; the mean position of the sun was at 262°, and the mean position of Mars at 163°32', and the anomaly of the eccentric circle was 43°52'.

With that before us let the eccentric circle *ABC* be described. Let *D* be its centre, *ADC* its diameter, *A* the apogee, *C* the perigee, and let

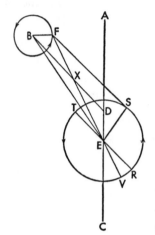

$$\text{ecc. } DE = 1{,}460,$$
where *AD* = 10,000.

Now let

$$\text{arc } AB = 43°52'.$$

Now with *B* as centre and radius *BF* of 500, whereof *AD* is 10,000, let the epicycle be described. Let

$$\text{angle } DBF = \text{angle } ADB;$$

and let *BD*, *BE*, *BF*, and *FE* be joined. Moreover, let *RST* the great orbital circle of the Earth be described around centre *E* with its diameter *RET* parallel to *BD*; and on the diameter let *R* be the apogee of the planet's regular movement of parallax, and *T* the perigee. Now let the Earth be at *S*; and in accordance with the regular anomaly of parallax as computed, let

$$\text{arc } RS = 98°28'.$$

Let *FE* be extended in the straight line *FEV*, which will cut *BD* at point *X* and the convex arc of the orbital circle of the Earth at *V*, where the true apogee of parallax is.

Accordingly, in triangle *BDE* [159^b] two sides are given:

$$DE = 1{,}460,$$
where, *BD* = 10,000;

and they comprehend angle *BDE*. And

$$\text{angle } ADB = 43°52';$$

now

$$\text{angle } BDE = 180° - 43°52' = 136°8'.$$

Hence it will be shown that

$$\text{base } BE = 11{,}097$$

and

$$\text{angle } DBE = 5°13'.$$

But by hypothesis

angle *DBF* = angle *ABD*;

by addition

angle *EBF* = 49°5',

and is comprehended by the given sides *EB* and *BF*. On that account

angle *BEF* = 2°

and

side *FE* = 10,776,

where *DB* = 10,000.

Accordingly

angle *DXE* = 7°13',

because

angle *DXE* = angle *XBE* + angle *XEB*,

the interior and opposite angles. Angle *DXE* is the subtractive additosubtraction, the difference by which angle *ADB* exceeds angle *XED*, and by which the mean position of Mars exceeds the true. Now the mean position is reckoned as 163°32'; therefore the true position is to the west at 156°19'. But its position appears to be at 191°28' to those viewing it from *S*. Therefore its parallax, or commutation, is 35°9' eastward. Therefore it is clear that

angle *EFS* = 35°9'.

Now as *RT* is parallel to *BD*,

angle *DXE* = angle *REV*,

and similarly

arc *RV* = 7°13'.

Thus, by addition,

arc *VRS* = 105°41',

which is the corrected anomaly of parallax, and hence angle *VES* exterior to triangle *FES* is given. Hence, as being interior and opposite

angle *FSE* = 70°32',

where 2 rt. angles = 180°.

But as the angles of the triangle are given, the ratio of the sides is given too: therefore

FE = 9,428

and

ES = 5,727

where diameter of circle circumscribing triangle = 10,000.

Accordingly

[160ª] *ES* ≒ 6,580,

where *EF* = 10,776

and *BD* = 10,000;

and that is approximately the same as Ptolemy's findings. But by addition

$$ADE = 11,460,$$

and by subtraction

$$EC = 8,540.$$

And at the lowest apsis of the eccentric circle the epicycle adds the 500 which it subtracts at A the highest apsis, so that the remainder at the highest apsis is 10,960, and the sum at the lowest apsis is 9,040. Accordingly, in so far as the radius of the orbital circle of the Earth is 1P, Mars will have a greatest distance of 1P39'57" at its apogee, a least distance of 1P22'26", and a mean distance of 1P39'11". So too in the case of Mars the movements, magnitudes, and distances have been explicated in a fixed ratio by means of the movement of the Earth.

20. On the Planet Venus

Now that we have set out the movements of the three higher planets Saturn, Jupiter, and Mars which circle around the Earth, it is time to speak of the planets which the Earth circles around. And first of Venus, which admits an easier and clearer demonstration of its movement than does Mercury, if only the necessary observations of some positions are not wanting; since if its greatest distances, *i.e.*, at morning and at evening, in either direction from the mean position of the sun are found equal to one another, then we have as certain that the highest or lowest apsis of the eccentric circle of Venus is at the midpoint between these two positions of the sun. The apsides are distinguished from one another by the fact that such equal (angular) elongations are smaller when they take place around the apogee and greater when they take place around the perigee. Finally at its other positions we perceive through the differences by which the angular elongations exceed one another how far distant the orb of Venus is from the highest or lowest apsis and also what its eccentricity is, according as these things have been passed on to us by Ptolemy with great clarity, so that there is no need to repeat them separately, except in so far as things from Ptolemy's observations are applicable to our hypothesis of terrestrial mobility.

He took as his first observation one made by the mathematician Theo of Alexandria in the 16th year of Hadrian, he tells us, on the 21st day of the month Pharmuthi, at the first hour of the following night; and that was in the year of Our Lord 132 on the evening of the 8th day before the Ides of March. And Venus was seen at its greatest evening distance of $47^1/_4°$ from the mean position of the sun, [160^b] while the mean position of the sun was by calculation at 337°41' in the sphere of the fixed stars. With this observation he compared one of his own which he said he made in the 4th year of Antoninus on the 12th day of the month Thoth at daybreak, *i.e.*, in the year of Our Lord 142 on the early morning of the third day before the Kalends of August.

He says that at this time the greatest morning elongation of Venus was equal to the previous elongation and was 47°15' from the mean position of the sun, which was at 119° in the sphere of the fixed stars and which on the previous date had been at 337°41'. Now it is manifest that midway between these mean positions are the apsides diametrically opposite one another at $48^{1}/_{3}$° and $228^{1}/_{3}$°. When the $6^{2}/_{3}$° of the precession of the equinoxes has been added to both of them, they will fall upon 25° of Taurus and of Scorpio according to Ptolemy, and the diametrically opposite highest and lowest apsides of Venus must be at those positions.

Once more for the further confirmation of the thing, he assumed another observation made by Theo in the 4th year of Hadrian at morning twilight on the 20th day of the month Athyr, which was in the year of Our Lord 119 on the morning of the fourth day before the Ides of October, at which time Venus was again found at a great distance of 47°32' from the mean position of the sun at 181°13'. With that he connected his own observation made in the 21st year of Hadrian, which was the year of Our Lord 136, on the 9th day of the month Mechyr by the Egyptian calendar but by the Roman calendar the 8th day before the Kalends of January, at the first hour of the following night, and the evening distance was found to be 47°32' from the mean position of the sun at 265°25'. But in the preceding observation made by Theo the mean position of the sun was at 191°13'. Again the apsides fall midway between these positions, at 48°20' and at 228°20' approximately, where the apogee and the perigee must be. And they are distant from the equinoxes at 25° of Taurus and at 25° of Scorpio, and Ptolemy separated them by two other observations as follows.

The first observation was made by Theo in the 13th year of Hadrian on the 3rd day of the month Epiphi, but in the year of Our Lord 129 at early morning on the 12th day before the Kalends of January, and he found the farthest morning elongation of Venus to be 44°48', while the sun by its mean movement was at $48^{10}/_{12}$°, and Venus was apparent at 4° of the sphere of the fixed stars. Ptolemy himself made the other observation in the 21st year of Hadrian on the 2nd day of the month [161ᵃ] Tybi by the Egyptian calendar, which was by the Roman calendar the year of Our Lord 136 on the 5th day before the Kalends of January at the 1st hour of the following night, while the sun by its mean movement was at 228°54', from which Venus had a greatest evening elongation of 47°16' and was itself apparent at $276^{1}/_{6}$°. Hence the apsides are distinguished from one another, that is to say, the highest is put at $48^{1}/_{3}$°, where the shorter wanderings of Venus are, and the lowest at $228^{1}/_{3}$°, where the greater wanderings are—as was to be demonstrated.

21. WHAT THE RATIO OF THE DIAMETERS OF THE ORBITAL CIRCLE OF THE EARTH AND OF VENUS IS

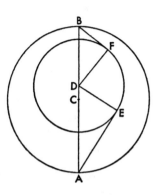

Furthermore from these last two observations the ratio of the diameters of the orbital circles of the Earth and Venus will be apparent. For let AB the orbital circle of the Earth be described around centre C. Let ACB be its diameter through both apsides; and on ACB let D be taken as the centre of the orbital circle of Venus which is eccentric to circle AB. Now let A be the position of the apogee; and when the Earth is there, the centre of the orbital circle of Venus is at its greatest distance, while AB is the line of mean movement of the sun—$48^1/_3°$ at A and $228^1/_3°$ at B. Now let the straight lines AE and BF be drawn touching the orbital circle of Venus at points E and F, and let DE and DF be joined. Accordingly, since as at the centre

$$\text{angle } DAE = 44^4/_5°,$$

and

$$\text{angle } AED = 90°,$$

then triangle DAE will have its angles given and hence its sides:

$$DE = {}^1/_2 \text{ ch. } 2\ DAE = 7{,}046,$$
$$\text{where } AD = 10{,}000.$$

In the same way in the right triangle BDF

$$\text{angle } DBF = 47^1/_3°,$$

and

$$\text{ch. } DF = 7{,}346,$$
$$\text{where } BD = 10{,}000.$$

Accordingly

$$BD = 9{,}582,$$
$$\text{where } DF = DE = 7{,}046.$$

Hence by addition

$$ACB = 19{,}582$$

and

$$AC = {}^1/_2 ACB = 9{,}791;$$

and by subtraction

$$CD = 209.$$

Accordingly, in so far as

$$[161^b]\ AC = 1^P,$$

$$DE = 43^{1}/_{6}',$$

and

$$CD \doteqdot 1^{1}/_{4}';$$

and

$$DE = DF \doteqdot 7{,}193$$

and

$$CD \doteqdot 213,$$

where $AC = 10{,}000$.

And that was to be demonstrated.

22. On the Twofold Movement of Venus

But by the argument from two of Ptolemy's observations, Venus does not have a simple regular movement around D. He made the first observation in the 18th year of Hadrian on the 2nd day of the month Pharmuthi by the Egyptian calendar, but by the Roman calendar it was the year of Our Lord 134 at early morning on the 12th day before the Kalends of March. For at that time the sun by its mean movement was at $318^{10}/_{12}°$; and Venus, which was apparent in the morning at $275^{1}/_{4}°$ of the ecliptic, had reached a farthest limit of elongation of $43°35'$.

He made the second in the 3rd year of Antoninus on the 4th day of the month Pharmuthi by the Egyptian calendar, which by the Roman calendar was in the year of Our Lord 140 on the evening of the 12th day before the Kalends of March. And at that time the mean position of the sun was at $318^{10}/_{12}°$; and Venus was at a greatest evening elongation of $48^{1}/_{3}°$ and was visible at $7^{10}/_{12}°$ in longitude.

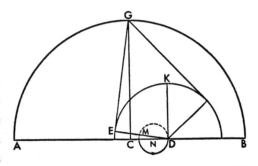

With that set out, let point G be taken as the position of the Earth in the same terrestrial orbital circle, so that arc AG is a quadrant of a circle—the quadrant which measures how far the sun diametrically opposite at both observations according to its mean movement was seen to be west of the apogee of the eccentric circle of Venus. Let GC be joined, and let DK be drawn parallel to GC. Let GE and GF be drawn touching the orbital circle of Venus; and let DE, DF, and DG be joined. Accordingly, since,

$$\text{angle } EGC = 43°35',$$

which was the morning elongation at the time of the first observation, and since

$$\text{angle } CGF = 48^{1}/_{3}°,$$

which was the evening elongation at the time of the second observation;

angle EGF = angle EGC + angle $CGF = 91^{11}/_{12}°$;

and accordingly

angle $DGF = ^1/_2 EGF = 45°47^1/_2'$;

and by subtraction

angle $CGD = 2°23'$.

But

angle $DCG = 90°$.

Accordingly, as the angles of triangle CGD are given, the ratios of the sides are given too; and

$$CD = 416,$$

where $CG = 10,000$.

Now it has already been shown that the distance between the centres was 208; and now the distance has become approximately twice as great. Accordingly, if CD is bisected at point M, similarly

[162a] $DM = 208$,

the total variation in this approach and withdrawal. Again if DM is bisected at N, it will be seen to be the mean and regular point in this movement.

Hence, as in the case of the three higher planets, the movement of Venus happens to be compounded of two regular movements, either by reason of the epicycle of an eccentric circle, as above, or by any other of the aforesaid modes. This planet however is somewhat different from the others in the order and commensurability of its movements; and, as I opine, there will be an easier and more convenient demonstration by means of the eccentric circle of an eccentric circle. In this way let us take N as centre and DN as radius and describe a small circle, on which (the centre of) the orbital circle of Venus is borne and moved around according to the law that whenever the Earth falls upon diameter ACB, on which the highest and lowest apsides of the eccentric circle are, the centre of orbital circle of the planet will always be at least distance, i.e., at point M; and when the Earth is at its mean apsis, i.e., at G, the centre of the orbital circle of the planet will reach point D and the greatest distance CD. Hence you are given to understand that at the time when the Earth has made one orbital circuit, the centre of the orbital circle of the planet has made two revolutions around centre N in the same direction as the Earth, i.e., eastward. For according to such an hypothesis in the case of Venus, all the regular and apparent movements agree with the observations, as will be shown later. Now all this which has so far been demonstrated concerning Venus is found to be consonant with our times, except that the eccentricity has decreased approximately one sixth, so that what before was 416 is now 350, as many observations teach us.

23. ON THE EXAMINATION OF THE MOVEMENT OF VENUS

In this connection I have taken two positions observed very accurately, the first by Timochares in the 13th year of Ptolemy Philadelphus the 52nd year after the death of Alexander in the early morning [162ᵇ] of the 18th day of Mesori the eight month by the Egyptian calendar; and it was recorded that Venus was seen to have occupied the position of the fixed star which is westernmost of the four stars in the left wing of Virgo and is sixth in the description of the sign; its longitude $151^{1}/_{2}°$, its northern latitude $1^{1}/_{6}°$; and it is of third magnitude. Accordingly the position of Venus was made manifest in this way; and the mean position of the sun was by calculation at 194°23'.

With this as an example, let the figure be drawn with point A still at 48°20':

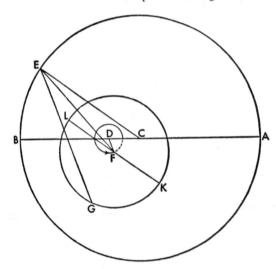

arc AE = 146°3',

and by subtraction

arc BE = 33°57'.

Angle CEG = 42°53',

which is the angular distance of the planet from the mean position of the sun. Accordingly, since

line CD = 312,

where CE = 10,000,

and

angle BCE = 33°57';

in triangle CDE

angle CED = 1°1',

and

base DE = 9,743.

But

angle CDF = 2BCE = 67°54';

and

angle BDF = 180° – 67°54' = 112°6'.

And, as the exterior angle of triangle CDE,

angle BDE = 33°57'.

Hence it is clear that

angle EDF = 144°4',

DF = 104,

where DE = 9,743.

So in triangle DEF

angle DEF = 20',

and by addition

$$\text{angle } CEF = 1°21',$$

and

$$\text{side } EF = 9,831.$$

But it has already been shown that

$$\text{angle } CEG = 42°53';$$

accordingly, by subtraction,

$$\text{angle } FEG = 41°32';$$

and, as radius of the orbital circle,

$$FG = 7,193,$$

$$\text{where } EF = 9,831$$

Accordingly, since in triangle EFG angle FEG and the ratios of the sides are given, the remaining angles are given too. And

$$[163^a] \text{ angle } EFG = 72°5'.$$

$$\text{Arc } KLG = 180° + \text{angle } EFG = 252°5',$$

measured from the highest apsis of the orbital circle. And so we have shown that in the 13th year of Ptolemy Philadelphus on the 18th day of the month Mesori the anomaly of parallax of Venus was 252°5'.

We ourselves made observations of a second position of Venus in the year of Our Lord 1529 on the 4th day before the Ides of March, 1 hour after sunset and at the beginning of the 8th hour after midday. We saw the moon begin to occult Venus at the midpoint of the dark part between the horns, and the occultation lasted till the end of the hour or a little later, until the planet was seen to emerge towards the west on the other side at the midpoint of the gibbosity of the horns. Accordingly it is clear that at the middle of the hour or thereabouts, the centres of the moon and Venus were in conjunction, and we had a full view at Frauenburg. Venus was still in her evening increase (of elongation) and this side of the point of tangency of the orbital circle with a line from the Earth. Accordingly from the birth of Christ there have been 1529 Egyptian years 87 days $7^1/_2$ hours by apparent time but by equal time 7 hours 34 minutes, and the mean position of the sun considered simply had reached 332°11'; and the precession of the equinoxes was 27°24'. The regular movement of the moon was 33°57' away from the sun, the regular movement of anomaly was 205°1', and the movement in latitude was 71°59'. Hence it is reckoned that the true position of the moon was at 10°, but measured from the equinox it was at 7°24' of Taurus with a northern latitude of 1°13'. But since the 15° of Libra were rising, on that account the lunar parallax in longitude was 48°32', and so the apparent position (of the moon) was at 6°26' of Taurus; but its longitude in the sphere of the fixed stars was 9°11' with a northern latitude of 41'; and the apparent position of Venus was at an evening distance of 37°1' from the mean position of the sun, and the distance of the Earth from the highest apsis of Venus was 76°9' to the west.

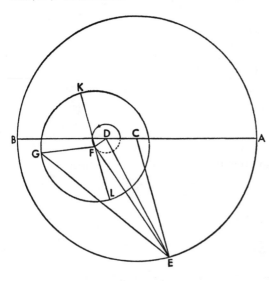

Now let the figure be drawn again according to the previous mode of construction, except that

angle *ECA* = 76°9',

and

angle *CDF* = 2*ECA* = 152°18';

and

ecc. *CD* = 246,

as it is found today; and

DF = 104,

where *CE* = 10,000. Therefore in triangle *CDE*, by subtraction

angle *DCE* = 103°51'

and is comprehended by the given sides. From that it will be shown that

angle *CED* = 1°15',

and

base *DE* = 10,056,

and

angle [163b] *CDE* = 74°54'.

But

angle *CDF* = 2*ACE* = 152°18'.

And

angle *EDF* = angle *CDF* – angle *CDE* = 77°24'.

So again in triangle *DEF* two sides are given;

DF = 104,

where *DE* = 10,056;

and they comprehend the given angle *EDF*. Moreover,

angle *DEF* = 35',

and

base *EF* = 10,034.

Hence by addition

angle *CEF* = 1°50'.

Furthermore,

angle *CEG* = 37°1',

which measures the apparent distance of the planet from the mean position of the sun. Now

angle FEG = angle CEG – angle CEF = 35°11'.

Similarly in triangle EFG two sides are given:

$$EF = 10,034,$$

where $FG = 7,193$

and the angle at E is given: hence too the remaining angles are calculable;

angle $EGF = 53^1/_2°$

and

angle EFG = 91°19',

which is the distance of the planet from the true perigee of its orbital circle.

But since diameter KFL has been drawn parallel to CE, so that K is the apogee of the regular movement and L is the perigee, and since

angle EFL = angle CEF;

then

angle LFG = angle EFG – angle EFL = 89°29';

and

arc KG = 180° – 89°29' = 90°31',

which is the planet's anomaly of parallax as measured from the highest apsis of regular movement of the orbital circle—and that is what we were investigating at the time of this observation of ours.

But at the time of the observation made by Timochares the anomaly was 252°5'; accordingly during the years between there was a movement of 198°26' besides the 1115 complete revolutions. Now the time from the 13th year of Ptolemy [164a] Philadelphus at early morning on the 18th day of the month Mesori to the year of Our Lord 1529 on the 4th day before the Ides of March $7^1/_2$ hours after midday was 1800 Egyptian years 236 days 40 minutes (of a day) approximately.

Accordingly when we have multiplied the movement of 1115 revolutions and 198°26' by 365 days, and divided the product by 1800 years 226 days 40 minutes, we shall have an annual movement of 225°1'45"3'"40"".

Once more the distribution of this through 365 days leaves a daily movement of 36'59"28'", which were added to the table which we set out above.

24. ON THE POSITIONS OF THE ANOMALY OF VENUS

Now from the first Olympiad to the 13th year of Ptolemy Philadelphus at early morning of the 18th day of the month Mesori there are 503 Egyptian years 228 days 40 minutes (of a day), during which time the movement was reckoned to be 290°39'. But if 290°39' is subtracted from 252°5' and 360° is borrowed, the remainder will be 321°26', the position of the movement at the beginning of the first Olympiad.

The remaining positions are in proportion to the movement and time so often spoken of: 81°52' at the beginning of the years of Alexander; 70°26' at the beginning of the years of Caesar; and 126°45' at the beginning of the years of Our Lord.

25. On Mercury

It has been shown how Venus is bound up with the movement of the Earth and in what ratio of circles the regularity of its movement is concealed. Mercury remains, and without fail will also submit to the principle assumed, although it has more complicated wanderings than Venus or any of the aforesaid planets. It has been established experimentally by ancient observations that in the sign of Libra, Mercury has its least angular elongations from the sun and has *greater* elongations in the opposite sign, as is right. But Mercury does not have its *greatest* elongations in this position but in some other positions higher and beyond, as in Gemini and Aquarius, particularly at the time of Antoninus according to Ptolemy; and that occurs in the case of no other planet. When the ancient mathematicians, who supposed the reason for this [164b] to be the immobility of the Earth and the movement of Mercury in its great epicycle along an eccentric circle, had noticed that one simple eccentric circle could not account satisfactorily for these appearances, not only did they grant that the movement on the eccentric circle was not around its own centre but around a foreign centre, but they were also compelled to admit that this same eccentric circle carrying the epicycle moved along another small circle, as they admitted the moon's eccentric circle did. And so there were three centres, namely that of the eccentric circle carrying the epicycle, that of the small circle, and that of the circle which the moderns call the equant. They passed over the first two circles and acknowledged that the epicycle did not move regularly except around the centre of the equant, which was the most foreign to the true centre, to its ratio, and to both the centres already extant. But they judged that the appearances of this planet could be saved by no other scheme, as Ptolemy declares at great length in his *Composition*.

But in order that this last planet may be freed from the liability to injury and disparagement and that the regularity of its movement in relation to the mobility of the Earth may be no less clear than in the case of the other preceding planets; we shall assign to it too a circle eccentric to an eccentric circle instead of the epicycle which the ancients assumed, but in a way different from that of Venus. And nevertheless an epicycle does move on the eccentric circle, but the planet is not borne on its circumference but up and down along its diameter: that can take place through regular circular movements, as was set forth above in connection with the precession of the equinoxes. And it is not surprising, since Proclus in his commentary on the *Elements* of Euclid admits that a straight line can be described by many movements, by all of which movements its appearance will be demonstrable.

But in order that the hypothesis may be grasped more perfectly, let *AB* be the great orbital circle of the Earth with centre *C* and diameter *ACB*. And on *ACB* between points *B* and *C* let *D* be taken as a centre and with one-third *CD* as radius let the small circle *EF* be described, so that *F* is its greatest distance from *C*, and *E* its least. Let *HI*

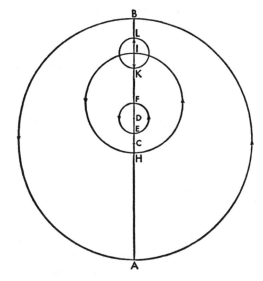

the orbital circle of Mercury be described around centre *F*; and then with *I* the highest apsis as centre let the epicycle which the planet traverses be added. Let *HI* be the orbital circle which is eccentric to an eccentric circle and carries the epicycle. When the figure has been drawn in this way, all those things will fall in order on the straight line *AHCEDFKILB*.

Meanwhile let the planet be set up at *K*, *i.e.*, at the least distance *KF* from centre *F*. [165ª] Now with that point established as the starting-point of the revolutions of Mercury, let it be understood that the centre *F* makes two revolutions for every one of the Earth, and in the same direction too, *i.e.*, eastward; similarly too the planet on *LK*, but along the diameter, up and down with respect to the centre of circle *HI*.

For it follows from this that whenever the Earth is at *A* or *B*, the centre of the orbital circle of Mercury is at *F*, which is the position farthest away from *C*; but whenever the Earth is at the middle quadrants, the centre is at *E* which is the position nearest (to *C*), and this in a manner contrary to that of Venus. Moreover, according to this law Mercury traversing the diameter of epicycle *KL* is nearest to the centre of the orbital circle carrying the epicycle, *i.e.*, is at *K*, when the Earth falls upon the diameter *AB*; and when the Earth is at its mean positions, the planet will be at *L* the most distant position. In this way there take place the two twin revolutions of the centre of the orbital circle on the circumference of the small circle *EF* and of the planet along the diameter *LK* which are equal to one another and commensurable with the annual movement of the Earth.

But meanwhile let the epicycle or line *FI* be moved by its own proper movement along orbital circle *HI*, and let its centre move regularly, completing one revolution simply and with respect to the sphere of the fixed stars in approximately 88 days. But by that movement, whereby it outruns the movement of the Earth and which we call the movement of parallax, it has with respect to the Earth one revolution in 116 days, as can be derived more exactly [165^b] from the table of mean movements. Hence it follows that Mercury by its own proper movement does not always describe the same circumference of a circle, but in proportion to its distance from the centre of its orbital circle it describes a circumference of greatly varying magnitude, least at point K, greatest at L, and middling around I, in practically the same way as was viewed in the case of the lunar epicycle on an epicycle. But what the moon did on the circumference, Mercury does on the diameter by means of a reciprocal movement which is nevertheless compounded of regular movements. We showed above in connection with the precession of the equinoxes how that takes place. But we shall bring to bear some other things concerning this, farther down in connection with latitudes. And this hypothesis is sufficient for all the appearances which are seen to be Mercury's, as is manifest from the history of the observations of Ptolemy and others.

26. On the Positions of the Highest and Lowest Apsides of Mercury

For Ptolemy observed Mercury in the 1st year of Antoninus after sunset on the 20th day of the month Epiphi, while the planet in the evening was at its greatest angular distance from the mean position of the sun. Now at this time there had been 137 Christian years 188 days $42^{1}/_{2}$ minutes (of a day) at Cracow, and accordingly the mean position of the sun according to our calculation was at 63°50', and the planet observed through the instrument was at 7° of Cancer, he says. But after the subtraction of the precession of the equinoxes, which at that time was 6°40', it was shown that the position of Mercury was at 90°20' from the beginning of Aries in the sphere of the fixed stars, and its greatest angular elongation from the mean position of the sun was $26^{1}/_{2}$°.

He made a second observation in the 4th year of Antoninus on the early morning of the 19th day of the month Phamenoth when 140 years 67 days 12 minutes (of a day) approximately had passed since the beginning of the years of Christ, and the mean position of the sun was at 303°19'. Now Mercury was apparent through the instrument at $13^{1}/_{2}$° of Capricorn, but it was at approximately 276°49' from the fixed beginning of Aries, and accordingly the greatest morning distance was similarly $26^{1}/_{2}$°. Accordingly since the limits of elongation on either side of the mean position of the sun are equal, it is necessary that the apsides of Mercury be either way at the midpoint between these positions, i.e., between 226°49' and 90°20'. And they are 3°34' and 183°34' diametrically opposite, where the highest and the lowest apsides [166^a] of Mercury must be.

As in the case of Venus, the apsides are distinguished through two observations. He made the first in the 19th year of Hadrian on the early morning of the 15th day of the month Athyr, while the mean position of the sun was at 182°38'. The greatest morning distance of Mercury from the sun was 19°3', since the apparent position of Mercury was at 163°35'. And in the same 19th year of Hadrian, which was the year of Our Lord 135, at dusk of the 19th day of the month Pachon by the Egyptian calendar, Mercury was found by the aid of the instrument at 27°43' in the sphere of the fixed stars, while the sun by its mean movement was at 4°28'. Again it was shown that the greatest evening distance of the planet was 23°15'—which is greater than the previous distance—whence it was clear enough that the apogee of Mercury at that time could be only at approximately 183$^1/_2$°—as was to be taken note of.

27. How Great the Eccentricity of Mercury Is and What the Commensurability of Its Circles Is

Moreover through this the distance between the centres and the magnitudes of the orbital circles are demonstrated simultaneously, For let AB be the straight line passing through A the highest and B the lowest apsis of Mercury, and also the diameter of the great circle, whose centre is D. And with D taken as centre let the orbital circle of the planet be described. Therefore let lines AE and BF be drawn touching the orbital circle, and let DE and DF be joined.

Accordingly, since at the first of the two preceding observations the greatest morning distance of the planet was seen to be 19°3',

$$\text{angle } CAE = 19°3'.$$

But at the second observation the greatest evening distance was seen to be 23$^1/_4$°. Accordingly, as both of the right triangles AED and BFD have their angles given, [166^b] the ratios of their sides will be given too, so that, as radius of the orbital circle,

$$ED = 32,639,$$

where $AD = 100,000$.

But

$$FD = 39,474,$$

where $BD = 100,000$.

But according as

$$FD = ED$$

as radius of the orbital circle,

$$FD = 32,639,$$

where $AD = 100,000$;

and by subtraction

$$DB = 82,685$$

Hence

$$AC = \tfrac{1}{2}AB = 91,342;$$
and by subtraction
$$CD = 8,658,$$
which is the distance between the centres. And the radius of the orbital circle of Mercury will be 21'26", where $AC = 1^P = 60'$; and
$$CD = 5'4'';$$
and
$$DF = 35,733$$
and
$$CD = 9,479,$$
where $AC = 100,000$
as was to be demonstrated.

But these magnitudes also do not stay everywhere the same; but are quite different from those found in connection with the mean apsides, as the apparent morning and evening longitudes observed at those positions and recorded by Theo and Ptolemy teach us. For Theo observed the evening limit of Mercury in the 14th year of Hadrian on the 18th day of the month Mesori after sunset; and at 129 years 216 days 45 minutes (of a day) after the birth of Christ, while the mean position of the sun was $93\tfrac{1}{2}°$, i.e., approximately at the mean apsis of Mercury. Now the planet was seen through the instrument to be $3\tfrac{10}{12}°$ to the west of Basiliscus in Leo; and on that account its position was $119\tfrac{3}{4}°$ and its greatest evening distance was $26\tfrac{3}{4}°$.

Ptolemy reported that the second limit was observed by him in the 2nd year of Antoninus on the 21st day of the month Mesori at early morning, at which time there had been 138 Christian years 219 days 12 minutes, and the mean position of the sun was similarly 93°39'; [167ª] and the greatest morning distance of Mercury from that was found to be $20\tfrac{1}{4}°$. For Mercury was visible at $73\tfrac{2}{5}°$ in the sphere of the fixed stars.

Therefore let $ACDB$ which is the diameter of the great orbital circle (of the Earth) and which passes through the apsides of Mercury, be again drawn, as before; and at point C let CE the line of mean movement of the sun be erected at right angles. Let point F be taken between C and D; and let the orbital circle of Mercury be described around F. Let the straight lines EH and EG touch this small circle; and let FG, FH, and EF be joined.

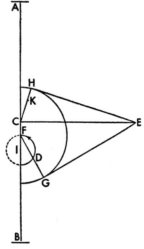

Now once more our problem is to find point F, and what ratio radius FG has to AC. For since

$$\text{angle } CEG = 26^1/_4°,$$

and

$$\text{angle } CEH = 20^1/_4°;$$

accordingly by addition

$$\text{angle } HEG = 46^1/_2°.$$

And

$$\text{angle } HEF = {}^1/_2 HEG = 23^1/_4°.$$

And so by subtraction

$$\text{angle } CEF = 3°.$$

For that reason the sides of the right triangle CEF are given:

$$CF = 524$$

and

$$FE = 10,014,$$
$$\text{where } CE = AC = 10,000.$$

But it has been shown already that

$$CD = 948,$$

while the Earth is at the highest or lowest apsis of the planet. Hence DF will be the excess (of CD over CF) and the diameter of the small circle which the centre of the orbital circle of Mercury describes.

$$DF = 424,$$

and

$$\text{radius } IF = 212.$$

Hence by addition

$$CFI = 736.$$

Similarly, as in triangle HEF

$$\text{angle } H = 90°,$$

and

$$\text{angle } HEF = 23^1/_4° :$$

hence

$$FH = 3,947,$$
$$\text{where } EF = 10,000.$$

But

$$FH = 3,953,$$
$$\text{where } EF = 10,014$$
$$\text{and } CE = 10,000.$$

Now it has been shown above that

$$FK = 3{,}573.$$

Therefore by subtraction

$$HK = 380,$$

which is the greatest difference in the planet's distance from F the centre of its orbital circle; and this greatest difference is found when the planet is between its highest or lowest apsis and its mean apsis. On account of this varying distance from F the centre of its orbital circle, the planet describes unequal circles in proportion to the varying distances—the least distance being 3,573, the greatest 3,953, and the mean 3,763—as was to be demonstrated.

28. WHY THE ANGULAR DIGRESSIONS OF MERCURY AT AROUND 60° FROM THE PERIGEE APPEAR GREATER THAN THOSE AT THE PERIGEE

Hence too it will seem less surprising that Mercury has greater angular digressions at a distance of 60° from the perigee than when at the perigee, since they are also greater than the ones which we have already demonstrated; consequently it was held by the ancients that in one revolution [167^b] of the Earth, Mercury's orb was twice very near to the Earth.

For let the construction be made such that

$$\text{angle } BCE = 60°.$$

On that account

$$\text{angle } BIF = 120°$$

For F is put down as making two revolutions for one of E the Earth. Therefore let EF and EI be joined. Accordingly, since it has been shown that

$$CI = 736,$$

where $EC = 10{,}000,$

and

$$\text{angle } ECI = 60°;$$

hence in triangle ECI

$$\text{base } EI = 9{,}655;$$

and

$$\text{angle } CEI \doteqdot 3°47',$$

which is the difference between angle ACE and angle CIE. But

$$\text{angle } ACE = 120°.$$

Accordingly

$$\text{angle } CIE = 116°13'.$$

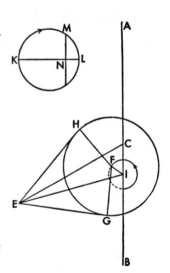

But also

$$\text{angle } FIB = 120°,$$

since by construction

$$\text{angle } FIB = 2 \ ECI;$$

and

$$\text{angle } CIF = 180° - 120° = 60°;$$

[and

$$\text{angle } BIE = 63°47']$$

hence by subtraction

$$\text{angle } EIF = 56°13'.$$

But it was shown that

$$IF = 212,$$
$$\text{where } EI = 9,655;$$

and EI and IF comprehend the given angle EIF. Hence it is inferred that

$$\text{angle } FEI = 1°4';$$

and by subtraction

$$\text{angle } CEF = 2°44',$$

which is the difference between the centre of the orbital circle of the planet and the mean position of the sun; and

$$\text{side } EF = 9,540.$$

Now let GH the orbital circle of Mercury be described around centre F; and from E let EG be drawn touching the orbital circle, and let FG and FH be joined.

We must first examine how great the radius FG or FH is under these circumstances; and we shall do that as follows:

For let a small circle be taken, whose

$$\text{diameter} = 380,$$
$$\text{where } AC = 10,000.$$

And let it be understood that the planet on straight line FG or FH approaches and recedes from centre F along that diameter or along a line equal to it, as we set forth above in connection with the precession of the equinoxes. And in accordance with our hypothesis, wherein angle BCE intercepts 60° of the circumference, let

$$\text{arc } KM = 120°;$$

and let MN be drawn at right angles to KL. And since

$$MN = {}^1/_2 \text{ ch. } 2ML = {}^1/_2 \text{ ch. } 2 \ KM,$$

then

$$LN = 95,$$

which is one quarter of the diameter—as is shown by [168ª] Euclid's *Elements*, XIII, 12 and V, 15. Accordingly,

$$KN = \frac{3}{4} KL = 285.$$

Line *KN* and the least distance of the planet added together make the distance sought for this position, *i.e.*,

$$FG = FH = 3,858,$$
$$\text{where } AC = 10,000$$
$$\text{and } EF = 9,540.$$

Wherefore two sides of right triangle *FEG* or *FEH* have been given: so angle *FEG* or *FEH* will also be given. For

$$FG = FH = 4,044,$$
$$\text{where } EF = 10,000;$$

and

$$FG = FH = \text{ch. } 23°52',$$

so that by addition

$$\text{angle } GEH = 47°44'.$$

But at the lowest apsis only $46\frac{1}{2}°$ is seen, and at the mean apsis similarly $46\frac{1}{2}°$. Accordingly the elongation here becomes $1°14'$ greater, not because the orbital circle of the planet is nearer to the Earth than it was at the perigee, but because the planet is here describing a greater circle than there. All these things are consonant with both present and past observations, and follow from the regular movements.

29. EXAMINATION OF THE MEAN MOVEMENTS OF MERCURY

For it is found by the ancient observations that in the 21st year of Ptolemy Philadelphus in the morning twilight of the 19th day of the month Thoth by the Egyptian calendar, Mercury was apparent on the straight line passing through the first and second of the stars in the forehead of Scorpio and was two lunar diameters distant to the east but was separated from the first star by one lunar diameter to the north. Now it is known that the position of the first star is $209\frac{2}{3}°$ in longitude and $1\frac{1}{3}°$ in northern latitude; and the position of the second is $209°$ in longitude and $1\frac{5}{6}°$ in southern latitude. From that it was concluded that the position of Mercury was $110\frac{2}{3}°$ in longitude and approximately $1\frac{5}{6}°$ in northern latitude. Now there were 59 years 17 days 45 minutes (of a day) since the death of Alexander; and the mean position of the sun according to our calculation was $228°8'$; and the morning distance of the star was $17°28'$ and was still increasing, as was noted during the four following days. Hence it was certain that the planet had not yet arrived at the farthest morning limit or at the point of tangency of its orbital circle, but was still moving in the lower part of the circumference nearer to the Earth. But since the highest apsis was at $183°20'$, there were $44°48'$ to the mean position of the sun.

[168^b] Therefore again let *ACB* be the diameter of the great orbital circle, as above; and from centre *C* let *CE* the line of mean movement of the sun be drawn, in such fashion that

angle *ACE* = 44°48'.

And let there be described around centre *I* the small circle on which the centre *F* of the eccentric circle is home. And since by hypothesis

angle *BIF* = 2 angle *ACE*,

let

angle *BIF* = 89°36'.

And let *EF* and *EI* be joined.

Accordingly, in triangle *ECI* two sides have been given:

CI = 736^1/$_2$,

where *CE* = 10,000.

And sides *CI* and *CE* comprehend the given angle *ECI*. And

angle *ECI* = 180° − angle *ACE* = 135°12'; side

EI = 10,534;

and

angle *CEI* = 2°49',

which is the excess of angle *ACE* over angle *EIC*. Therefore too

angle *CIE* = 41°59'.

But

angle *CIF* = 180° − angle *BIF* = 90°24'.

Therefore by addition

angle *EIF* = 132°23';

and angle *EIF* is comprehended by the given sides *EI* and *IF* of triangle *EFI*, and

side *EI* = 10,534

and

side *IF* = 211^1/$_2$,

where *AC* = 10,000,

Hence

angle *FEI* = 50';

and

side *EF* = 10,678

And by subtraction

329

angle $CEF = 1°59'$.

Now let the small circle LM be taken; and let

diameter $LM = 380$,

where $AC = 10,000$.

And in accordance with the hypothesis let

arc $LN = 89°36'$.

Let chord LN also be drawn; and let NR be drawn perpendicular to LM. Accordingly, since

sq. $LN = $ rect. LM, LR;

that ratio being given,

side $LR \doteq 189$,

where diameter $LM = 380$.

That straight line, *i.e.*, LR, measures the distance of the planet from F the centre of its orbital circle at the time when line EC has completed angle ACE. Accordingly, by the addition of this [169ª] line to the least distance

$$189 + 3{,}573 = 3{,}672,$$

which is the distance at this position.

Accordingly, with the centre F and radius 3,762, let a circle be described; and let EG be drawn cutting the convex circumference at point G, in such a way that

angle $CEG = 17°28'$,

which is the apparent angular elongation of the planet from the mean position of the sun. Let FG be joined; and let FK be drawn parallel to CE. Now

angle $FEG = $ angle $CEG - $ angle $CEF = 15°29'$.

Hence in triangle EFG two sides have been given;

$$EF = 10{,}678,$$

and

$$FG = 3{,}762,$$

and

angle $FEG = 15°29'$:

whence it will be clear that

angle $EFG = 33°46'$.

Now, since

angle $EFK = $ angle CEF,

angle $KFG = $ angle $EFG - $ angle $RFK = 31°48'$;

and

arc $KG = 31°48'$,

which is the distance of the planet from K the mean perigee of its orbital circle.

arc $KG + 180° = 211°48'$,

which was the mean movement of the anomaly of parallax at the time of this observation—as was to be shown.

30. ON THREE MODERN OBSERVATIONS OF THE MOVEMENTS OF MERCURY

The ancients have directed us to this method of examining the movement of this planet, but they were favoured by a clearer atmosphere at a place, where the Nile—so they say—does not give out vapours as the Vistula does among us. For nature has denied that convenience to us who inhabit a colder region, where fair weather is rarer; and furthermore on account of the great obliquity of the sphere it is less frequently possible to see Mercury, as its rising does not fall within our vision at its greatest distance from the sun when it is in Aries or Pisces, and its setting in Virgo and Libra is not visible; and it is not apparent in Cancer or Gemini at evening or early morning, and never at night, except when the sun has receded through the greater part of Leo. On this account the planet has made us take many detours and undergo much labour in order to examine its wanderings. On this account we have borrowed three positions from those which have been carefully observed at Nuremburg.

The first observation was taken by Bernhard Walther, a pupil of Regiomontanus, in the year of Our Lord 1491 on the 9th of September, the fifth day before the Ides, 5 equal hours after midnight, by means of an astrolabe brought into relation with the Hyades. And he saw Mercury at $13^1/_2°$ [169^b] of Virgo with a northern latitude of $1^5/_6°$; and at that time the planet was at the beginning of its morning occultation, while during the preceding days its morning (elongation) had decreased continuously. Accordingly there were 1491 Egyptian years 258 days $12^1/_2$ minutes (of a day) since the beginning of the years of Our Lord; the simple mean position of the sun was at from the spring equinox but in 26°47' of Virgo, wherein the position of Mercury was approximately $13^1/_2°$.

The second was taken by Johann Schöner in the year of Our Lord 1504 on the 5th day before the Ides of January $6^1/_2$ hours after midnight, when 10° of Scorpio was in the middle of the heavens over Nuremburg; and the planet was apparent at $3^1/_3°$ of Capricorn with a northern latitude of 45'. Now by our calculation the mean position of the sun away from the spring equinox was at 27°7' of Capricorn and a morning Mercury was 23°42' to the west of that.

The third observation was taken by this same Johann Schöner in the same year 1504 on the 15th day before the Kalends of April, at which time he found Mercury at $26^1/_{10}°$ of Aries with a northern latitude of approximately 3°, while 25° of Cancer was in the middle of the heavens over Nuremburg—as seen through an astrolabe brought into relation with the Hyades, at $12^1/_2$ hours after midday, at which time the mean position of the sun away from the spring equinox was at 5°39' of Aries, and an evening Mercury was 21°17' away from the sun.

Accordingly from the first position to the second, there are 12 Egyptian years 125 days 3 minutes (of a day) 45 seconds, during which time the simple movement of the sun was 120°14', and Mercury's movement of anomaly of parallax was 316°1'. During the second interval there were 69 days 31 minutes 45 seconds the simple mean position of the sun was 68°32', and Mercury's mean anomaly of parallax was 216°.

Accordingly we wish to examine the movements of Mercury during our time by means of these three observations, and I think we must grant that the commensurability of the circles has remained from Ptolemy's time to now, since in the case of the other planets the good authorities who preceded us are not found to have been mistaken here. If we have the position of the apsis of the eccentric circle together with these observations, nothing further should be desired in the case of the apparent movement of this planet. Now we have taken the position of the highest apsis as $211^1/_2$°, *i.e.*, at $28^1/_2$° of Scorpio; for it was not possible to take it as less without prejudice to the observations. And so we shall have the anomaly of the eccentric circle—I mean [170ª] the distance of the mean movement of the sun from the apogee—as 298°15' at the first terminus, as 58°29' at the second, and as 127°1' at the third.

Therefore let the figure be constructed as before, except that

angle ACE = 61°45',
which measures the westward distance of the line of mean movement of the sun from the apogee at the time of the first observation; and then the rest according to the hypothesis. And since

$$IC = 736^1/_2,$$

where AC = 10,000;
and angle ECI in triangle ECI is also given;
then

angle CEI = 3°35',

and

side IE = 10,369,
where EC = 10,000;

and

$$IF = 211^1/_2.$$

So in triangle EFI also there are two sides having a given ratio; and since by construction
angle BIF = 2 angle ACE;
angle BIF = $123^1/_2$°,

and

angle $CIF = 180° - 123^1/_2° = 56^1/_2°$.

Therefore by addition

angle $EIF = 114°40'$.

Accordingly

angle $IEF = 1°5'$,

and

side $EF = 10,371$.

Hence

angle $CEF = 2^1/_2°$.

But in order that we may know how greatly the orbital circle, whose centre is F, is increased by the movement of approach and withdrawal from the apogee or perigee, let a small circle be drawn and quadrisected by the diameters LM, NR at centre O. And let

angle $POM = 2$ angle $ACE = 123^1/_2°$;

and from point P let PS be drawn perpendicular to LM. Accordingly by the ratio given,

$OP : OS = LO : OS = 10,000 : 8,349 = 190 : 105$.

Whence

$$LS = 295,$$

where $[170^b] AC = 10,000$;

and LS measures the farther removal of the planet from centre F. As the least distance is 3,573,

$$LS + 3,573 = 3,868,$$

which is the present distance.

And with 3,868 as radius and F as centre, let circle HG be drawn. Let EG be joined; and let EF be extended in the straight line EFH. Accordingly, it has been shown that

angle $CEF = 2^1/_2°$,

and by observation

angle $GEC = 13^1/_4°$,

which is the morning distance of the planet from the mean sun. Therefore by addition

angle $FEG = 15^3/_4°$.

But in triangle EFG

$EF : EG = 10,371 : 3,868$;

and angle EFG is also given; that shows us that

angle $EGF = 49°8'$.

Hence

angle $GFH = 64°53'$,

as it is the exterior angle; and

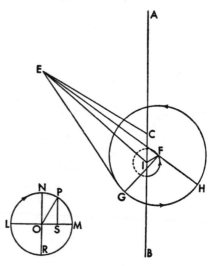

$$360° - \text{angle } GFH = 295°7',$$

which is the true anomaly of parallax. And

$$295°7' + \text{angle } CEF = 297°37',$$

the mean and regular anomaly of parallax—which is what we were looking for. And

$$297°37' + 316°1' = 253°38',$$

which is the regular anomaly of parallax at the second observation—and we shall show that this number is certain and is consonant with the observations.

For let us make

$$\text{angle } ACE = 58°29'$$

in accordance with the second movement of anomaly of the eccentric circle. Then also in triangle CEI two sides are given:

$$IC = 736,$$

where $EC = 10,000$;

and IC and EC comprehend angle ECI, and

$$\text{angle } ECI = 121°31';$$

accordingly

$$\text{side } EI = 10,404$$

and

$$\text{angle } CEI = 3°28'.$$

Similarly, since in triangle EIF

$$\text{angle } EIF = 118°3',$$

and

$$\text{side } IF = 211^{1}/_{2}°,$$
$$\text{where } IE = 10,404:$$
$$\text{side } EF = 10,505,$$

and

$$\text{angle } IEF = 61'.$$

And so by subtraction

$$\text{angle } FEC = 2°27',$$

which is the additive additosubtraction of the eccentric circle; and the addition of angle FEC to the mean movement of parallax makes the true movement to be 256°5'.

Now also in the epicycle of approach [171a] and withdrawal let us take

$$\text{angle } LOP = 2 \text{ angle } ACE = 116°58'.$$

Then too, as in right triangle OPS

$$OP : OS = 1,000 : 455;$$
$$OS = 85,$$

where $OP = OL = 190$.

And by addition

$$LOS = 276.$$

The addition of *LOS* to the least distance of 3,573 makes 3,849.

With 3,849 as radius let circle *HG* be described around centre *F*, so that the apogee of parallax is at point *H* from which the planet has the westward distance of 103°55' of arc *HG*, which measures the difference between a full revolution and the 256°5' of the movement of corrected parallax. And on that account

$$\text{angle } EFG = 180° - 103°55' = 76°5'.$$

So again in triangle *EFG* two sides are given:

$$FG = 3,849,$$
$$\text{where } EF = 10,505.$$

On that account

$$\text{angle } FEG = 21°19';$$

and

$$\text{angle } CEG = \text{angle } FEG + \text{angle } CEF = 23°46'.$$

That is the apparent distance between *C* the centre of the great orbital circle and *G* the planet; and it differs very little from the observation.

All this will be further confirmed by the third example, wherein we have set down that

$$\text{angle } ACE = 127°1',$$

or

$$\text{angle } BCE = 180° - 127°1' = 52°59'.$$

Hence it is shown that

$$\text{angle } CEI = 3°31',$$

and

$$\text{side } IE = 9,575,$$
$$\text{where } EC = 10,000.$$

And since by construction

$$\text{angle } EIF = 49°28',$$

and the sides comprehending angle *EIF* are given:

$$FI = 211^1/_2,$$
$$\text{where } EI = 9,575;$$
$$\text{side } EF = 9,440,$$

and

$$\text{angle } IEF - 59'.$$

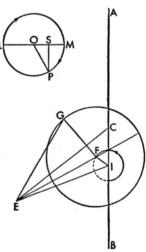

Angle *FEC* = angle *IEC* − 59' = 2°32', which is the subtractive additosubtraction of the anomaly of the eccentric circle. When the 2°32' has been added to the mean anomaly of parallax, which we reckoned as 109°38' after adding the 216° of the second (movement of anomaly), the sum will be the 112°10' of the true anomaly of parallax.

Now in the epicycle let

$$\text{angle } LOP = 2 \text{ angle } ECI = 105°58';$$

here too by the ratio $PO : OS$,

$$OS = 52,$$

so that by addition

$$LOS = 242.$$

Now the least distance is 3,573; and

$$3,573 + 242 = 3,815,$$

which is the corrected distance.

With 3,815 as radius and F as centre, let the circle be described, in which the highest apsis of parallax is H, which is on the straight line made by extending line EFH. And in proportion to the true anomaly of parallax [171^b] let

$$\text{arc } HG = 112°10';$$

and let GF be joined. Therefore

$$\text{angle } GFE = 180° - 112°10' = 67°50',$$

which is comprehended by the given sides:

$$GF = 3,815$$

and

$$EF = 9,440.$$

Hence

$$\text{angle } FEG = 23°50'.$$

Now angle CEF is the additosubtraction; and

$$\text{angle } CEG = \text{angle } FEG - \text{angle } CEF = 21°18',$$

which is the apparent angular distance between the evening planet and the centre of the great orbital circle. And that is approximately the distance found by observation.

Therefore these three positions which are in agreement with observations testify indubitably that the position of the highest apsis, of the eccentric circle is the one which we assumed at $211^1/_2°$ in the sphere of the fixed stars for our time; and that what follows is also certain, namely, that the regular anomaly of parallax was 297°37' at the first position, 253°38' at the second, and 109°38' at the third position; and that is what we were looking for.

But at the ancient observation made in the 21st year of Ptolemy Philadelphus in the early morning of the 19th day of Thoth, the 1st month of the Egyptian calendar the position of the highest apsis of the eccentric circle was 182°20' according to Ptolemy, and the position of regular anomaly of parallax was 211°47'. Now the time between this latest and that ancient observation amounts to 1768 Egyptian years 200 days 33 minutes (of a day), during which time the highest apsis of the eccentric circle moved 28°10' in the sphere of the fixed stars, and the movement of parallax was

257°51' besides the 5,570 complete [172ª] revolutions—as approximately 63 periods are completed in 20 years and that amounts to 5,544 periods in 1,760 years and 26 revolutions in the remaining 8 years 200 days. Similarly in the 1,768 years 200 days 33 minutes there are in addition to the 5,570 revolutions 257°51', which is the distance between the position observed in ancient times and the one observed by us. That agrees with the numbers which we set out in the tables. Now when we have compared the 28°10' with the time during which the apogee of the eccentric circle has moved, it will be seen to have moved 1° per 63 years, if only the movement were regular.

31. On Determining the Former Positions of Mercury

Accordingly there have been 1504 Egyptian years 87 days 48 minutes (of a day) from the beginning of the years of Our Lord to the hour of the last observation, during which time Mercury's movement of anomaly of parallax was 63°14'—not counting the complete revolutions. When 63°14' has been subtracted from 109°38', it will leave 46°24' as the position of the movement of anomaly at the beginning of the years of Our Lord.

Again between that time and the beginning of the first Olympiad there are 775 Egyptian years 12¹/₂ days, during which the movement was calculated to be 95°3' besides the whole revolutions.

If 95°3' is subtracted from the position at the beginning of the years of our Lord and one revolution is borrowed, 311°21' will be left as the position at the time of the first Olympiad.

Moreover between this and the death of Alexander there are 451 years 247 days, and by computation the position is 213°3'.

32. On Another Explanation of Approach and Withdrawal

But before we leave Mercury, let us survey another method no less credible than the former, by which that approach and withdrawal can take place and can be understood.

For let *GHKP* be a circle quadrisected at centre *F*, and around centre *F* let *LM* a small homocentric circle be inscribed. And again, with *L* as centre and radius *LFO* equal to *FG* or *FH*, let another circle *OR* be described.

Now let it be postulated that this whole configuration of circles [172ᵇ] together with its sections *GFR*

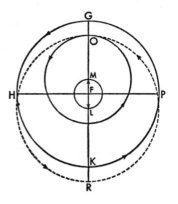

and *HFP* moves eastward around centre *F* in a daily movement of approximately 2°7'—namely, a movement as great as that whereby the movement of parallax of the planet exceeds the movement of the Earth in the ecliptic—away from the apogee of the eccentric circle of the planet; and the planet meanwhile furnishes the remaining movement away from *G* along *OR* the proper circle of parallax; and this movement is approximately equal to the terrestrial movement. Let it also be assumed that in this same annual revolution the centre of *OR*, the orbital circle which carries the planet, is borne by a movement of libration along a diameter *LFM* which is twice as great as the one which we laid down at first; and let it move back and forth, as was said above.

With this as the set-up, we have placed the Earth by its mean movement in a position corresponding to the apogee of the eccentric circle of the planet; and at that time the centre of the orbital circle carrying the planet, at *L*; and the planet itself, at point *O*. The planet then being at its least distance from *F* will describe by its whole movement its least circle, whose radius is *FO*. Consequently, when the Earth is in the neighbourhood of the mean apsis, the planet falling upon point *H* at its greatest distance from *F* will describe its greatest arcs, namely in proportion to the circle having *F* as centre. For at that time *OR* the deferent coincides with circle *GH* on account of the unity of centre at *F*. Hence as the Earth advances in the direction of the perigee and the centre of the orbital circle *OR* towards *M* the other extreme, the orbital circle itself is placed beyond *GK*, and the planet at *R* again is at its least distance from *F*, and the same things occur as in the beginning. For here the three revolutions—namely, that of the Earth through the apogee of the eccentric orbital circle of Mercury, the libration of the centre along diameter *LM*, and the movement of the planet in the same direction away from line *FG*—are equal to one another, and only the movement of the sections *GH* and *KP* away from the apsis of the eccentric circle is different from those revolutions, as we said.

And so in the case of this planet nature has sported a wonderful variety, but one which she has confirmed by a perpetual, certain, and unchanging order. And we should note here that the planet does not traverse the middle spaces of quadrants *GH* and *KP* without any irregularity in longitude, provided that the diversity of the centres, which comes into play, necessarily makes some additosubtraction; but the instability of that centre prevents that. For if, for example, the centre abided at *L* and the planet proceeded from *O*, then it would admit greatest irregularity at *H* in proportion [173ᵃ] to the eccentricity *FL*. But it follows from the assumptions that the planet proceeding from *O* begins and promises to cause the irregularity which is due to *FL* the distance between the centres; but as the mobile centre approaches the midpoint *F*, more and more is taken away from the promised irregularity, and it is made void to such an extent that it wholly vanishes at the mean sections *H* and *P*, where you would expect it to be greatest. And nevertheless, as we acknowledge, when the planet is made small beneath the rays of the sun,[1] it undergoes occultation; and when the planet is rising or setting in the morning or evening, it is not discernible on the curves of the circle. And we are unwilling to pass over this method which is no less rational than the former, and which will come into open use in connection with the movements in latitude.

33. ON THE TABLES OF THE ADDITOSUBTRACTIONS OF THE FIVE WANDERING STARS

These things have been demonstrated concerning the regular and apparent movements of Mercury and the other planets and set out in numbers. And by their example the way to calculating differences of movement for certain other positions will be clear. And for this use we have made ready separate tables for each planet: six column, thirty rows, ascending by triads of 3°, as is usual. The first two columns will contain the common numbers—both of the anomaly of the eccentric circle and of the parallax. The third, the sums of the additosubtractions of the eccentric circle—I mean the total differences occurring between the regular and irregular movements of the orbital circles. The fourth, the proportional minutes—and they go up to 60'—by which the parallaxes are increased or diminished on account of the greater or lesser distance of the Earth. The fifth, the additosubtractions which are the parallaxes occurring at the highest apsis of the eccentric circle and which arise from the great orbital circle. The sixth and last, their excesses over the parallaxes which take place at the lowest apsis of the eccentric circle. And the tables are as follows:

---

[1] *i.e.*, when the planet is in conjunction with the sm.

ADDITIONS-AND-SUBTRACTIONS OF SATURN

Common Numbers		Additosubtractions of the eccentric circle		Proportional Minutes	Parallaxes of the orbital circle of the Earth		Excesses over the parallax of the lowest apsis	
Deg.	Deg.	Deg.	Min.	Min.	Deg.	Min.	Deg.	Min.
3	357	0	20	0	0	17	0	2
6	354	0	40	0	0	34	0	4
9	351	0	58	0	0	51	0	6
12	348	1	17	0	1	3	0	8
15	345	1	36	1	1	23	0	10
18	342	1	55	1	1	40	0	12
21	339	2	13	1	1	56	0	14
24	336	2	31	2	2	11	0	16
27	333	2	49	2	2	26	0	18
30	330	3	6	3	2	42	0	19
33	327	3	33	3	2	56	0	21
36	324	3	39	4	3	10	0	23
39	321	3	55	4	3	25	0	24
42	318	4	10	5	3	38	0	26
45	315	4	25	6	3	52	0	27
48	312	4	39	7	4	5	0	29
51	309	4	52	8	4	17	0	31
54	306	5	5	9	4	28	0	33
57	303	5	17	10	4	38	0	34
60	300	5	29	11	4	49	0	35
63	297	5	41	12	4	59	0	36
66	294	5	50	13	5	8	0	37
69	291	5	59	14	5	17	0	38
72	288	6	7	16	5	24	0	38
75	285	6	14	17	5	31	0	39
78	282	6	19	18	5	37	0	39
81	279	6	23	19	5	42	0	40
84	276	6	27	21	5	46	0	41
87	273	6	29	22	5	50	0	42
90	270	6	31	23	5	52	0	42
93	267	6	31	25	5	52	0	43
96	264	6	30	27	5	53	0	44
99	261	6	28	29	5	53	0	45
102	258	6	26	31	5	51	0	46
105	255	6	22	32	5	48	0	46
108	252	6	17	34	5	45	0	45
111	249	6	12	35	5	40	0	45
114	246	6	6	36	5	36	0	44
117	243	5	58	38	5	29	0	43
120	240	5	49	39	5	22	0	42
123	237	5	40	41	5	13	0	41
126	234	5	28	42	5	3	0	40
129	231	5	16	44	4	52	0	39
132	228	5	3	46	4	41	0	37
135	225	4	48	47	4	29	0	35
138	222	4	33	48	4	15	0	34
141	219	4	17	50	4	1	0	32
144	216	4	0	51	3	46	0	30
147	213	3	42	52	3	30	0	28
150	210	3	24	53	3	13	0	26
153	207	3	6	54	2	56	0	24
156	204	2	46	55	2	38	0	22
159	201	2	27	56	2	21	0	19
162	198	2	7	57	2	2	0	17
165	195	1	46	58	1	42	0	14
168	192	1	25	59	1	22	0	12
171	189	1	4	59	1	2	0	9
174	186	0	43	60	0	42	0	7
177	183	0	22	60	0	21	0	4
180	180	0	0	60	0	0	0	0

ADDITIONS-AND-SUBTRACTIONS OF JUPITER

Common Numbers		Addito-subtractions of the eccentric circle		Proportional Minutes		Parallaxes of the orbital circle of the Earth		Excesses over the parallax of the lowest apsis	
Deg.	Deg.	Deg.	Min.	Min.	Sec.	Deg.	Min.	Deg.	Min.
3	357	0	16	0	3	0	28	0	2
6	354	0	31	0	12	0	56	0	4
9	351	0	47	0	18	1	25	0	6
12	348	1	2	0	30	1	53	0	8
15	345	1	18	0	45	2	19	0	10
18	342	1	33	1	3	2	46	0	13
21	339	1	48	1	23	3	13	0	15
24	336	2	2	1	48	3	40	0	17
27	333	2	17	2	18	4	6	0	19
30	330	2	31	2	50	4	32	0	21
33	327	2	44	3	26	4	57	0	23
36	324	2	58	4	10	5	22	0	25
39	321	3	11	5	40	5	47	0	27
42	318	3	23	6	43	6	11	0	29
45	315	3	35	7	48	6	34	0	31
48	312	3	47	8	50	6	56	0	34
51	309	3	58	9	53	7	18	0	36
54	306	4	8	10	57	7	39	0	38
57	303	4	17	12	0	7	58	0	40
60	300	4	26	13	10	8	17	0	42
63	297	4	35	14	20	8	35	0	44
66	294	4	42	15	30	8	52	0	46
69	291	4	50	16	50	9	8	0	48
72	288	4	56	18	10	9	22	0	50
75	285	5	1	19	17	9	35	0	52
78	282	5	5	20	40	9	47	0	54
81	279	5	9	22	20	9	59	0	55
84	276	5	12	23	50	10	8	0	56
87	273	5	14	25	23	10	17	0	57
90	270	5	15	26	57	10	24	0	58
93	267	5	15	28	33	10	25	0	59
96	264	5	15	30	12	10	33	1	0
99	261	5	14	31	43	10	34	1	1
102	258	5	12	33	17	10	34	1	2
105	255	5	10	34	50	10	33	1	3
108	252	5	6	36	21	10	29	1	3
111	249	5	1	37	47	10	23	1	3
114	246	4	55	39	0	10	15	1	3
117	243	4	49	40	25	10	5	1	2
120	240	4	41	41	50	9	54	1	1
123	237	4	32	43	18	9	41	1	0
126	234	4	23	44	46	9	25	1	0
129	231	4	13	46	11	9	8	0	59
132	228	4	2	47	37	8	56	0	58
135	225	3	50	49	2	8	27	0	57
138	222	3	38	50	22	8	5	0	55
141	219	3	25	51	46	7	39	0	53
144	216	3	13	53	6	7	12	0	50
147	213	2	59	54	10	6	43	0	47
150	210	2	45	55	15	6	13	0	43
153	207	2	30	56	12	5	41	0	39
156	204	2	15	57	0	5	7	0	35
159	201	1	59	57	37	4	32	0	31
162	198	1	43	58	6	3	56	0	27
165	195	1	27	58	34	3	18	0	23
168	192	1	11	59	3	2	40	0	19
171	189	0	53	59	36	2	0	0	15
174	186	0	35	59	58	1	20	0	11
177	183	0	17	60	0	0	40	0	6
180	180	0	0	60	0	0	0	0	0

ADDITIONS-AND-SUBTRACTIONS OF MARS

Common Numbers		Addito-subtractions of the eccentric circle		Proportional Minutes		Parallaxes of the orbital circle of the Earth		Excesses over the parallax of the lowest apsis	
Deg.	Deg.	Deg.	Min.	Min.	Sec.	Deg.	Min.	Deg.	Min.
3	357	0	32	0	0	1	8	0	8
6	354	1	5	0	2	2	16	0	17
9	351	1	37	0	15	3	24	0	25
12	348	2	8	0	28	4	31	0	33
15	345	2	39	0	42	5	38	0	41
18	342	3	10	0	57	6	45	0	50
21	339	3	41	1	13	7	52	0	59
24	336	4	11	1	34	8	58	1	8
27	333	4	41	2	1	10	5	1	16
30	330	5	10	2	31	11	11	1	25
33	327	5	38	3	2	12	16	1	34
36	324	6	6	3	32	13	22	1	43
39	321	6	32	4	3	14	26	1	52
42	318	6	58	4	37	15	31	2	2
45	315	7	22	5	16	16	35	2	11
48	312	7	47	6	2	17	39	2	20
51	309	8	10	6	50	18	42	2	30
54	306	8	32	7	39	19	45	2	40
57	303	8	53	8	30	20	47	2	50
60	300	9	12	9	27	21	49	3	5
63	297	9	30	10	25	22	50	3	11
66	294	9	47	11	28	23	48	3	22
69	291	10	3	12	33	24	47	3	34
72	288	10	19	13	38	25	44	3	46
75	285	10	32	14	46	26	40	3	59
78	282	10	42	16	4	27	35	4	11
81	279	10	50	17	24	28	29	4	24
84	276	10	56	18	45	29	21	4	36
87	273	11	5	20	8	30	12	4	50
90	270	11	7	21	32	31	0	5	5
93	267	11	8	22	58	31	45	5	20
96	264	11	8	24	32	32	30	5	35
99	261	11	7	26	7	33	13	5	51
102	258	11	5	27	43	33	53	6	7
105	255	11	1	29	21	34	30	6	25
108	252	10	56	31	2	35	3	6	45
111	249	10	45	32	46	35	34	7	4
114	246	10	33	34	41	35	59	7	25
117	243	10	11	36	16	36	21	7	46
120	240	10	7	38	1	36	37	8	11
123	237	9	51	39	46	36	49	8	34
126	234	9	33	41	30	36	55	8	59
129	231	9	13	43	12	36	45	9	24
132	228	8	50	44	50	36	25	9	49
135	225	8	27	46	26	35	59	10	17
138	222	8	2	48	1	35	30	10	47
141	219	7	36	49	35	34	24	11	15
144	216	7	7	51	2	33	3	11	45
147	213	6	37	52	25	32	26	12	12
150	210	6	5	53	38	31	5	12	35
153	207	5	34	54	50	30	5	12	54
156	204	5	0	56	0	28	8	13	7
159	201	4	25	57	6	26	21	13	28
162	198	3	49	57	54	23	28	13	47
165	195	3	12	58	22	20	21	12	12
168	192	2	35	58	50	16	51	10	59
171	189	1	57	59	11	13	51	9	40
174	186	1	18	59	44	8	51	6	28
177	183	0	39	60	0	4	32	3	28
180	180	0	0	60	0	0	0	0	0

ADDITIONS-AND-SUBTRACTIONS OF VENUS

Common Numbers		Addito-subtractions of the eccentric circle		Proportional Minutes		Parallaxes of the orbital circle of the Earth		Excesses over the parallax of the lowest apsis	
Deg.	Deg.	Deg.	Min.	Min.	Sec.	Deg.	Min.	Deg.	Min.
3	357	0	6	0	0	1	15	0	1
6	354	0	13	0	0	2	30	0	2
9	351	0	19	0	10	3	45	0	3
12	348	0	25	0	39	4	59	0	5
15	345	0	31	0	58	6	13	0	6
18	342	0	36	1	20	7	28	0	7
21	339	0	42	1	39	8	42	0	9
24	336	0	48	2	23	9	56	0	11
27	333	0	53	2	59	11	10	0	12
30	330	0	59	3	38	12	24	0	13
33	327	1	4	4	18	13	37	0	14
36	324	1	10	5	3	14	50	0	16
39	321	1	15	5	45	16	3	0	17
42	318	1	20	6	32	17	16	0	18
45	315	1	25	7	22	18	28	0	20
48	312	1	29	8	18	19	40	0	21
51	309	1	33	9	31	20	52	0	22
54	306	1	36	10	48	22	3	0	24
57	303	1	40	12	8	23	14	0	26
60	300	1	43	13	32	24	24	0	27
63	297	1	46	15	8	25	34	0	28
66	294	1	49	16	35	26	43	0	30
69	291	1	52	18	0	27	52	0	32
72	288	1	54	19	33	28	57	0	34
75	285	1	56	21	8	30	4	0	36
78	282	1	58	22	32	31	9	0	38
81	279	1	59	24	7	32	13	0	41
84	276	2	0	25	30	33	17	0	43
87	273	2	0	27	5	34	20	0	45
90	270	2	0	28	28	35	21	0	47
93	267	2	0	29	58	36	20	0	50
96	264	2	0	31	28	37	17	0	53
99	261	1	59	32	57	38	13	0	55
102	258	1	58	34	26	39	7	0	58
105	255	1	57	35	55	40	0	1	0
108	252	1	55	37	23	40	49	1	4
111	249	1	53	38	52	41	36	1	8
114	246	1	51	40	19	42	18	1	11
117	243	1	48	41	45	42	59	1	14
120	240	1	45	43	10	43	35	1	18
123	237	1	42	44	37	44	7	1	22
126	234	1	39	46	6	44	32	1	26
129	231	1	35	47	36	44	49	1	50
132	228	1	31	49	6	45	4	1	36
135	225	1	27	50	12	45	10	1	41
138	222	1	22	51	17	45	5	1	47
141	219	1	17	52	33	44	51	1	53
144	216	1	12	53	48	44	22	2	0
147	213	1	7	54	28	43	36	2	6
150	210	1	1	55	0	42	34	2	13
153	207	0	55	55	57	41	12	2	19
156	204	0	49	56	47	39	20	2	34
159	201	0	43	57	33	36	58	2	27
162	198	0	37	58	16	33	58	2	27
165	195	0	31	58	59	30	14	2	27
168	192	0	25	59	39	25	42	2	16
171	189	0	19	59	48	20	20	1	56
174	186	0	13	59	54	14	7	1	26
177	183	0	7	59	58	7	16	0	46
180	180	0	0	60	0	0	16	0	0

ADDITIONS-AND-SUBTRACTIONS OF MERCURY

Common Numbers		Addito-subtractions of the eccentric circle		Proportional Minutes		Parallaxes of the orbital circle of the Earth		Excesses over the parallax of the lowest apsis	
Deg.	Deg.	Deg.	Min.	Min.	Sec.	Deg.	Min.	Deg.	Min.
3	357	0	8	0	3	0	44	0	8
6	354	0	17	0	12	1	28	0	15
9	351	0	26	0	24	2	12	0	23
12	348	0	34	0	50	2	56	0	31
15	345	0	43	1	43	3	41	0	38
18	342	0	51	2	42	4	25	0	45
21	339	0	59	3	51	5	8	0	53
24	336	1	8	5	10	5	51	1	1
27	333	1	16	6	41	6	34	1	8
30	330	1	24	8	29	7	15	1	16
33	327	1	32	10	35	7	57	1	24
36	324	1	39	12	50	8	38	1	32
39	321	1	46	15	7	9	18	1	40
42	318	1	53	17	26	9	59	1	47
45	315	2	0	19	47	10	38	1	55
48	312	2	6	22	8	11	17	2	2
51	309	2	12	24	31	11	54	2	10
54	306	2	18	26	17	12	31	2	18
57	303	2	24	29	17	13	7	2	26
60	300	2	29	31	39	13	41	2	34
63	297	2	34	33	59	14	14	2	42
66	294	2	38	36	12	14	46	2	51
69	291	2	43	38	29	15	17	2	59
72	288	2	47	40	45	15	46	3	8
75	285	2	50	42	58	16	14	3	16
78	282	2	53	45	6	16	40	3	24
81	279	2	56	46	59	17	4	3	32
84	276	2	58	48	50	17	27	3	40
87	273	2	59	50	36	17	48	3	48
90	270	3	0	52	2	18	6	3	56
93	267	3	0	53	43	18	23	4	3
96	264	3	1	55	4	18	37	4	11
99	261	3	0	56	14	18	48	4	19
102	258	2	59	57	14	18	56	4	27
105	255	2	58	58	1	19	2	4	34
108	252	2	56	58	40	19	3	4	42
111	249	2	55	59	14	19	3	4	49
114	246	2	53	59	40	18	59	4	54
117	243	2	49	59	57	18	53	4	58
120	240	2	44	60	0	18	42	5	2
123	237	2	39	59	49	18	27	5	4
126	234	2	34	59	35	18	8	5	6
129	231	2	28	59	19	17	44	5	9
132	228	2	22	58	59	17	17	5	9
135	225	2	16	58	32	16	44	5	6
138	222	2	10	57	56	16	7	5	3
141	219	2	3	56	41	15	25	4	59
144	216	1	55	55	27	14	38	4	52
147	213	1	47	54	55	13	47	4	41
150	210	1	38	54	25	12	52	4	26
153	207	1	29	53	54	11	51	4	10
156	204	1	19	53	23	10	44	3	53
159	201	1	10	52	54	9	34	3	33
162	198	1	0	52	33	8	20	3	10
165	195	0	51	52	18	7	4	2	43
168	192	0	41	52	8	5	43	2	14
171	189	0	31	52	3	4	19	1	43
174	186	0	21	52	2	2	54	1	9
177	183	0	10	52	2	1	27	0	35
180	180	0	0	52	2	0	0	0	0

34. HOW THE POSITIONS IN LONGITUDE OF THE FIVE
PLANETS ARE CALCULATED

[178ᵇ] Therefore by means of the tables drawn up in this way by us we shall calculate without any difficulty the positions in longitude of the five wandering stars. There is approximately the same method of computation in all of them, though the three outer planets differ slightly from Venus and Mercury in this respect.

Therefore let us speak of Saturn, Jupiter, and Mars first. In their case the calculation is such that the mean movements—that is, the simple movement of the sun and the movement of parallax of the planet—are sought for any given time by the method described above. Next, the position of the highest apsis of the eccentric circle is subtracted from the simple position of the sun, and the movement of parallax is subtracted from the remainder; the first remainder is the anomaly of the eccentric circle of the planet. We shall look it up among the common numbers in one of the first two columns of the table, and correspondingly in the third column we shall take the additosubtraction of the eccentric circle, and the proportional minutes in the following column. We shall add this additosubtraction to the movement of anomaly of parallax and subtract it from the anomaly of the eccentric circle, if the number whereby we entered (the table) was found in the first column; and conversely we shall subtract it from the anomaly of the eccentric circle—if the number was found in the second column. The sum or remainder will be the corrected anomaly of parallax or the corrected anomaly of the eccentric circle—the proportional minutes being reserved for a use we shall speak of soon. Then we shall look up this corrected anomaly (of parallax) in the first two columns of common numbers; and from the corresponding place in the fifth column we shall take the additosubtraction arising from the movement of parallax, together with its excess found in the last column; and of that excess we shall take the proportional part in accordance with the number of proportional minutes; and we shall always add this proportional part to the additosubtraction. The sum will be the true parallax of the planet; and is to be subtracted from the corrected anomaly of parallax, if the (corrected anomaly) is less than a semicircle, or added, if greater than a semicircle. For in this way we shall have the true and apparent distance of the planet westward from the mean position of the sun; and when we have subtracted that distance from the mean position of the sun, the remainder will be the sought position of the planet [179ᵃ] in the sphere of the fixed stars, and the addition of the precession of the equinoxes will determine the position of the planet in relation to the spring equinox.

In the case of Venus and Mercury we shall use the distance from the highest apsis to the mean position of the sun as the anomaly of the eccentric circle; and by means of this anomaly we shall correct the movement of parallax and the anomaly of the eccentric circle, as was said already. But if the additosubtraction of the eccentric circle and the corrected parallax are of the same quality or species (*i.e.*, are both additive or both subtractive), they are simultaneously added to or subtracted from the mean position of the sun. But if they are of different species, the lesser is subtracted from the greater; and by means of the remainder there will take place that which we have just mentioned, according to the additive or subtractive property of the greater number; and the final result will be the position which we are looking for.

35. On the Stations and Retrogradations of the Five Wandering Stars

Moreover, the knowledge of where and when the stations, retrogradations, and returns take place and how great they are seems also to pertain to the account of movement in longitude. The mathematicians, especially Apollonius of Perga, have dealt a good deal with them; but they have done so under the assumption of only one irregular movement, namely, that whereby the planets are moved with respect to the sun and which we have called the parallax due to the great orbital circle of the Earth.

For if the circles of the planets—whereon all the planets are borne with unequal periods of revolution but in the same direction, *i.e.*, towards the east—are homocentric with the great orbital circle of the Earth, and some planet on its own orbital circle and within the great orbital circle, such as Venus or Mercury, has greater velocity than the movement of the Earth has; *and if a straight line drawn from the Earth cuts the orbital circle of the planet in each a way that half the segment comprised within the orbital circle has the same ratio to the line which extends from our point of vision the Earth to the lower and convex are of the intersected orbital circle, as does the movement of the Earth to the velocity of the planet then, if a point is made at the extremity of this line drawn to the arc which is at the perigee of the circle of the planet, the point will separate the retrogradation from the progression, so that when the planet is at that position, it will have the appearance of stopping.*

Similarly in the case of the three outer planets which have a movement slower than the velocity [179b] of the Earth, *if a straight line drawn through our point of vision cuts the great orbital circle, in such a way that half the segment comprised within the orbital circle has the same ratio to the line which extends from the planet to our point of vision located on the nearer and convex surface of the orbital circle, as does the movement of the planet to the velocity of the Earth; then the planet when in that position will present to our vision the appearance of stopping.*

But if half the segment comprised within the circle, as was said, *has a greater ratio to the remaining external segment than the velocity of the Earth has to the velocity of Venus or Mercury, or than the movement of any of the three upper planets has to the velocity of the*

Earth; then the planet will progress eastward; but if the ratio is less, then it will retrograde westward.

In order to demonstrate all this, Apollonius took a certain lemma, which was in accord with the hypothesis of the immobility of the Earth but which none the less squares with our principle of terrestrial mobility and which for that reason we too shall employ. And we can enunciate it in this form: *if the greater side of a triangle is so cut that one of the segments is not less than the adjoining side, then this segment will have a greater ratio to the remaining segment than the angles on the side cut, taken in reverse order will have to one another.*

For let *BC* be the greater side of triangle *ABC*; and if on side *BC*

$$CD < AC,$$

then I say that

$$CD : BD > \text{angle } ABC : \text{angle } BCA.$$

Now it is demonstrated as follows. Let the parallelogram *ADCE* be completed; and *BA* and *CE* extended will meet at point *E*. Accordingly since

$$AE < AC,$$

the circle described with centre *A* and radius *AE* will pass through *C* or beyond it. Now let *GEC* be the circle, and let it pass through *C*. Since

$$\text{trgl. } AEF > \text{sect. } AEG,$$

while

$$\text{trgl. } AEC < \text{sect. } AEC;$$

then

$$\text{trgl. } AEF : \text{trgl. } AEC > \text{sect. } AEG : \text{sect. } AEC.$$

But

$$\text{trgl. } AEF : \text{trgl. } AEC = \text{base } FE : \text{base } EC.$$

Therefore

$$FE : EC > \text{angle } FAE : \text{angle } EAG.$$

But

$$FE : EC = CD : DB.$$

And

$$\text{angle } FAE = \text{angle } ABC;$$

and

$$\text{angle } EAC = \text{angle } BCA.$$

Accordingly

[180ª] $CD : DB > \text{angle } ABC : \text{angle } ACB.$

Now it is manifest that the ratio will be much greater if it is not assumed that

$$CD = AC = AE$$

but that

$$CD > AE.$$

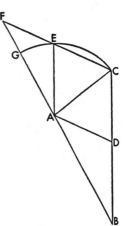

Now let *ABC* be the circle of Venus or Mercury around centre *D*; and let the Earth *E* outside the circle be movable around the same centre *D*. From *E* our point of vision let the straight line *ECDA* be drawn through the centre of the circle; and let *A* be the position farthest from the Earth, and *C* the nearest. And let *DC* be put down as having a greater ratio to *CE* than the movement of the point of vision has to the velocity of the planet. Accordingly it is possible to find a line *EFB* such that half *BF* has the same ratio to *FE* that the movement of the point of vision has to the movement of the planet. For let line *EFB* be moved away from centre *D* and be decreased along *FB* and increased along *EF*, until we meet with what is demanded.

I say that *when the planet is set up at point F, it will present to us the appearance of stopping; and that whatever size of the arc we take on either side of F, we shall find the planet progressing, if the arc is taken in the direction of the apogee, and retrograding, if in as direction of the perigee.*

For first let the arc *FG* be taken in the direction of the apogee: let *EGK* be extended, and let *BG*, *DG*, and *DF* be joined. Accordingly since in triangle *BGE* segment *BF* of the greater side *BE* is greater than *BG*, then

$$BF : EF > \text{angle } FEG : \text{angle } GBF.$$

Furthermore,

$$1/_2 BF : FE > \text{angle } FEG : 2 \text{ angle } GBF,$$

i.e.,

$$1/_2 BF : FE > \text{angle } FEG : \text{angle } GDF.$$

But

$$1/_2 BF : FE = \text{movement of Earth : movement of planet.}$$

Therefore

$$\text{angle } FEG : \text{angle } GDF < \text{velocity of Earth : velocity of planet.}$$

Now let

$$\text{angle } FEL : \text{angle } FDG = \text{movement of Earth : movement of planet.}$$

Therefore

$$\text{angle } FEL > \text{angle } FEG.$$

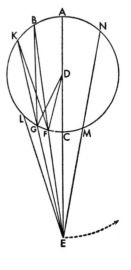

Accordingly, during the time in which the planet traverses arc *GF* of the orbital circle, our line of sight [180ᵇ] will be thought to have traversed during that time the contrary space between line *EF* and line *EL*. It is manifest that in the same time in which to our sight arc *GF* transports the planet westward in accordance with the smaller angle *FEG*, the passage of the Earth drags it back eastward in accordance with the greater angle *FEL*, so that the planet will go on increasing its angular distance eastward by angle *GEL* and will not seem to have come to a stop yet.

Now it is manifest that the opposite of this can be shown by the same means. If in the same diagram we put down that

$$\frac{1}{2}GK : GE = \text{movement of Earth : velocity of planet;}$$

and if we take arc GF in the direction of the perigee and away from straight line EK, and join KF and make triangle KEF, where

$$GE > EF;$$

then

$$KG : GE < \text{angle } FEG : \text{angle } FKG.$$

Thus too

$$\frac{1}{2}KG : GE < \text{angle } FEG : 2 \text{ angle } FKG,$$

i.e.,

$$\frac{1}{2}KG : GE < \text{angle } FEG : \text{angle } GDF,$$

conversely to what was shown before. And it is inferred by the same means that

$$\text{angle } GDF : \text{angle } FEG < \text{velocity of planet : velocity of line of sight.}$$

Accordingly, when angle GDF has been made greater, so that the angles have the same ratio, then the planet will complete a greater movement westwards than progression demands.

Hence it is also manifest that if we make

$$\text{arc } FC = \text{arc } CM$$

the second station will be at point M; and if line EMN is drawn,

$$\frac{1}{2}MN : ME = \frac{1}{2}BF : FE = \text{velocity of Earth : velocity of planet;}$$

and accordingly points M and F will designate the two stations and will determine the whole arc FCM as retrogressive and the remainder of the circle as progressive.

Moreover, it follows that at certain distances

$$DC : CE > \text{velocity of Earth : velocity of planet;}$$

and it will not be possible to draw another straight line in the ratio (which the velocity of the Earth has to the velocity of the planet); and the planet will not seem to stop or to retrograde. For since it was assumed that in triangle DEG

$$DC < EG;$$
$$\text{angle } CEG : \text{angle } CDG < DC : CE$$

but

$$DC : CE > \text{velocity of Earth : velocity of planet.}$$

Therefore also

$$\text{angle } CEG : \text{angle } CDG < \text{velocity of Earth : velocity of planet.}$$

Where that occurs, the planet will progress; [181a] and we shall not find anywhere in the orbital circle of the planet an arc through which it seems to retrograde. All this concerning Venus and Mercury, which are inside the great orbital circle (of the Earth).

We can demonstrate this concerning the three outer planets by the same method and with the same diagrams—merely by reversing the names, so that we put down

ABC as the great orbital circle of the Earth and as the circuit of our point of vision and the planet at *E*, whose movement in its own orbital circle is less than the speed of our point of vision in the great orbital circle. The rest of the demonstration will proceed as before.

36. HOW THE TIMES, POSITIONS, AND ARCS OF THE RETROGRADATIONS ARE DETERMINED

Now if the orbital circles which bear the wandering stars were homocentric with the great orbital circle, it would be easy to establish that which the demonstrations promise, as the ratio of the velocity of the planet to the velocity of the point of vision would always be the same. But the orbital circles are eccentric, and hence their movements appear as irregular. For that reason it will be necessary for us to assume irregular and corrected movements everywhere as the differences of velocity and to employ them in the demonstrations, and not the simple and regular movements, except when the planet happens to be at its mean longitudes, the only place where it seems to be carried in its orbital circle with a mean movement.

Now we shall show this in the case of Mars, so that the retrogradations of the other planets may become clearer by means of this example. For let *ABC* be the great orbital circle, on which our point of vision revolves; and let the planet be at point *E*. From the planet let the straight line *ECDA* be drawn through the centre of the orbital circle; and let *EFB* also be drawn. Half of chord *BF*—*i.e.*, chord *GF*—will have the ratio to line *EF* which the varying velocity of the planet has to the velocity of the line of sight, whereby it exceeds the planet. Our problem is to find arc *FC* of half the retrogradation, or *ABF*, so as to know at what distance from its farthest position from *A* the planet becomes stationary and what the angle comprehended by *FEC* is. For by means of this we shall foretell the time and position of such an affection of the planet.

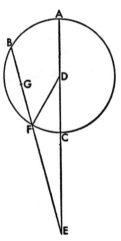

Now let the planet be placed at the mean apsis of the eccentric circle, where the movements of longitude and of anomaly differ very little from the regular movements. Therefore in the case of the planet Mars, since its mean movement (that is half of the line *BF*) [181^b] is 1P8°7", the motion of parallax, which is the relation of our vision to the mean movement of the star, consists of one part and is the straight line *EF*. Hence

$$EB = 3\text{P}16'14"$$

and likewise

$$\text{rect. } BE, EF = 3\text{P}16'14".$$

Now we have shown that

$$DA = 6,580,$$

where $DE = 10,000$,

and DA is the radius of the orbital circle.

But

$$DA = 39\text{P}29',$$
$$= \text{where } DE = 60\text{P};$$

and

$$AE : EC = 99\text{P}29' : 20\text{P}31'.$$

And

$$\text{rect. } AE, EC = \text{rect. } BE, EF = 2,041\text{P}4'.$$

Accordingly by reduction

$$2,041\text{P}4' \div 3\text{P}16'14'' = 624\text{P}4',$$

and similarly

$$\text{side } EF = 24\text{P}58'52'',$$

where $DE = 60\text{P}$.

But

$$EF = 4,163$$

where $DE = 10,000$

and $DF = 6,580$.

Accordingly, as the sides of triangle DEF are given,

$$\text{angle } DEF = 27°15',$$

which is the angular retrogradation of the planet, and

$$\text{angle } CD\!F = 16°50',$$

which is the angular anomaly of parallax. Accordingly, since the planet, when first stationary, appeared on line EF; and the planet, when opposite the sun, on line EC; if the planet is not moved eastward, the 16°15' of arc CF will comprehend the 27°25' of angle AEF found to be the retrogradation; but according to the ratio set forth of the velocity of the planet to the velocity of our line of sight, 16°5' corresponds to the section of the anomaly of parallax and approximately 19°6'39" corresponds (to the section of the anomaly) of longitude of the planet. Now

$$27°15' - 19°6'39'' = 8°8',$$

which is the distance from the other station to the solar opposition—and there are approximately $36^1/_2$ days during which the anomaly in longitude is 19°6'39"—and hence the total retrogradation is 16°16' in 73 days. These things which have been demonstrated for the mean longitudes of the eccentric circle can be similarly demonstrated for other positions—the planet being credited with an always varying velocity, according as its position demands, as we said.

Hence in Saturn, Jupiter, and Mars the same way of demonstration is open, provided we take the point of sight instead of the planet and the planet instead of the point of sight. Now the reverse of what occurs in the orbital circles which the Earth encloses occurs in the orbital circles which enclose the Earth; and let that be enough, so that we won't have to repeat the same old song. Nevertheless, since the variable movement of the planet with respect to the point of sight and to the ambiguity of the stationary points—of which the theorem of Apollonius does not relieve us—give no little difficulty; I do not know whether it would not be better to investigate the stations simply and in connection with the nearest position, by the method whereby we investigate by means of the known numbers of their movements the conjunction of the planet, when opposite the sun, with the line of mean movement of the sun, or the conjunction of any of the planets. And we shall leave that to your pleasure.

BOOK SIX

[182ª] We have indicated to the best of our ability what power and effect the assumption of the revolution of the Earth has in the case of the apparent movement in longitude of the wandering stars and in what a sure and necessary order it places all the appearances. It remains for us to occupy ourselves with the movements of the planets by which they digress in latitude and to show how in this case too the selfsame mobility of the Earth exercises its command and prescribes laws for them here also. Moreover this is a necessary part of the science, as the digressions of these planets cause no little variation in the rising and setting, apparitions and occultations, and the other appearances of which there has been a general exposition above. And their true positions are said to be known only when their longitude together with their latitude in relation to the ecliptic has been established. Accordingly by means of the assumption of the mobility of the Earth we shall do with perhaps greater compactness and more becomingly what the ancient mathematicians thought to have demonstrated by means of the immobility of the Earth.

1. GENERAL EXPOSITION OF THE DIGRESSION IN LATITUDE OF THE FIVE WANDERING STARS

The ancients found in all the planets two digressions in latitude answering to their twofold irregularity in longitude—one digression taking place by reason of the eccentricity of the orbital circles, and the other in accordance with the epicycles. In place of the epicycles, as has been often repeated, we have taken the single great orbital circle of the Earth—not that the orbital circle has some inclination with respect to the plane of the ecliptic fixed once and forever, since they are the same, but that the orbital circles of the planets are inclined to this plane [182ᵇ] with a variable obliquity, and this variability is regulated according to the movement and revolutions of the great orbital circle of the Earth.

But since the three higher planets, Saturn, Jupiter, and Mars, move longitudinally under different laws from those under which the remaining two do, so also they differ not a little in their latitudinal movement. Accordingly, the ancients first examined where and how great their farthest northern limits in latitude were. Ptolemy found the limits in the case of Saturn and Jupiter around the beginning of Libra, but in the case of Mars around the end of Cancer near the apogee of the eccentric circle. But in our time we found this northern limit in the case of Saturn at 7° of Scorpio, in the case of Jupiter at 27° of Libra, in the case of Mars at 27° of Leo, according as the apogees have been changing around down to our time; for the inclinations and the cardinal points of latitude follow upon the movement of those orbital circles. At corrected or apparent distances of 90° between these limits, they seem to be making no digression in latitude, wherever the Earth happens to be at that time. Therefore, when they are at these mean longitudes, they are understood to be at the common section of their orbital circles with the ecliptic, just as the moon was at the ecliptic sections. Ptolemy calls these points the nodes: the ascending node, after which the planet enters upon northern latitudes; and the descending node, after which the planet crosses over into southern latitudes— not that the great orbital circle of the Earth, which always remains the same in the plane of the ecliptic, gives them any latitude; but every digression in latitude is measured from the nodes and varies greatly in positions different from the nodes. And according as the Earth approaches other positions, where the planets are seen to be opposite the sun and *acronycti*, the planets always move with a greater digression than in any other position of the Earth: in the northern semicircle to the north, and in the southern to the south, and with greater variation than the approach or withdrawal of the Earth demands. By that happening, it is known that the inclination of their orbital circles is not fixed, but that it changes in a certain movement of libration commensurable with the revolutions of the great circle of the Earth, as will be said a little farther on.

Now Venus and Mercury seem to digress somewhat differently but under a fixed law which has been observed to hold at the mean, highest, and lowest apsides. For at the mean longitudes, namely when the line of the mean movement of the sun is at a quadrant's distance from their highest or lowest apsis, and the planets as evening or morning stars are themselves at a distance of a quadrant of their orbital circle from the same line of mean movement of the sun; [183ª] the ancients found that the planets had not digressed from the ecliptic, and hence the ancients understood them to be at that time the common section of their separate orbital circles and the ecliptic. This section passes through their apogees and perigees; and accordingly when they are higher or lower than the Earth, they then make manifest digressions—the greatest digressions at their greatest distances from the Earth, *i.e.*, at the evening apparition or at the morning occultation, when Venus is farthest north, and Mercury farthest south. And conversely at a position nearer to the Earth, when they undergo occultation in the evening or emerge in the morning, Venus is to the south, and Mercury to the north. Vice versa, when the Earth is at the position opposite to this and at the other mean apsis namely, when the anomaly of the eccentric circle is 270°—Venus is apparent at its greater southern distance from the Earth, and Mercury is to the north, and at a nearer position of the Earth Venus is to the north, and Mercury to the south. At the solstice of the Earth at the apogee of these planets, Ptolemy found that the latitude of Venus the morning star was northern, and of Venus the evening star, southern; and inversely in the case of Mercury: southern when the morning star, and northern when the evening star. These relations are similarly reversed at the opposite position of the perigee, so that

Venus Lucifer is seen in the south, and Venus Vesperugo in the north; but Mercury as morning star in the north, and Mercury as evening star in the south. And the ancients found that at both these positions the northern digression of Venus was always greater than the southern, and that the southern digression of Mercury was greater than the northern.

Taking this as an occasion, the ancients reasoned out a twofold latitude for this position, and a threefold latitude universally. They called the first latitude, which occurs at the mean longitudes, the inclination; the second, which occurs at the highest and lowest apsides, the obliquation; and the third one, which occurs in conjunction with the second, the deviation: it is always northern in the case of Venus and southern in the case of Mercury. Between these four limits the latitudes are mixed with one another, and alternately increase and decrease and yield mutually; and we shall give the right causes for all that.

2. HYPOTHESES OF THE CIRCLES ON WHICH THE PLANETS ARE MOVED IN LATITUDE

Accordingly in the case of these five planets we must assume that their orbital circles are inclined to the plane of the ecliptic—the common section being through the diameter of the ecliptic—by a variable but regular inclination, [183^{b}] since in Saturn, Jupiter, and Mars the angle of section receives a certain libration around that section as around an axis, like the libration which we demonstrated in the case of the precession of the equinoxes, but simple and commensurable with the movement of parallax. The angle of section is increased and decreased by this libration within a fixed period, so that, whenever the Earth is nearest to the planet, *i.e.*, to the planet in opposition to the sun, the greatest inclination of the orbital circle of the planet occurs; at the contrary position the least inclination; at the mean position, the mean inclination: consequently, when the planet is at its farthest limit of northern or southern latitude, its latitude appears much greater at the nearness of the Earth than at its greatest distance from the Earth. And although this irregularity can be caused only by the unequal distances of the Earth, in accordance with which things nearer seem greater than things farther away; nevertheless there is a rather great difference between the excess and the deficiency of these planetary latitudes: and that cannot take place unless the orbital circles too have a movement of libration with respect to their obliquity. But, as we said before, in the case of things which are undergoing a libration, we must take a certain mean between the extremes.

In order that this may be clearer, let *ABCD* be the Earth's great orbital circle in the plane of the ecliptic with centre *E*; and let *FGKL* the orbital circle of the planet be inclined to *ABCD* in a mean and permanent declination whereof *F* is the northern limit in latitude, *K* the southern, *G* the descending node of section, and *BED* the

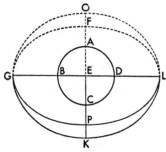

common section, which is extended in the straight lines *GB* and *DL*. And those four termini do not change, except along with the movement of the apsides. Let it be understood however that the movement of the planet in longitude takes place not on the plane of circle *FG* but on *OP,* another circle which is inclined to and homocentric with *FG.* These two circles cut one another in that same [184ᵃ] straight line *GBDL.* Therefore, while the planet is being borne on orbital circle *OP,* it meanwhile falls upon plane *FK* by the movement of libration, goes beyond plane *FK* in either direction, and on that account makes the latitude appear variable.

For first let the planet be at point *O* at its greatest northern latitude and at its position nearest to the Earth in *A;* then the latitude of the planet will increase in proportion to *OGF* the angle of greatest inclination of orbital circle *OGP.* This movement (of libration) is a movement of approach and withdrawal, because by hypothesis it is commensurable with the movement of parallax: if then the Earth is at *B,* point *O* will coincide with *F,* and the latitude of the planet will appear less in the same position than before; and it will be much less if the Earth is at point *C.* For *O* will cross over to the farthest and most diverse part of its libration, and will leave only as much latitude as is in excess over the subtractive libration of the northern latitude, namely over the angle equal to *OGF.* Hence the latitude of the planet around *F* in the north will increase throughout the remaining semicircle *CDA,* until the Earth returns to the first point *A,* from which it set out. There will be the same way of progress for the meridian planet set up around point *K*—the movement of the Earth starting from *C.* But if the planet, in opposition to the sun or hidden by it, is at one of the nodes *G* or *L,* even though at that time the orbital circles *FK* and *OP* have their greatest inclination to one another, on that account no planetary latitude is perceptible, namely because the planet is at the common section of the orbital circles. From that, I judge, it is easily understood how the northern latitude of the planet decreases from *F* to *G;* and the southern latitude increases from *G* to *K,* but vanishes totally at *L* and becomes northern. And this is the way with those three higher planets.

Venus and Mercury differ from them no little in their latitudes, as in longitude, because they have the common sections of the orbital circles located through the apogee and perigee. Now their greatest inclinations at the mean apsides become changeable by a movement of libration, as in the case of the higher planets; but they undergo furthermore a libration dissimilar to the first. Nevertheless both librations are commensurable with the revolutions of the Earth, but not in the same way. For the first libration has the following property; when there has been one revolution of the Earth with respect to the apsides of the planets, there have been two revolutions of the movement of libration having as an immobile axis the section through the apogee and the perigee, which we spoke of; so that whenever the line of mean movement of the sun is at the perigee or apogee of the planets, the greatest angle of section occurs; while the least angle occurs at the mean longitudes. [184b] But the second libration supervening upon this one differs from it in that, by possessing a movable axis, it has the following effect: namely, that when the Earth is located at a mean longitude, the planet of Venus or Mercury is always on the axis, *i.e.*, at the common section of this libration, but shows its greatest deviation when the Earth is in line with its apogee or perigee—Venus always being to the north, as was said, and Mercury to the south; although on account of the former simple inclination they should at this time be lacking latitude.

For example, when the mean movement of the sun is at the apogee of Venus and Venus is in the same position, it is manifest that, in accordance with the simple inclination and the first libration, Venus, being at the common section of its orbital circle with the plane of the ecliptic, would at that time have had no latitude; but the second libration, which has its section or axis along the transverse diameter of the eccentric orbital circle and cuts at right angles the diameter passing through the highest and lowest apsis, adds its greatest deviation to the planet. But if at this time Venus is in one of the other quadrants and around the mean apsides of its orbital circle, then the axis of this libration will coincide with the line of mean movement of the sun, and Venus itself will add to the northern obliquity the greatest deviation, which it subtracts from the southern obliquity and leaves smaller. In this way the libration of deviation is made commensurate with the movement of the Earth.

In order that these things may be grasped more easily, let *ABCD* be drawn again as the great orbital circle. Let *FGK* be the orbital circle of Venus or Mercury: it is eccentric to circle *ABC* and inclined to it in accordance with the equal inclination *FGK*. Let

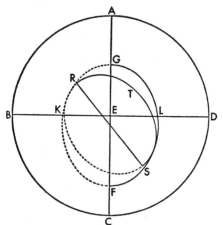

FG be the common section of these two circles through *F* the apogee of the orbital circle and *G* the perigee. First for the sake of an easier demonstration let us put down *GKF* the inclination of the eccentric orbital circle as simple and fixed, or, if you prefer, as midway between the greatest and the least inclination, except that the common section *FG* [185a] changes according to the movement of the perigee and apogee. When the Earth is on this common section, *i.e.*, at *A* or *C*, and the planet is on the same line, it is manifest that at that time the planet would have no latitude, since all latitude is sideways in the semicircles *GKF* and *FLG*, whereon the planet effects its northern or southern approaches, as has been said, in proportion to the inclination of circle *FKG* to the plane of the ecliptic. Now some call this digression of the planet the obliquation; others, the reflexion. But when the Earth is at *B* or *D*, *i.e.*, at the mean apsides of the planet, there will be the same latitudes *FKG* and *GFL* above and below; and they call them the declinations. And so these latitudes differ nominally rather than really from the former latitudes, and even the names are interchanged at the middle positions. But since the angle of inclination of these circles is found to be greater in the obliquation than in the declination, the ancients understood this as taking place through a certain libration, curving itself around section *FG* as an axis, as was said above. Accordingly since the angle of section is known in both cases, it will be easy to understand from the difference between them how great the libration from least to greatest inclination is.

Now let there be understood another circle, the circle of deviation, which is inclined to circle *GKFL* and homocentric in the case of Venus but eccentric to the eccentric circle in the case of Mercury, as will be said later: And let *RS* be their common section as axis of libration, an axis movable in a circle, in such fashion that when the Earth is at *A* or *B*, the planet is at the farthest limit of deviation, wherever that is, as at point *T*; and as far as the Earth has advanced from *A*, so far away from *T* let the planet be understood to have moved, while the inclination of the circle of deviation decreases, so that when the Earth bas measured the quadrant *AB*, the planet should be understood as having arrived at the node of this latitude, *i.e.*, at *R*. But as at this time the planes coincide at the mean movement of libration and are tending in different directions, the remaining semicircle of the deviation, which before was southerly, becomes northern; and as Venus passes into this semicircle, Venus avoids the south and seeks the north again, never to seek the south by this libration, just as Mercury, by crossing in the opposite direction, stays in the south; and Mercury also differs from Venus in that its libration takes place not in a circle homocentric with an eccentric circle but in a circle eccentric to an eccentric circle.

We employed an epicycle instead of this eccentric circle in demonstrating the irregularity in the movement in longitude. But since there we were considering longitude without latitude, and here latitude [185ᵇ] without longitude, and as one and the same revolution comprehends and brings them to pass equally; it is clear enough that it is one and the same movement and one and the same libration which can cause both irregularities and be eccentric and have an inclination at the same time; and that there is no other hypothesis besides this which we have just spoken of and will say more about below.

3. How Great the Inclinations of the Orbital Circles of Saturn, Jupiter, and Mars Are

After setting out our hypothesis for the digressions of the five planets, we must descend to the things themselves and discern singulars; and first how great the inclinations of the single circles are. We measure these inclinations against the great circle which passes through the poles of the circle having the inclination and is at right angles to the ecliptic; the transits in latitude are observed in relation to this great circle. For when we have apprehended these (inclinations), the way of learning the latitudes of each planet will be disclosed. Beginning once more with the three higher planets, we find that according to Ptolemy the digression of Saturn in opposition to the sun at the farthest limits of southern latitude was 3°5', the digression of Jupiter 2°7', that of Mars 7°; but in opposite positions, namely when they were in conjunction with the sun, the digression of Saturn was 2°2', that of Jupiter 1°5', and that of Mars only 5', so that it almost touched the ecliptic—according as it is possible to mark the latitudes from the observations which he took in the neighbourhood of their occultations and apparitions.

Let that be kept before us; and in the plane which is at right angles to the ecliptic and through its centre, let *AB* be the common section (of the plane) with the ecliptic, and *CD* the common section (of the plane) with any of the three eccentric circles

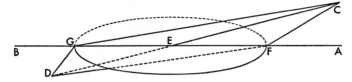

through the greatest northern and southern limits. Moreover, let *E* be the centre of the ecliptic, and *FEG* the diameter of the great orbital circle of the Earth. Now let *D* be the southern latitude and *C* the northern; and let *CF*, *CG*, *DF*, and *DG* be joined.

But the ratios of *EG* the great orbital circle of the Earth to *ED* the eccentric circle of the planet at any of their given positions have already been demonstrated above in the cases of the single (planets). But the positions of greatest latitudes have been given by the observations. Therefore, since angle *BGD*, the angle of greatest southern latitude and an exterior angle of triangle *EGD*, has been given, the interior and opposite angle *GED*, the angle of greatest southern inclination of the eccentric circle to the plane of the ecliptic, will also be given by what has been shown concerning plane triangles.

Similarly we shall demonstrate the least inclination by mean of the least southern latitude, namely by means of angle [186ᵃ] *EFD*. Since in triangle *EFD* the ratio of side *EF* to side *ED* is given together with angle *EFD*, we shall have *GED* given, the exterior angle and angle of least southern inclination: hence from the difference between both declinations we shall have the total libration of the eccentric circle in relation to the ecliptic. Moreover against these angles of inclination we shall measure the opposite northern latitudes, that is to say, angles *AFC* and *EGC*; and if they agree with the observations, it will be a sign that we have not erred at all.

Now as our example we shall take Mars, which has a greater digression in latitude than any of the others. Ptolemy marked the greatest southern latitude as being approximately 7° in the case of the perigee of Mars, and the greatest northern latitude as 4°20' at the apogee. But as we have assumed that

$$\text{angle } BGD = 6°50',$$

we shall find that correspondingly

$$\text{angle } AFC \fallingdotseq 4°30'.$$

For since

$$EG : ED = 1^P : 1^P22'26''$$

and since

$$\text{angle } BCD = 6°50';$$
$$\text{angle } DEG \fallingdotseq 1°51',$$

which is the angle of greatest southern inclination.
And since

$$EF : CE = 1^P : 1^P39'57'',$$

and

$$\text{angle } CEF = \text{angle } DEG = 1°51',$$

it follows that, as angle *CFA* is the exterior angle which we spoke of,

$$\text{angle } CFA = 4^1/_2°,$$

when the planet is in opposition to the sun.

Similarly, in the opposite position where it is in conjunction with the sun, if we assume that

$$\text{angle } DFE = 5',$$

then, since sides *DE* and *EF* and angle *EFD* are given,

$$\text{angle } EDF = 4',$$

and, as exterior angle,

$$\text{angle } DEG \fallingdotseq 9',$$

which is the angle of least inclination. And that will show us that

$$\text{angle } CGE = 6',$$

which is the angle of northern latitude. Therefore, by the subtraction of the least inclination from the greatest,

$$1°5' - 9' = 1°42'$$

which is the libration of this inclination, and

$$^1/_2(1°42') \fallingdotseq 50^1/_2'.$$

In the case of the other two, Jupiter and Saturn, there is a similar method for dis-covering the angles of the inclinations together with the latitudes; for the greatest incli-nation of Jupiter is 1°42', and the least 1°18'; [186ᵇ] so that its total libration does not comprehend more than 24'. Now the greatest inclination of Saturn is 2°44', and the least 2°16'; and the libration between them is 19'. Hence by means of the least angles of inclination, which occur at the opposite position, when the planets are hidden beneath the sun, their digressions in latitude away from the ecliptic will be exhibited: that of Saturn as 2°3' and that of Jupiter as 1°6'—as were to be shown and reserved for the tables to be drawn up below.

4. ON THE EXPOSITION OF THE OTHER LATITUDES IN PARTICULAR AND IN GENERAL

Now that these things have been shown, the latitudes of these three planets will be made clear in general and in particular. For as before, let *AB* the line through the far-thest limits of digression be the common section of the plane perpendicular to the ecliptic. And let the northern limit be at *A*; and let *CD*, which cuts *AB* in point *D*, be

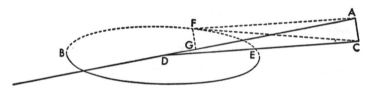

the perpendicular common section of the orbital circle of the planet. And with *D* as centre let *EF* the great orbital circle of the Earth be described. From the opposition, which is at *E*, let any known arc, such as *EF*, be measured, and from *F* and from *C*, the position of the planet, let the perpendiculars *CA* and *FG* be drawn to *AB*; and let *FA* and *FC* be joined.

We are first looking to see how great *ADC* the angle of inclination of the eccentric cir-cle is, with this set-up. Now it has been shown that the inclination was greatest when the Earth was at point *E*. Moreover it has been made clear that the total libration is commen-surate with the revolution of the Earth on circle *EF* in relation to the diameter *BE*, as the nature of libration demands. Therefore on account of arc *EF* being given, the ratio of *ED* to *EG* will be given; and that is the ratio of the total libration to which angle *ADC* has just now decreased. For that reason angle *ADC* is given in this case. Accordingly triangle *ADC* has all its angles given together with its sides. But since by the foregoing, *CD* has a given ratio to *ED*, the ratio of *CD* to the remainder *DG* is given. Accordingly the ratios of *CD* and *AD* to *GD* are given. And hence the remainder *AG* is given. Hence too *FG* is given; for

$$FG = \frac{1}{2} \text{ ch. } 2 \ EF.$$

Therefore as two sides of the right triangle AGF have been given, side AF is given, and the ratio of AF to AC. Finally as two sides of right triangle ACF [187a] have been given, angle AFC will be given; and that is the angle of apparent latitude, which we were looking for.

Once more we shall take Mars as our example of this. Let its limit of greatest southern latitude be around A, which is approximately at its lowest apsis. Now let the position of the planet be at C, where—as has been demonstrated—the angle of inclination was greatest, *i.e.*, 1°50', when the Earth was at point E. Now let us put the Earth at point F and the movement of parallax at 45° in accordance with arc EF: therefore

$$\text{line } FG = 7,071,$$
$$\text{where } ED = 10,000,$$

and

$$GE = 10,000 - 7,071 = 2,929,$$

which is the remainder of the radius. Now it has been shown that

$$\frac{1}{2} \text{ libration of angle } ADC = 50\frac{1}{2}';$$

and half of the libration has the following ratio of increase and decrease in this case,

$$DE : GE = 50\frac{1}{2}' : 15'.$$

Now at present

$$\text{angle } ADC = 1°50' - 15' = 1°35',$$

which is the angle of inclination. On that account triangle ADC will have its sides and angles given; and since it has been shown above that

$$CD = 9,040,$$
$$\text{where } ED = 6,580;$$
$$FG = 4,653,$$
$$AD = 9,036,$$

and by subtraction

$$AEG = 4,383,$$

and

$$AC = 249\frac{1}{2}.$$

Accordingly, in right triangle AFG, since

$$\text{perpendicular } AG = 4,383$$

and

$$\text{base } FG = 4,653$$
$$\text{side } AF = 6,392.$$

Thus finally in triangle ACF, whereof

$$\text{angle } CAF = 90°$$

and sides AC and AF are given,

$$\text{angle } ACF = 2°15',$$

which is the angle of apparent latitude in relation to the Earth placed at *F*. We shall apply similar reasoning in the case of Saturn and Jupiter.

5. On the Latitudes of Venus and Mercury

Venus and Mercury remain, and their transits in latitude will be demonstrated, as I said, by means of three simultaneous and complicated latitudinal divagations. [187ᵇ] In order that they may be discerned separately, we shall begin with the one which the ancients call declination, as if from a simpler handling of it. And it happens to the declination alone to be sometimes separate from the others; and that occurs around the mean longitudes and around the nodes in accordance with the exact movements in longitude when the Earth has moved through a quadrant of a circle from the apogee or perigee of the planet. For when the Earth is very near, a northern or southern latitude of 6°22' is found in the case of Venus, and 4°5' in the case of Mercury; but at the greatest distance from the Earth, 1°2' in the case of Venus; and in the case of Mercury, 1°45'. Thereby the angles of inclination at this position are made manifest by means of the tables of additosubtractions which have been drawn up; and for Venus in that position at its greatest distance from the Earth the latitude is 1°2', and at its least distance 6°22', and on either side (of the mean latitude) the arc of the circle (through the poles of the orbital circle and perpendicular to the plane of the ecliptic) is approximately 2¹/₂°; but in the case of Mercury the 1°45' at its greatest distance and the 4°5' at its least demand 6¹/₄° as the (total) arc of its circle: consequently the angle of inclination of the circles of Venus is 2°30', and that of Mercury is 6¹/₄°, whereof four right angles are equal to 360°. By means of these (angles) the particular latitudes of declination can be unfolded, as we shall demonstrate, and first in the case of Venus.

For in the plane of the ecliptic and through the centre of the perpendicular plane, let *ABC* be the common section (of the two planes) and *DBE* the common section (of the perpendicular plane) with the plane of the orbital circle of Venus. And let *A* be the

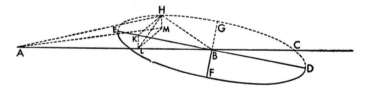

centre of the Earth, *B* the centre of the orbital circle of the planet, and *ABE* the angle of inclination of the orbital circle to the ecliptic. Let circle *DFEG* be described around *B*, and let diameter *FBG* be drawn perpendicular to diameter *DE*. Now let it be understood that the plane of the circle is so related to the assumed perpendicular plane that lines in the plane of the circle which are drawn at right angles to *DE* are parallel to one another and to the plane of the ecliptic; and in the plane of the circle line *FBG* alone has been drawn.

Now our problem is to find out, by means of the given straight lines *AB* and *BC* together with angle *ABE* the given angle of inclinations, how far distant in latitude the planet is, when, for example, [188ª] it is 45° distant from *E* the point nearest to the Earth; and, following Ptolemy, we have chosen this position so that it may become apparent whether the inclination of the orbital circle adds any difference in longitude to Venus or Mercury. For such differences should be most visible around the positions midway between the limits *D*, *F*, *E*, and *G*, because the planet when situated at these four limits has the same longitude as it would have without declination, as is manifest of itself.

Therefore, as was said, let us assume that
$$\text{arc } EH = 45°;$$
and let *HK* be drawn perpendicular to *BE*, and *KL* and *HM* perpendicular to the plane of the ecliptic; and let *HB*, *LM*, *AM*, and *AH* be joined. We shall have the right parallelogram *LKHM*, as *HK* is parallel to the plane of the ecliptic. For angle *LAM* comprehends the additosubtraction in longitude; and angle *HAM* comprehends the transit in latitude, since *HM* also falls perpendicular upon the same plane of the ecliptic. Accordingly, since
$$\text{angle } HBE = 45°;$$
$$HK = \frac{1}{2} \text{ ch. 2 } HE = 7{,}071,$$
$$\text{where } EB = 10{,}000.$$

Similarly in triangle *KBL*
$$\text{angle } BKL = 2\frac{1}{2}°$$
and
$$\text{angle } BLK = 90°,$$
and
$$\text{side } BK = 7{,}071,$$
$$\text{where } BE = 10{,}000;$$
hence
$$\text{side } KL = 308$$
and
$$\text{side } BL = 7{,}064.$$
But since, by what was shown above,
$$AB : BE \doteqdot 10{,}000 : 7{,}193;$$
then
$$HK = 5{,}086,$$
and

$$HM = KL = 221,$$

and

$$BL = 5,081;$$

hence, by subtraction,

$$LA = 4,919.$$

Moreover, as in triangle ALM side AL is given,

$$LM = HK,$$

and

$$\text{angle } ALM = 90°;$$

then

$$\text{side } AM = 7,075$$

and

$$\text{angle } MAL = 45°57',$$

which is the additosubtraction or great parallax of Venus according to calculation. Similarly, as in triangle MAH

$$\text{side } AM = 7,075$$

and

$$\text{side } MH = KL;$$
$$\text{angle } MAH = 1°47',$$

which is the angular declination in latitude.

And if it is not boring to examine what difference in the longitude of Venus is caused by this inclination, let us take triangle ALH, as we understand side LH to be the diagonal of parallelogram $LKHM$. For

$$LH = 5,091,$$
$$\text{where } AL = 4,919$$

and

$$\text{angle } ALK = 90° :$$

hence

$$\text{side } AH = 7,079.$$

Accordingly, as the ratio of the sides is given,

$$\text{angle } HAL = 45°59'.$$

But it has been shown that

$$\text{angle } MAL = 45°57';$$

therefore there is a difference of only 2', as was to be shown.

Again, in the case of Mercury, [188b] with a similar scheme of declination we shall demonstrate the latitudes with the help of a diagram similar to the foregoing: wherein

$$\text{arc } EH = 45°,$$

so that again

$$HK = KB = 7,071,$$

where side $AB = 10,000$.

Accordingly, as can be gathered from the differences in longitude which have already been demonstrated, in this case

$$BK = KH = 2,975,$$

where radius $BH = 3,953$

and $AB = 9,964$.

And since it has been shown that

angle of inclination $ABE = 6°15'$,

where 4 rt. angles = 360°;

accordingly, as the angles of right triangle BKL are given,

base $KL = 304$

and

perpendicular $BL = 2,778$.

And so by subtraction

$$AL = 7,186.$$

But also

$$LM = HK = 2,795;$$

accordingly, as in triangle ALM

angle $L = 90°$

and sides AL and LM have been given;

side $AM = 7,710$

and

angle $LAM = 21°16'$,

which is the additosubtraction calculated.

Similarly, since in triangle AMH side AM has been given,

side $MH = KL$

and

angle $M = 90°$,

which is comprehended by sides AM and MH;

angle $MAH = 2°16'$,

which is the latitude sought for. But if we wish to inquire how much is due to the true and the apparent additosubtraction, let us take LH the diagonal of the parallelogram: we deduce from the sides (of the parallelogram) that

$$LH = 2,811.$$

And

$$AL = 7,186.$$

Hence

angle *LAH* = 21°23',

which is the additosubtraction of apparent movement and has an excess of approximately 7' over the previously reckoned difference, (angle *LAM*), as was to be shown.

6. ON THE SECOND TRANSIT IN LATITUDE OF VENUS AND MERCURY ACCORDING TO THE OBLIQUATION OF THEIR ORBITAL CIRCLES IN THE APOGEE AND THE PERIGEE

That is enough on the transit in latitude of these planets, which occurs around the mean longitudes of their orbital circles: we have said that these latitudes are called the declinations. Now we must speak of those latitudes which occur at the perigee and apogee and to which the third digression, the deviation, is conjoined—not as the latitudes occur in the three higher planets, but as follows, in order that the third digression may be more easily separated and discerned by reason. For Ptolemy observed that these latitudes appeared greatest at the time when the planets were on the straight lines from the centre of the Earth which touch the orbital circles; and that occurs, [189ᵃ] as we said, at their greatest morning and evening distances from the sun. He found that the northern latitudes of Venus were $1/3$° greater than the southern, but that the southern latitudes of Mercury were approximately $1/2$° greater than the northern. But, wishing to reduce the difficulty and labour of calculations, he took in accordance with a certain mean ratio $2^1/_2$° in different directions of latitude; the latitude themselves subtend these degrees in the circle perpendicular to the ecliptic and around the Earth, against which circle the latitudes are measured—especially as he did not think the error would on that account be very great, as we shall soon show. But if we take only $2^1/_2$° as the equal digression on each side of the ecliptic and exclude the deviation for the time being, until we have determined the latitudes of the obliquations, our demonstrations will be simpler and easier. Accordingly we must first show that this latitudinal digression is greatest around the point of tangency of the eccentric circle, where the additosubtractions in longitude are also greatest.

For let there be drawn the common section of the plane of the ecliptic and the plane of the eccentric circle of Venus or Mercury—the common section through the apogee and the perigee; and on it let *A* be taken as the position of the Earth and *B* as

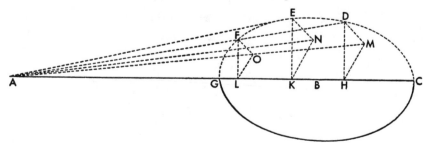

the centre of eccentric circle *CDEFG* which is inclined to the ecliptic, so that straight lines drawn anywhere at right angles to *CG* comprehend angles equal to the obliquation; and let *AE* be drawn tangent to the circle, and *AD* as cutting it somewhere. Moreover, from points *D*, *E*, and *F* let *DH*, *EK*, and *FL* be drawn perpendicular to line *CG*, and *DM*, *EN*, and *FO* perpendicular to the underlying plane of the ecliptic; and let *MH*, *NK*, and *OL* be joined, and also *AN* and *AOM*; for *AOM* is a straight line, since its three points are each in two planes—namely, in the plane of the ecliptic and in the plane *ADM* perpendicular to the plane of the ecliptic.

Accordingly since in the present obliquation the angles *HAM* and *KAN* comprehend the additosubtractions of these planets; and the angles *DAM* and *EAN* are the digressions in latitude: [189^{b}] I say, first, that angle *EAN*, the angle situated at the point of tangency, where the additosubtraction in longitude is also approximately greatest, is the greatest of all the angles of latitude.

For since angle *EAK* is greater than any of the others,

$$KE : EA > HD : DA$$

and

$$KE : EA > LF : FA.$$

But

$$EK : EN = HD : DM = LF : FO.$$

For, as we said,

$$\text{angle } EKN = \text{angle } HDM = \text{angle } LFO;$$

and

$$\text{angle } M = \text{angle } N = \text{angle } O = 90°.$$

Therefore

$$NE : EA > MD : DA$$

and

$$NE : EA > DF : FA;$$

and again

$$\text{angle } DMA = \text{angle } ENA = \text{angle } OFA = 90°.$$

Accordingly

$$\text{angle } EAN > \text{angle } DAM,$$

and angle *EAN* is greater than each of the other angles constructed in this way. Whence it is manifest that among the differences occurring between the additosubtractions and arising from the obliquation in longitude, the difference which is determined at point *E* in the greatest transit is the greatest. For

$$HD : HM = KE : KN = LF : FO,$$

on account of their subtending equal angles (in similar triangles). And since these lines are in the same ratio as the differences between them,

$$EK - KN : EA > HD - HM : AD$$

and

$$EK - KN : EA > LF - FO : AF.$$

Hence it is also clear that the additosubtractions in longitude of the segments of the eccentric circle will have the same ratio to the transits in latitude as the greatest additosubtraction in longitude has to the greatest transit in latitude, since

$$KE : EN = LF : FO = HD : DM,$$

—as was set before us to be demonstrated.

7. How Great the Angles of Obliquation of Venus and Mercury Are

Having first noted all that, let us see how great an angle is comprehended by the obliquation of the planes of either planet; and let us repeat what was said before: each planet has 5° between its greatest and least distance (in latitude), so that for the most part they become more northern or southern at contrary times and in accordance with their position on the orbital circle, for when the transit or manifest difference of Venus makes a digression greater or less than 5° through the apogee or perigee of the eccentric circle, the transit of Mercury however is more or less at $^1/_2$°.

[190ᵃ] Accordingly as before, let ABC be the common section of the ecliptic and the eccentric circle; and let the orbital circle of the planet be described around centre

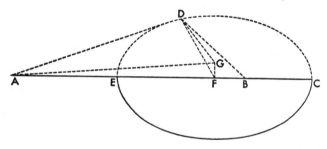

B oblique to the plane of the ecliptic in the way set forth. Now from the centre of the Earth let straight line AD be drawn touching the orbital circle at point D; and from D let DF be drawn perpendicular to CBE and DG perpendicular to the underlying plane of the ecliptic; and let BD, FG, and AG be joined. Moreover, let it be assumed that

angle $DAG = 2^1/_2$°,

where 4 rt. angles = 360°,

which is half the difference in latitude set forth for each planet.

Our problem is to find how great the angle of obliquation between the planes is, i.e., to find angle DFG.

Accordingly, since in the case of the planet Venus it has been shown that

the greater distance, at the apogee = 10,208,

where radius = 7,193,

369

and
$$\text{the lesser, at the perigee} = 9{,}792,$$

and
$$\text{the mean distance} = 10{,}000,$$

which Ptolemy decided to assume in this demonstration, as he wished to avoid labour and difficulty and to make an epitome; for where the extremes do not cause any great difference, it is better to use the mean. Accordingly,
$$AB : BD = 10{,}000 : 7{,}193,$$

and
$$\text{angle } ADB = 90°.$$

Therefore
$$\text{side } AD = 6{,}947.$$

Again, since
$$\text{angle } DAG = 2\tfrac{1}{2}°$$

and
$$\text{angle } AGD = 90°;$$

accordingly, the angles of triangle AGD are given, and
$$\text{side } DG = 303,$$
$$\text{where } AD = 6{,}947.$$

Thus also, two sides DF and DG have been given, and
$$\text{angle } DGF = 90°;$$

hence
$$\text{angle } DFG = 3°29',$$

which is the angle of inclination or obliquation. But since the excess of angle DAF over angle FAG comprehends the difference made by the parallax in longitude, hence that difference is to be determined by the measurement of those magnitudes. For it has been shown that
$$AD = 6{,}947$$

and
$$DF = 4{,}997,$$
$$\text{where } DG = 303.$$

Now
$$\text{sq. } AD - \text{sq. } DG = \text{sq. } AG,$$

and
$$\text{sq. } FD - \text{sq. } DG = \text{sq. } GF;$$

therefore
$$AG = 6{,}940$$

and

$$FG = 4,988.$$

But

$$FG = 7,187,$$
where $AG = 10,000$,

and

$$\text{angle } FAG = 45°57';$$

and

$$DF = 7,193,$$
where $AD = 10,000$,

and

$$\text{angle } DAF \doteqdot 46°.$$

Therefore at the greatest obliquation the additosubtraction of the parallax is deficient by approximately 3'. [190^b] Now it was made clear that at the mean apsis the angle of inclination of the orbital circles was $2^1/_2°$; but here it has increased by approximately 1°, which the first movement of libration—of which we have spoken—has added to it.

There is a similar demonstration in the case of Mercury. For the greatest distance of the orbital circle from the Earth is 10,948, where the radius of the orbital circle is 3,573; the least is 9,052; and the mean between these is 10,000. Moreover

$$AB : BD = 10,000 : 3,573.$$

Therefore

$$\text{side } AD = 9,340;$$

and since

$$BD : BF = AB : AD;$$

therefore

$$DF = 3,337.$$

And since

$$\text{angle } DAG = 2^1/_2°,$$

which is the angle of latitude;

$$DG = 407,$$
where $DF = 3,337$.

And so in triangle DFG the ratio of these two sides is given, and

$$\text{angle } G = 90°;$$

hence

$$\text{angle } DFG \doteqdot 7°.$$

And that is the angle of inclination or obliquation between the orbital circle of Mercury and the plane of the ecliptic. But it has been shown that around the mean longitudes or quadrants the angle of inclination was 6°15'. Therefore 45' have now been added to it by the movement of libration.

There is a similar argument in picking out the additosubtractions and their differences, after it has been shown that

$$DG = 407,$$
$$\text{where } AD = 9,340$$
$$\text{and } DF = 3,337$$

Accordingly,

$$\text{sq. } AD - \text{sq. } DG = \text{sq. } AG,$$

and

$$\text{sq. } DF - \text{sq. } DG = \text{sq. } FG.$$

Therefore

$$AG = 9,331$$

and

$$FG = 3,314;$$

hence it is inferred that

$$\text{angle } GAF = 20°48',$$

which is the additosubtraction, and

$$\text{angle } DAF = 20°56',$$

which is approximately 8' greater than the angle proportionate to the obliquation. It still remains for us to see if such angles of obliquation and the latitudes in accordance with the greatest and least distance of the orbital circle are found to be in conformity with those gathered from observation.

Wherefore once more with the same diagram; and first at the greatest distance of the orbital circle of Venus, let

$$AB : BD = 10,208 : 7,193.$$

And since

$$\text{angle } ADB = 90°;$$
$$DF = 5,102.$$

[191ª] But it has been found that

$$\text{angle } DFG = 3°29',$$

which is the angle of obliquation; hence

$$\text{side } DG = 309,$$
$$\text{where } AD = 7,238.$$

Accordingly,

$$DG = 427,$$
$$\text{where } AD = 10,000;$$

whence it is concluded that at the greatest distance from the Earth

$$\text{angle } DAG = 2°27'.$$

But at the least distance

$$AB = 9{,}792,$$
where radius of orbital circle = 7,193.

And

$$AD = 6{,}644,$$
which is perpendicular to the radius; and similarly, since
$$BD : DF = AB : AD,$$
$$DF = 4{,}883.$$

But

$$\text{angle } DFG = 3°28';$$
therefore

$$DG = 297,$$
where $AD = 6{,}644$.
And as the sides of the triangle have been given,
$$\text{angle } DAG = 2°34'.$$
But neither 3' nor 4' is large enough to be measured by means of an astrolabe; therefore that which was considered to be the greatest latitude of obliquation of the planet Venus is correct.

Again, let the greatest distance of the orbital circle of Mercury be taken, *i.e.*, let
$$AB : AD = 10{,}948 : 3{,}573;$$
consequently, by demonstrations similar to the foregoing, we still infer that
$$AD = 9{,}452$$
and

$$DF = 3{,}085.$$
But here too we have it recorded that
$$\text{angle } DFG = 7°,$$
which is the angle of obliquation; hence
$$DG = 376,$$
where $DF = 3{,}085$
and $DA = 9{,}452$.
Accordingly, as the sides of right triangle DAG are given,
$$\text{angle } DAG \fallingdotseq 2°17',$$
which is the greatest digression in latitude. But at the least distance
$$AB : BD = 9{,}052 : 3{,}573;$$
therefore

$$AD = 8{,}317$$
and

$$DF = 3{,}283.$$
Now since by reason of this same (obliquation)

$$DF : DG = 3,283 : 400,$$

where $AD = 8,317$;

whence

angle $DAG = 2°45'$.

Accordingly there is a difference of at least 13' between the $2^1/_2°$ of the digression in latitude according to the mean ratio and the digression at the apogee; and at the most a difference of 15' between the mean digression and that at the perigee. And in making our calculations according to the mean ratio we shall use $^1/_4°$ as the difference; for it is not sensibly diverse from the observed differences.

Having demonstrated these things and also that the greatest additosubtractions in longitude have the same ratio to the greatest transit in latitude as the additosubtractions in the remaining sections of the orbital circle have to the particular transits in latitude, we shall have at hand the numbers of all the latitudes, which occur on account of the obliquation of the orbital circle of Venus and Mercury. But we have calculated only those latitudes which occur midway between the apogee and the perigee, as we said; and it was shown that the greatest of these latitudes is $2^1/_2°$, and the greatest [191^b] additosubtraction in the case of Venus is 46° and that in the case of Mercury about 22°. And in the tables of irregular movements we have already placed the additosubtractions opposite the particular sections of the orbital circles. Accordingly in the case of each of the two planets we shall take from the $2^1/_2°$ a part proportionate to the excess of the greatest additosubtraction over each of the lesser additosubtractions; we shall inscribe it in the table to be drawn up below with all its numbers; and in this way we shall have unfolded all the particular latitudes of the obliquations which occur when the Earth is at their highest apsis and at their lowest—just as we set forth the latitudes of the declinations in the case of the mean quadrants and mean longitudes. The latitudes which occur between these four limits can be unfolded by the subtle art of mathematics with the help of the proposed hypothesis of circles but not without labour. Now Ptolemy— who is compendious wherever he can be so—seeing that each of these aspects of latitude as a whole and in all its parts increased and decreased proportionally, like the latitude of the moon, accordingly took twelve parts of it, since their greatest latitude is 5° and that number is a twelfth part of 60, and made proportional minutes out of them, to be used not only in the case of these two planets but also in that of the three higher planets, as will be made clear below.

8. On the Third Aspect of the Latitude of Venus and Mercury, Which They Call the Deviation

Now that these things have been set forth, it still remains to say something about the third movement in latitude, which is the deviation. The ancients, who held the

Earth down at the centre of the world, believed that the deviation took place by reason of the inclination of an eccentric circle which has an epicycle and which revolves around the centre of the Earth—the deviation occurring most greatly when the epicycle is at the apogee or perigee and being always $1/6°$ to the north in the case of Venus and $3/4°$ to the south in the case of Mercury, as we said before. It is not however sufficiently clear whether they meant the inclination of the orbital circles to be equal and always the same: for their numbers indicate that, when they order a sixth part of the proportional minutes to be taken as the deviation of Venus, and three parts out of four as that of Mercury. That does not hold, unless the angle of inclination always remains [192a] the same, as is demanded by the ratio of the minutes, which they take as their base. But if the angle remains the same, it is impossible to understand how the latitude of the planets suddenly springs back from the common section into the same latitude which it had just left, unless you say that takes place in the manner of refraction of light, as in optics. But here we are dealing with movement, which is not instantaneous but is by its own nature measured by time. Accordingly we must acknowledge that a libration such as we have expounded is present in those (circles) and makes the parts of the circle move over in different directions: And that necessarily follows, as the numbers differ $1/5°$ in the case of Mercury. That should seem less surprising, if in accordance with our hypothesis this latitude is variable and not wholly simple but does not produce any apparent error, as is to be seen in the case of all differences, as follows:

For in the plane perpendicular to the ecliptic let (*ABC*) be the common section (of the two planes), and in the common section let *A* be the centre of the Earth and *B* the centre of the circle *CDF* at greatest or least distance from the Earth and as it were through the poles of the inclined orbital circle. And when the centre of the orbital circle is at the apogee or

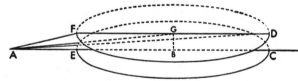

the perigee, *i.e.*, on line *AB*, the planet, wherever it is, is at its greatest deviation, in accordance with the circle parallel to the orbital circle; and *DF* is the diameter parallel to *CBE*, the diameter of the orbital circle. And *DF* and *CBE* are put down as the common sections of the planes perpendicular to plane *CDF*. Now let *DF* be bisected at *G*, which will be the centre of the parallel circle; and let *BG*, *AG*, *AD*, and *AF* be joined. Let us put down that

angle *BAG* = 10′,

as in the greatest deviation of Venus. Accordingly, in triangle *ABG*

angle *B* = 90°;

and we have the following ratio for the sides:

$$AB : BG = 10{,}000 : 29.$$

But
$$\text{line } ABC = 17{,}193;$$

and by subtraction
$$AE = 2{,}807;$$

and
$$^1/_2 \text{ ch. } 2 \ CD = \ ^1/_2 \text{ ch. } 2 \ EF = BG.$$

Accordingly
$$\text{angle } CAD = 6'$$

and
$$\text{angle } EAF \doteqdot 15'.$$

Now
$$\text{angle } BAG - \text{angle } CAD = 4',$$

while
$$\text{angle } EAF - \text{angle } BAG = 5';$$

and those differences can be neglected on account of their smallness. Accordingly, when the Earth is situated at its apogee or perigee, the apparent deviation of Venus will be slightly more or less than 10', [192^b] in whatever part of its orbital circle the planet is.

But in the case of Mercury when
$$\text{angle } BAG = 45',$$

and
$$AB : BG = 10{,}000 : 131,$$

and
$$ABC = 13{,}573,$$

and by subtraction
$$AE = 6{,}827;$$

then
$$\text{angle } CAD = 33°$$

and
$$\text{angle } EAF \doteqdot 70'.$$

Accordingly, angle CAD has a deficiency of 12', and angle EAF has an excess of 25'. But these differences are practically obliterated beneath the rays of the sun before Mercury emerges into our sight, wherefore the ancients considered only its apparent and as it were simple deviation. But if anyone wishes to examine with least labour the precise ratio of their passages when hidden beneath the sun, we shall show how that takes place, as follows. We shall take Mercury as our example, because it makes a more considerable deviation than Venus.

For let *AB* be the straight line in the common section of the orbital circle of the planet and the ecliptic, while the Earth—which is at *A*—is at the apogee or the

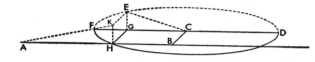

perigee of the planet's orbital circle. Now let us put down that

$$AB = 10,000$$

indifferently, as if the mean between greatest and least distance, as we did in the case of the obliquation. Now around centre *C* let there be described circle *DEF*, which is parallel to the eccentric orbital circle at a distance *CB*; and the planet on this parallel circle be understood as being at this time at its greatest deviation. Let *DCF* be the diameter of this circle, which is also necessarily parallel to *AB*; and *DCF* and *AB* are in the same plane perpendicular to the orbital circle of the planet. Therefore, for example, let

$$\text{arc } EF = 45°,$$

in relation to which we shall examine the deviation of the planet. And let *EG* be drawn perpendicular to *CF*, and *EK* and *GH* perpendicular to the underlying plane of the orbital circle. Let the right parallelogram be completed by joining *HK*; and let *AE*, *AK*, and *EC* also be joined.

Now in the greatest deviation of Mercury

$$BC = 131,$$
$$\text{where } AB = 10,000.$$

And

$$CE = 3,573;$$

and the angles of the right triangle *EGC* are given; hence

$$\text{side } EG = KH = 2526.$$

And since

$$BH = EG = CG,$$
$$AH = BA - BH = 7,474.$$

Accordingly, since in triangle *AHK*

$$\text{angle } H = 90°$$

and the sides of comprehending angle *H* are given;

$$\text{side } AK = 7,889.$$

But

$$\text{side } KE = CB = GH = 131.$$

Accordingly, since in triangle [193ª] *AKE* the two sides *AK* and *KE* comprehending the right angle *K* have been given, angle *KAE* is given, which answers to the deviation we were seeking for the postulated arc *EF* and differs very little from the angle observed. We shall do similarly in the case of Venus and the other planets; and we shall inscribe our findings in the subjoined table.

Having made this exposition, we shall work out proportional minutes for the deviations between these limits. For let *ABC* be the eccentric orbital circle of Venus or Mercury; and let *A* and *C* be the nodes of this movement in latitude, and *B* the limit of greatest deviation. And with *B* as centre let there be described the small circle *DFG*, with the diameter *DBF* across it, along which diameter the libration of the movement of deviation takes place. And since it has been laid down that when the Earth is at the apogee or the perigee of the eccentric orbital circle of the planet, the planet itself is at its greatest deviation, namely in point *F*, where at this time the circle carrying the planet touches the small circle. Now let the Earth be somewhere removed from the apogee or perigee of the eccentric circle of the planet;

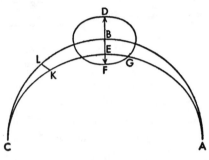

and in accordance with this movement let a similar arc *FG* be taken on the small circle. Let circle *AGC* be described, which bears the planet and will cut the small circle, and the diameter *DF* at point *E*; and let the planet be taken as being on this circle at point *K* in accordance with arc *EK* which is by hypothesis similar to arc *FG*; and let *KL* be drawn perpendicular to circle *ABC*.

Our problem is to find by means of *FG*, *EK*, and *BE* the magnitude of *KL*, *i.e.*, the distance of the planet from circle *ABC*. For since by means of arc *FG* arc *EG* will be given as a straight line hardly different from a circular or convex line, and *EF* will similarly be given in terms of the parts, whereof *BF* and the remainder *BE* will be given; for

$$BF : BE = \text{ch. } 2 \ CE : \text{ch. } 2 \ CK = BE : KL.$$

Accordingly if we put down *BF* and the radius of circle *CE* in terms of the same number, sixty, we shall have from them the number which picks out *BE*. When that number has been multiplied by itself and the product divided by sixty, we shall have *KL*, the minutes proportional to arc *EK*; and we shall inscribe them similarly in the fifth and last column of the table which follows:

LATITUDES OF SATURN, JUPITER, AND MARS

Common Numbers		SATURN				JUPITER				MARS				Proportional Minutes	
		Northern		Southern		Northern		Southern		Northern		Southern			
Deg.	Deg.	Deg.	Min.	Deg.	Min.	Deg.	Min.	Deg.	Min.	Deg.	Min.	Deg.	Min.	Deg.	Min.
3	357	2	3	2	2	1	6	1	5	0	6	0	5	59	48
6	354	2	4	2	2	1	7	1	5	0	7	0	5	59	36
9	351	2	4	2	3	1	7	1	5	0	9	0	6	59	6
12	348	2	5	2	3	1	8	1	6	0	9	0	6	58	36
15	345	2	5	2	3	1	8	1	6	0	10	0	8	57	48
18	342	2	6	2	3	1	8	1	6	0	11	0	8	57	0
21	339	2	6	2	4	1	9	1	7	0	12	0	9	56	48
24	336	2	7	2	4	1	9	1	7	0	13	0	9	54	36
27	333	2	8	2	5	1	10	1	8	0	14	0	10	53	18
30	330	2	8	2	5	1	10	1	8	0	14	0	11	52	0
33	327	2	9	2	6	1	11	1	9	0	15	0	11	50	12
36	324	2	10	2	7	1	11	1	9	0	16	0	12	48	24
39	321	2	10	2	7	1	12	1	10	0	17	0	12	46	24
42	318	2	11	2	8	1	12	1	10	0	18	0	13	44	24
45	315	2	11	2	9	1	13	1	11	0	19	0	15	42	12
48	312	2	12	2	10	1	13	1	11	0	20	0	16	40	0
51	309	2	13	2	11	1	14	1	12	0	22	0	18	37	36
54	306	2	14	2	12	1	14	1	13	0	23	0	20	35	12
57	303	2	15	2	13	1	15	1	14	0	25	0	22	32	36
60	300	2	16	2	15	1	16	1	15	0	27	0	24	30	0
63	297	2	17	2	16	1	17	1	17	0	29	0	25	27	12
66	294	2	18	2	18	1	18	1	18	0	31	0	27	24	24
69	291	2	20	2	19	1	19	1	19	0	33	0	29	21	24
72	288	2	21	2	21	1	21	1	21	0	35	0	31	18	24
75	285	2	22	2	22	1	22	1	22	0	37	0	34	15	24
78	282	2	24	2	24	1	24	1	24	0	40	0	37	12	24
81	279	2	25	2	26	1	25	1	25	0	42	0	39	9	24
84	276	2	27	2	27	1	27	1	27	0	45	0	42	6	24
87	273	2	28	2	28	1	28	1	28	0	48	0	45	3	12
90	270	2	30	2	30	1	30	1	30	0	51	0	49	0	0
93	267	2	31	2	31	1	31	1	31	0	55	0	52	3	12
96	264	2	33	2	33	1	33	1	33	0	59	0	56	6	24
99	261	2	34	2	34	1	34	1	34	1	2	1	0	9	9
102	258	2	36	2	36	1	36	1	36	1	6	1	4	12	12
105	255	2	37	2	37	1	37	1	37	1	11	1	8	15	15
108	252	2	39	2	39	1	39	1	39	1	15	1	12	18	18
111	249	2	40	2	40	1	40	1	40	1	19	1	17	21	21
114	246	2	42	2	42	1	42	1	42	1	25	1	22	24	24
117	243	2	43	2	43	1	43	1	43	1	31	1	28	27	12
120	240	2	45	2	45	1	44	1	44	1	36	1	34	30	0
123	237	2	46	2	46	1	46	1	46	1	41	1	40	32	37
126	234	2	47	2	48	1	47	1	47	1	47	1	47	35	12
129	231	2	49	2	49	1	49	1	49	1	54	1	55	37	36
132	228	2	50	2	51	1	50	1	51	2	2	2	5	40	6
135	225	2	52	2	53	1	53	1	53	2	10	2	15	42	12
138	222	2	53	2	54	1	52	1	54	2	19	2	26	44	24
141	219	2	54	2	55	1	53	1	55	2	29	2	38	47	24
144	216	2	55	2	56	1	55	1	57	2	37	2	48	48	24
147	213	2	56	2	57	1	56	1	58	2	47	3	4	50	12
150	210	2	57	2	58	1	58	1	59	2	51	3	20	52	0
153	207	2	58	2	59	1	59	2	1	3	12	3	32	53	18
156	204	2	59	3	0	2	0	2	2	3	23	3	52	54	36
159	201	2	59	3	1	2	1	2	3	3	34	4	13	55	48
162	198	3	0	3	2	2	2	2	4	3	46	4	36	57	0
165	195	3	0	3	2	2	2	2	5	3	57	5	0	57	48
168	192	3	1	3	3	2	3	2	5	4	9	5	23	58	36
171	189	3	1	3	3	2	3	2	6	4	17	5	48	59	6
174	186	3	2	3	4	2	4	2	6	4	23	6	15	59	36
177	183	3	2	3	4	2	4	2	7	4	27	6	35	59	48
180	180	3	2	3	5	2	4	2	7	4	30	6	50	60	0

LATITUDES OF VENUS AND MERCURY

Common Numbers		VENUS				MERCURY				Deviation of Venus		Deviation of Mercury		Proportional Minutes of the Deviation	
		Declination		Obliquation		Declination		Obliquation							
Deg.	Deg.	Deg.	Min.	Deg.	Min.	Deg.	Min.	Deg.	Min.	Deg.	Min.	Deg.	Min.	Deg.	Min.
3	357	1	2	0	4	0	7	1	45	0	5	0	33	59	36
6	354	1	2	0	8	0	7	1	45	0	11	0	33	59	12
9	351	1	1	0	12	0	7	1	45	0	16	0	33	58	25
12	348	1	1	0	16	0	7	1	44	0	22	0	33	57	14
15	345	1	0	0	21	0	7	1	44	0	27	0	33	55	41
18	342	1	0	0	25	0	7	1	43	0	33	0	33	54	9
21	339	0	59	0	29	0	7	1	42	0	38	0	33	52	12
24	336	0	59	0	33	0	7	1	40	0	44	0	34	49	43
27	333	0	58	0	37	0	7	1	38	0	49	0	34	47	21
30	330	0	57	0	41	0	8	1	36	0	55	0	34	45	4
33	327	0	56	0	45	0	8	1	34	1	0	0	34	42	0
36	324	0	55	0	49	0	8	1	30	1	6	0	34	39	15
39	321	0	53	0	53	0	8	1	27	1	11	0	35	35	53
42	318	0	51	0	57	0	8	1	23	1	16	0	35	32	51
45	315	0	49	1	1	0	8	1	19	1	21	0	35	29	41
48	312	0	46	1	5	0	8	1	15	1	26	0	36	26	40
51	309	0	44	1	9	0	8	1	11	1	31	0	36	23	34
54	306	0	41	1	13	0	8	1	8	1	35	0	36	20	39
57	303	0	38	1	17	0	8	1	4	1	40	0	37	17	40
60	300	0	35	1	20	0	8	0	59	1	44	0	38	15	0
63	297	0	32	1	24	0	8	0	54	1	48	0	38	12	20
66	294	0	29	1	28	0	9	0	49	1	52	0	39	9	55
69	291	0	26	1	32	0	9	0	44	1	56	0	39	7	38
72	288	0	23	1	35	0	9	0	38	2	0	0	40	5	39
75	285	0	20	1	38	0	9	0	32	2	3	0	41	3	57
78	282	0	16	1	42	0	9	0	26	2	7	0	42	2	34
81	279	0	12	1	46	0	9	0	21	2	10	0	42	1	28
84	276	0	8	1	50	0	10	0	16	2	14	0	43	0	40
87	273	0	4	1	54	0	10	0	8	2	17	0	44	0	10
90	270	0	0	1	57	0	10	0	0	2	20	0	45	0	0
93	267	0	5	2	0	0	10	0	8	2	23	0	45	0	10
96	264	0	10	2	3	0	10	0	15	2	25	0	46	0	40
99	261	0	15	2	6	0	10	0	23	2	27	0	47	1	28
102	258	0	20	2	9	0	11	0	31	2	28	0	48	2	34
105	255	0	26	2	12	0	11	0	40	2	29	0	48	3	57
108	252	0	32	2	15	0	11	0	48	2	29	0	49	5	39
111	249	0	38	2	17	0	11	0	57	2	30	0	50	7	38
114	246	0	44	2	20	0	11	1	6	2	30	0	51	9	55
117	243	0	50	2	22	0	11	1	16	2	30	0	51	12	20
120	240	0	59	2	24	0	12	1	25	2	29	0	52	15	0
123	237	1	8	2	26	0	12	1	35	2	28	0	53	17	40
126	234	1	18	2	27	0	12	1	45	2	26	0	54	20	39
129	231	1	28	2	29	0	12	1	55	2	23	0	55	23	34
132	228	1	38	2	30	0	12	2	6	2	20	0	56	26	40
135	225	1	48	2	30	0	13	2	16	2	16	0	57	29	41
138	222	1	59	2	30	0	13	2	27	2	11	0	57	32	51
141	219	2	11	2	29	0	13	2	37	2	6	0	58	35	53
144	216	2	25	2	28	0	13	2	47	2	0	0	59	39	25
147	213	2	43	2	26	0	13	2	57	1	53	1	0	42	0
150	210	3	3	2	22	0	13	3	7	1	46	1	1	45	4
153	207	3	23	2	18	0	13	3	17	1	38	1	2	47	21
156	204	3	44	2	12	0	14	3	26	1	29	1	3	49	43
159	201	4	5	2	4	0	14	3	34	1	20	1	4	52	12
162	198	4	26	1	55	0	14	3	42	1	10	1	5	54	9
165	195	4	49	1	42	0	14	3	48	0	59	1	6	55	41
168	192	5	13	1	27	0	14	3	54	0	48	1	7	57	14
171	189	5	36	1	9	0	14	3	58	0	36	1	8	58	25
174	186	5	52	0	48	0	14	4	2	0	24	1	9	59	12
177	183	6	7	0	25	0	14	4	4	0	12	1	9	59	36
180	180	6	22	0	0	0	14	4	5	0	0	1	10	60	0

9. On the Calculation of the Latitudes of the Five Wandering Stars

[195^b] Now this is the method of calculating the latitudes of the five wandering stars by means of these tables. For in the case of Saturn, Jupiter, and Mars we shall take the discrete, or corrected, anomaly of the eccentric circle among the common numbers: in the case of Mars, the anomaly as is; in that of Jupiter, after the subtraction of 20°; and in that of Saturn, after the addition of 50°. Accordingly we shall note the numbers which occur in the region of the 60's, in the proportional minutes placed in the last column. Similarly by means of the corrected anomaly of parallax we shall determine the proper number of each planet, corresponding to the latitude: the first and northern latitude, if the proportional minutes are in the first half of the column—which happens when the anomaly of the eccentric circle is less than 90° or more than 270°; the second and southern latitude, if the proportional minutes are in the second half of the column, *i.e.*, if the anomaly of the eccentric circle, whereby the table was entered upon, was more than 90° or less than 270°. Accordingly if we adjust one of these latitudes to its 60's, the result will be the distance north or south of the ecliptic in accordance with the denomination of the circles assumed.

But in the case of Venus and Mercury the three latitudes of declination, obliquation, and deviation, which are marked down separately, are to be taken first by means of the corrected anomaly of parallax, except that in the case of Mercury one tenth of the obliquation is to be subtracted, if the anomaly of the eccentric circle and its number are found in the first column of the table, or merely added, if in the second column of the table; and the remainder or sum is to be kept.

And we must discern whether their denominations are northern or southern, since if the corrected anomaly of parallax is in the apogeal semicircle, *i.e.*, is less than 90° or more than 270° and the anomaly of the eccentric circle is also less than a semicircle; or again, if the anomaly of parallax is in the perigeal arc, *i.e.*, is more than 90° and less than 270° and the anomaly of the eccentric circle is greater than a semicircle; the declination of Venus will be northern and that of Mercury southern. But if the anomaly of parallax is in the perigeal arc and the anomaly of the eccentric circle is less than a semicircle; [196^a] or if the anomaly of parallax is in the apogeal arc and the anomaly of the eccentric circle is more than a semicircle; conversely the declination of Venus will be southern and that of Mercury northern. But in the case of the obliquation, if the anomaly of parallax is less than a semicircle and the anomaly of the eccentric circle is apogeal; or if the anomaly of parallax is greater than a semicircle and the anomaly of the eccentric circle is perigeal; the obliquation of Venus will be to the north and that of Mercury to the south; and vice versa. But the deviations of Venus always remain northern and those of Mercury southern.

Then, corresponding to the corrected anomaly of the eccentric circle, the proportional minutes should be taken which are common to all the five planets, although they are ascribed to the three higher planets. These are assigned to the obliquation and lastly to the deviation. After this, when we have added 90° to the same anomaly of the eccentric circle, we shall once more take the sum and find the common proportional minutes which correspond to it and assign them to the latitude of declination. Having placed these things in this order, we shall adjust each of the three particular latitudes set forth to their proportional minutes; and the result will be the corrected latitude for the position and time, so that at last we may have the sum of the three latitudes of the two planets. If all the latitudes are of one denomination, they are added together; but if not, only the two are added which have the same denomination; and according as the sum is greater or less than the third latitude, which is different from there, there will be a subtraction; and the remainder will be the predominant latitude sought for.

Acknowledgments

This book would not have been possible without the help of a number of talented people who made different contributions at various stages of the book's development. Among those deserving special thanks are Michael Rosin, a consultant to Running Press, Gil King, and Mrs. Karen Sime, assistant to Professor Stephen Hawking.

Thanks are also due to several past and present members of the staff of Running Press: Carlo DeVito, Kathleen Greczylo, Kelly Pennick, Bill Jones, Deborah Grandinetti, and Sarah O'Brien.

About Stephen Hawking

Stephen Hawking is considered the most brilliant theoretical physicist since Einstein. He has also done much to popularize science. His book, *A Brief History of Time*, sold more than 10 million copies in 40 languages, achieving the kind of success almost unheard of in the history of science writing. His subsequent books, *The Universe in A Nutshell* and *The Future of Spacetime*, with Kip S. Thorne and others, have also been well received.

He was born in Oxford, England on January 8, 1942 (300 years after the death of Galileo). He studied physics at University College, Oxford, received his Ph.D. in Cosmology at Cambridge, and since 1979 has held the post of Lucasian Professor of Mathematics. The chair was founded in 1663 with money left in the will of the Reverend Henry Lucas, who had been the Member of Parliament for the University. It was first held by Isaac Barrow, and then in 1663 by Isaac Newton. It is reserved for those individuals considered the most brilliant thinkers of their time.

Professor Hawking has worked on the basic laws that govern the universe. With Roger Penrose, he showed that Einstein's General Theory of Relativity implied space and time would have a beginning in the Big Bang and an end in black holes. The results indicated it was necessary to unify General Relativity with Quantum Theory, the other great scientific development of the first half of the twentieth century. One consequence of such a unification that he discovered was that black holes should not be completely black but should emit radiation and eventually disappear. Another conjecture is that the universe has no edge or boundary in imaginary time.

Stephen Hawking has twelve honorary degrees and is the recipient of many awards, medals, and prizes. He is a Fellow of the Royal Society and a Member of the United States National Academy of Sciences. He continues to combine family life (he has three children and one grandchild) and his research into theoretical physics together with an extensive program of travel and public lectures.